AF683043

Multi-Body Dynamics

Conference Organizing Committee

P Brooks
University of Leeds

B Eickhoff
AEA Technology Rail

S Gupta
Mechanical Dynamics International UK Limited

P Kelly
Ford Werke AG

A Mostofi
Nuclear Electric Limited

H Rahnejat (Chair)
University of Bradford

IMechE Conference Transactions

I MECH E

International Conference on

Multi-Body Dynamics

New Techniques and Applications

10–11 December 1998
IMechE HQ, London, UK

Organized by the Automobile and Railway Divisions of the Institution of Mechanical Engineering (IMechE).

Co-sponsored by The Society of Automotive Engineers of Japan

IMechE Conference Transactions 1998–13

Professional Engineering Publishing

Published by Professional Engineering Publishing Limited for
The Institution of Mechanical Engineers, Bury St Edmunds and London, UK.

First Published 1998

This publication is copyright under the Berne Convention and the International Copyright Convention. All rights reserved. Apart from any fair dealing for the purpose of private study, research, criticism or review, as permitted under the Copyright, Designs and Patents Act, 1988, no part may be reproduced, stored in a retrieval system, or transmitted in any form or by any means, electronic, electrical, chemical, mechanical, photocopying, recording or otherwise, without the prior permission of the copyright owners. Reprographic reproduction is permitted only in accordance with the terms of licences issued by the Copyright Licensing Agency, 90 Tottenham Court Road, London W1P 9HE. *Unlicensed multiple copying of the contents of this publication is illegal.* Inquiries should be addressed to: The Publishing Editor, Professional Engineering Publishing Limited, Northgate Avenue, Bury St. Edmunds, Suffolk, IP32 6BW, UK. Fax: 01284 704006.

© The Institution of Mechanical Engineers 1998, unless otherwise stated.

ISSN 1356–1448
ISBN 1 86058 152 8

A CIP catalogue record for this book is available from the British Library.

Printed and bound in Great Britain

The Publishers are not responsible for any statement made in this publication. Data, discussion, and conclusions developed by authors are for information only and are not intended for use without independent substantiating investigation on the part of potential users. Opinions expressed are those of the Author and are not necessarily those of the Institution of Mechanical Engineers or its Publishers.

Related Titles of Interest

Title	Editor/Author	ISBN
Active Control of Vibration	Edited by C R Burrows and P S Keogh	0 85298 916 4
Handbook of Vehicle Design Analysis	John Fenton	0 85298 963 6
Multi-Body Dynamics: Monitoring and Simulation Techniques	Edited by H Rahnejat and R Whalley	1 86058 064 5
Multi-Body Dynamics Vehicles, Machines, and Mechanisms	H Rahnejat	1 86058 122 6
Vehicle Handling Dynamics	John R Ellis	0 85298 885 0
Vibrations in Rotating Machinery	IMechE Conference 1996–7	1 86058 009 2

For the full range of titles published by Professional Engineering Publishing contact:

Sales Department
Professional Engineering Publishing Limited
Northgate Avenue
Bury St Edmunds
Suffolk
IP32 6BW
UK

Tel: 01284 724384
Fax: 01284 718692

Contents

Powertrain and Rotor Dynamics

Structural and Mechanism Dynamics

Powertrain and Rotor Dynamics

C553/046/98

The NVH behaviour of vehicle-drivetrains – a difficult but very interesting challenge for acoustic engineers

J -W BIERMANN
Institute of Automotive Engineering, Aachen University of Technology, Germany

SUMMARY

The development of modern vehicles is governed more and more by design conflicts. For example, there are lightweight aspects, reduced fuel consumption on the one hand and improved NVH-behaviour and comfort on the other hand. To reach a well-balanced compromise a "profound" knowledge of the complex multi-body-system drivetrain/body will be needed in future. That is why there will be a demand for even more intense, i.e. more detailed test rig and simulation investigations in future.

1 INITIAL SITUATION

The development of new vehicle types is governed by a multitude of factors of different significance. Whereas over the past few years the emphasis was on criteria such as engine power, top speed and acceleration, priority factors today include safety, fuel consumption and comfort, not forgetting the omnipresent cost-reduction. Customers' demands have increased over recent years in many respects. For example, owners expect smaller vehicles such as the FORD Fiesta to be as comfortable as bigger cars like the FORD Mondeo or Scorpio.

In present vehicle development, characteristic target conflicts occur when designing special components. Vehicle designs which take account of lightweight aspects cause problems in NVH-behaviour due to the higher vibration. Nevertheless, the interior noise level does drop as a result of specific NVH-development work and fuel consumption decreases, too. The gain of around minus 5 dB(A) corresponds to a reduction of about 65% over the past 20 years. This is shown in Fig. 1.

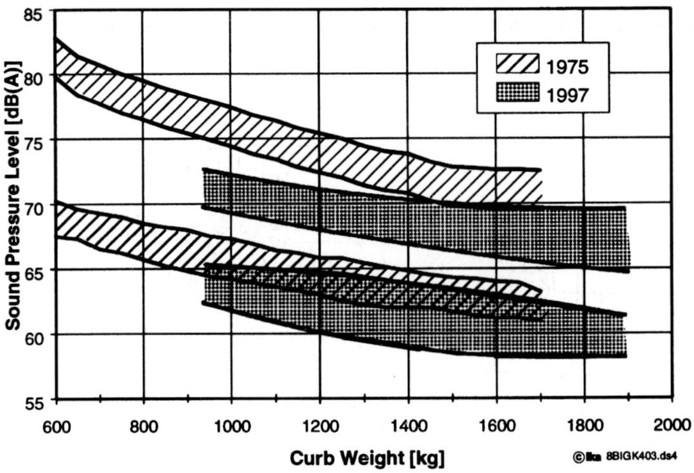

Fig. 1: Reduction in interior noise level

The work of acoustic engineers is becoming more and more complicated due to the reduction in the interior noise level. Noise which was previously imperceptible is now regarded as being annoying. This will have to be eliminated to meet the customers' future requirements.

2 PROBLEM DEFINITION AND SUGGESTIONS FOR IMPROVEMENTS

The interior noise in vehicles consists of emissions from many sources. Fig. 2 shows the main sources.

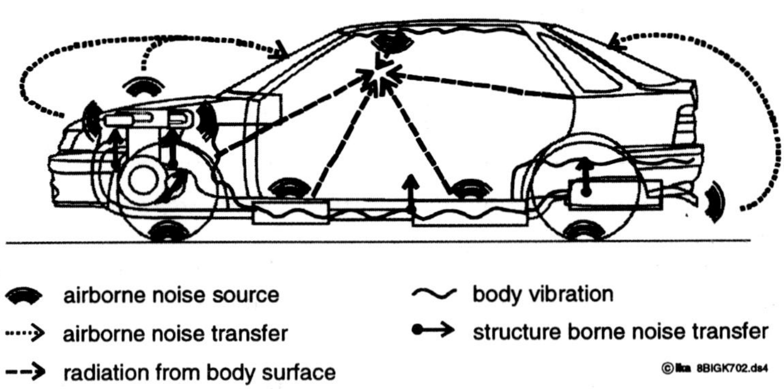

Fig. 2: Main sources of vehicle interior noise

The drivetrain is the most important of them. The body, on the other hand, acts like a transfer element with the panel surfaces radiating airborne noise into the interior.

The drivetrain is made up of several components such as the engine and transmission. These components in turn consist of many single parts. Fig. 3 provides an overview of the drivetrain assembly.

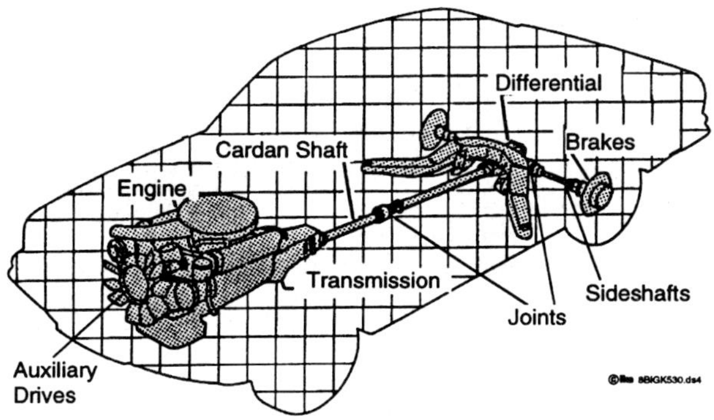

Fig. 3: Drivetrain system

The drivetrain is thus a complex multi-body-system, which vibrates in a multitude of bending and rotational modes, not least due to the excitation produced by dynamic operation. The NVH-phenomena resulting from these processes are relevant for both the interior comfort and total noise level. An overview of the main NVH-phenomena is given in Fig. 4.

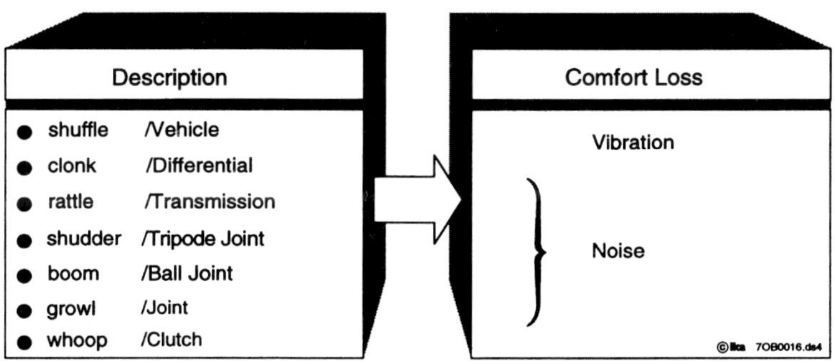

Fig. 4 Overview of drivetrain NVH-phenomena

Except for "Shuffle", a low-frequency vehicle vibration in the direction of travel, all of these phenomena are acoustic and are transferred into the interior directly as airborne noise or via the body as structure borne noise.

"Rattle-Noise" is caused by the impact of tooth faces in manual transmissions. The excitation is found in the frequency range of the second engine order in 4-cylinder engines. The main distributor of the "Clonk"-phenomenon is the differential (1). After a load cycle, i.e. a change in torque direction, the flanks of the toothed wheels strike against each other.

"Growl" and "Shudder" are caused by both side shafts, while clutch operation is the source of annoying "Whoop" noise (2). Although the engine noise is an acoustic measure of the performance characteristics of a vehicle, annoying noise such as "Boom" is also caused by the intake process.

The NVH-phenomena mentioned above are a great challenge to research and development departments in the automotive industry. The goal of reducing or preventing these problems can only be achieved with an exact knowledge of the main causes of and interdependencies between the various influencing variables. This means that initial measurements are required on special test rigs and in vehicles. In a second phase, mathematical simulation tools for the actual vibration behaviour are modelled on the basis of the results and information provided by the measured values. These tools are of enormous significance, particularly for future drivetrain developments, and are consequently a prime target of investigation.

The following brief explanation of some examples of special tasks and suggestions for tests clarifies this statement. The first example deals with gear rattle noise from manual transmissions. Fig. 5 and Fig. 6 show two test rigs which were developed and set up at the Institute of Automotive Engineering to investigate gear rattle.

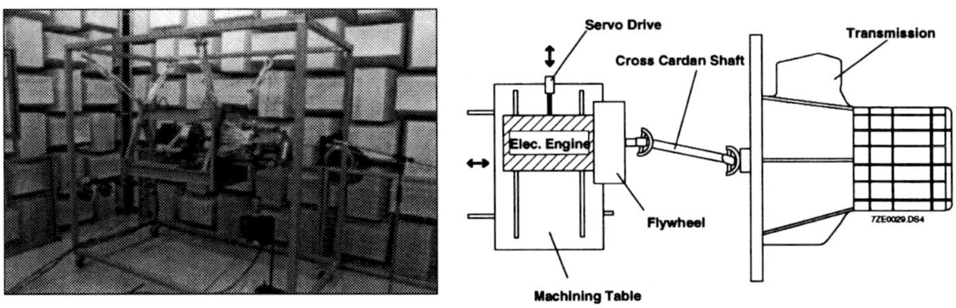

Fig. 5: IDLE Rattle Rig

The first testing device has recently been used as an aiming tool to investigate idle rattle in transmissions. The test rig is powered by a small electric engine. A modified cardan shaft is used in order to excite the second order of a 4-cylinder combustion-engine. The angular acceleration can be varied by the displacement angle of the cardan joint.

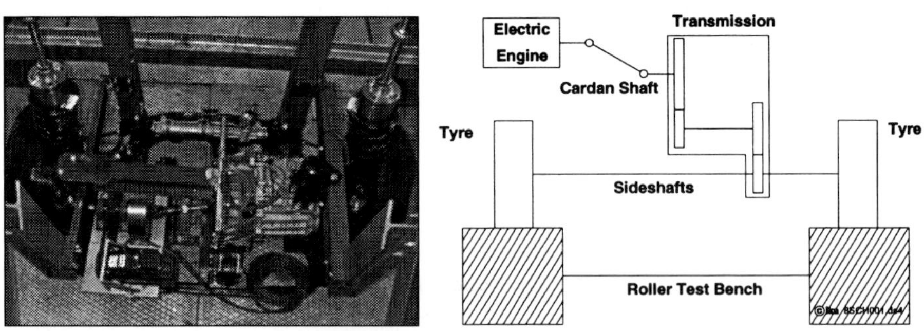

Fig. 6: IN GEAR Rattle Rig

The test rig in Fig. 6 is a current development. The set-up includes both the transmission and complete drivetrain with wheels. The test rig allows an investigation of the rattle behaviour in complete drivetrains at different speeds and torque levels (in gear rattle). This example clearly demonstrates the need for improved test equipment as mentioned above.

The second example illustrates the problem of the conflicting objectives of two design criteria. During sudden acceleration or deceleration, most drivetrains show a clearly perceivable impact-noise, so-called "Clonk", in combination with a subsequent low-frequency "Shuffle". These effects are intolerable from the customer's point of view, especially in higher vehicle classes. Simulation modelling is a good way to investigate this vibration process. Fig. 7 shows the simulation model, set up for those calculations at ika. All important parameters are taken into account (3). For programming, the itisim©-software is used.

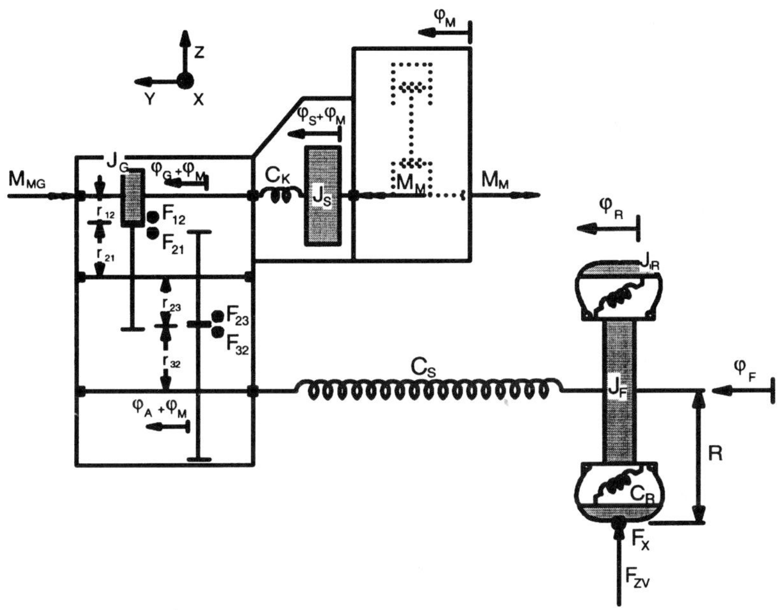

Fig. 7: Vehicle model (longitudinal dynamic)

As shown in the simulation example in Fig. 8, a clear improvement can be achieved with an electronic engine control which enables a damped torque delivery of the combustion engine (3).

The disadvantage of this measure is the reduction in agility. Agility means the engine response to a sudden gas pedal operation by the driver. This effect is unacceptable, especially in sports cars.

Thus, simulations combined with basic test-rig investigations allow an assessment of the potentials offered by several measures for this application. The most effective measure, in this case torque delivery control, was then designed as a balanced compromise.

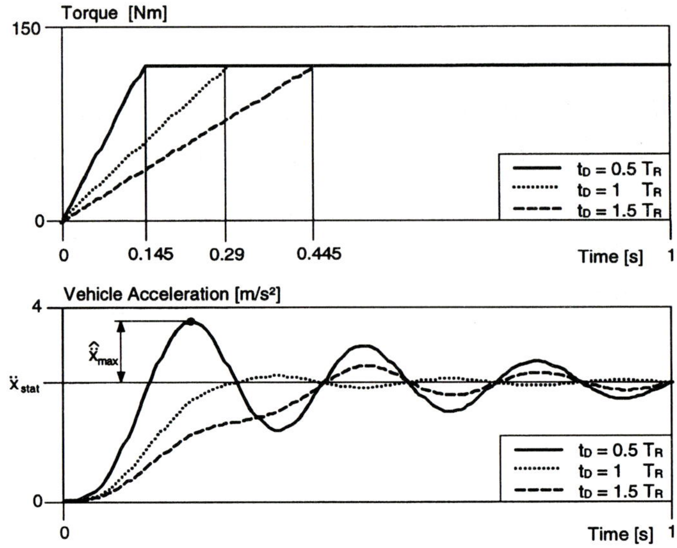

Fig. 8: Reduction of "Shuffle" vibration by Torque Delivery Control

This example shows quite clearly that conflicting development criteria can only be balanced on the basis of a "profound" knowledge, i.e. by means of extensive test-rig and simulation investigations.

3 RESULTS

The development of future vehicles is governed by increasing changes in boundary conditions. A "profound" knowledge of the complex vibration system "drivetrain" in combination with the vehicle body will be needed in future to reach selective, balanced compromises for conflicting objectives. Consequently, there will be a demand for even more intense, i.e. more detailed, test-rig and simulation investigations.

4 REFERENCES

(1) Biermann, J.-W., Hagerodt, B. — Investigation of Clonk Phenomenum in Vehicle Transmissions - Measurement, Modelling and Simulation. Multi Body Dynamics: Monitoring and Simulation Techniques. Conference Volume, Bradford, March 1997

(2) Kelly, P., Rahnejat, H. — Clutch Pedal Dynamic Noise and Vibration Investigation. Multi Body Dynamics: Monitoring and Simulation Techniques. Conference Volume, Bradford, March 1997

(3) Hagerodt, B. — Untersuchungen zu Lastwechselreaktionen frontgetriebener Pkw. Dissertation, Aachen, April 1998

© With Author 1998

C553/037/98

Simple modelling and analysis for crankshaft three-dimensional vibrations under firing conditions

H OKAMURA
Sophia University, Tokyo, Japan
T MORITA
NOM-MEGULASTIK Co Limited, Fujisawa, Japan

SYNOPSIS

The three-dimensional vibrations of an engine crankshaft system under running conditions were investigated by simple modeling and analysis by dynamic stiffness matrix method. In the analyses, we took account of the dynamic behavior of cylinder block, oil film, and torsional damper. To simplify the analyses, the crankshaft was idealized by a set of jointed structures consisting of simple round rods and simple beam blocks of rectangular cross-section. The main journal bearings were idealized by a set of linear springs and dash-pots, and the flywheel was idealized by FEM model. Then the dynamic stiffness matrix was derived in closed form for each constituent member.

To eliminate the complicated FEM analysis for a cylinder block of complicated structure[1,2], we derived the dynamic stiffness matrix from the inverse matrix of the compliance matrix. Here, the compliance matrix was derived in analytical form from the modal parameters obtained from a series of hammering tests. Finally, the dynamic stiffness matrix was constructed for the total engine system which consists of the crankshaft system and the cylinder block.

The three-dimensional vibrations of the crankshaft system and cylinder block surface vibrations near the main bearings under firing conditions were calculated for an automobile diesel engine, in which five kinds of solid pulleys, each with different masses and moments of inertia, and a torsional damper were attached to the crankshaft. The calculated results were compared with the experimental results.

1 Introduction

A crankshaft system is the most important moving part which has complicated vibration behavior. To develop a quiet and smooth running engine, early in the design stage, the engine designers should know the precise vibration behavior of the crankshaft system for the operating engine.

Several methods have been reported to analyze the vibration behavior of the total engine system, which includes the crankshaft, engine block(cylinder block), piston, and connecting rod. In most of these methods , the crankshaft system and cylinder block were idealized by a huge number of FEM models. Accordingly, huge computation times are required to solve a huge number of simultaneous equations of motion stepwise for very small time intervals.

Such analysis by FEM models seems to be impractical for a preliminary study in the engine design stage. To simplify the analyses, Okamura et al.[3] proposed a simple but efficient modeling method and a computation approach for the analysis of an engine crankshaft system. The modeling method and the analyses by the dynamic stiffness matrix method[DSMM] in the frequency domain are shown in reference[3].

The crankshaft system is supported by the oil film filled between the crank journals and the main bearings, a part of the cylinder block. To eliminate the complicated FEM analysis for a cylinder block, we derived the dynamic stiffness matrix DSM of the cylinder block by an approach which applied the experimental model analysis. Then we established the DSM for the total engine system in which the crankshaft system was assembled in the cylinder block as shown in Fig. 1. The three-dimensional vibration behavior was calculated for the rotating crankshaft system and for the cylinder block which remains stationary.

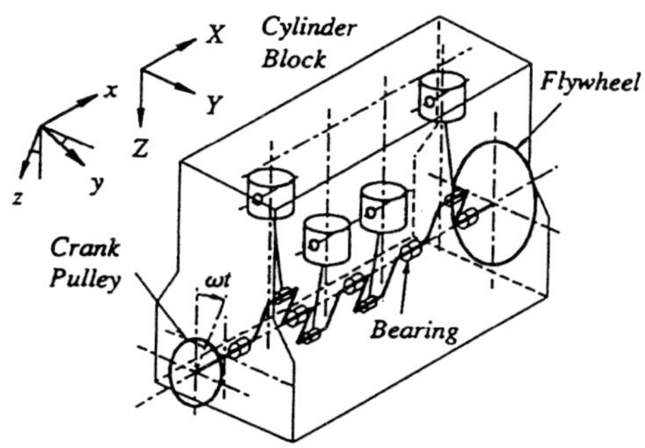

Fig. 1 Engine system

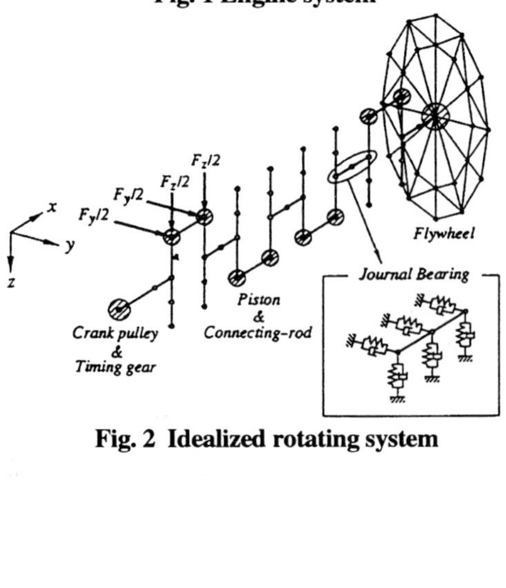

Fig. 2 Idealized rotating system

2 Modeling of crankshaft system

We applied a simple but effective modeling method[3,4,5], for a crankshaft system of a four-cylinder in-line engine under firing conditions, as shown in Fig. 1. That idealized model is shown in Fig.2. The $x-y-z$ axes are attached to the crankshaft, as shown in Figs. 1 and 2. The $X-Y-Z$ axes are attached to the cylinder block, as shown in Fig. 1.

The details of the modeling method can be summarized as follows.

- *The crankshaft body was* idealized by a set of jointed structures consisting of elements of round bars and beam blocks of rectangular cross section preserving each joint at its original position. *The front pulley and the crankshaft gear* were idealized by a set of lumped masses and moments of inertia attached at their centers of gravity.

- *The flywheel* was idealized by a set of FEM models consisting of triangular plate elements; since the flywheel was relatively thin, it has a relatively low stiffness in the x direction and about the y and z axes. However, the original mass and moments of inertia of the flywheel are preserved in the three directions of y, z, and θ_x, since the flywheel has sufficient rigidity in these directions.

- *The big end mass of the connecting rod* was idealized by a set of two equal rotating masses, attached at the two ends of the crankpin.

- The reciprocating masses consist of the masses of *the piston and the piston pin* and the mass of the small end of the connecting rod. To take account of the mean kinetic energy that would be induced by the reciprocating masses under running conditions[6], one quarter of the total reciprocating masses was attached at the two ends of the crankpin. Then we assumed that the inertia force due to the reciprocating masses is applied with the gas force directly at the two ends of the crankpin, as shown in Fig. 2.

- *The crank journal bearings* were idealized by three sets of linear springs and dash-pots in the y and z directions. They were attached at three points located equidistantly along the journal axis, to take account of the oil film pressure distribution along the x axis. The spring stiffness and dash-pot damping coefficients were calculated from a data book published by JSME[7] applying the Sommerfeld numbers of each bearing under operating conditions.

3 Calculation of crankshaft vibrations in rotating coordinate system

The vibration behavior of a rotating crankshaft system under firing conditions can be analyzed by using either (1) a stationary coordinate system attached to the engine body, or (2) a rotating coordinate system attached to the crankshaft body. For the former, since a crankshaft body has some non-isometric properties, e.g., each crank-throw has different bending stiffness in the two orthogonal $x-y$ and $x-z$ planes, time-variable properties can be introduced in the equations of motion. Therefore, the solution only can be obtained by numerical methods in the time domain. However for the latter, the equations of motion can be derived as a system of time-invariable properties, and the vibration behavior of the crankshaft system can be calculated directly in the frequency domain.

We can derive the equations of motion and the time-invariable characteristic matrices of the multidegrees of freedom for the crankshaft system referred to a rotating coordinate system.

3.1 Vectors of displacement and location

To consider the vibration behavior of a rotating system, one can idealize the rotating system as shown in Fig. 3. In the idealized system, the vibration system is attached to the $x-y-z$ coordinate system which is rotating about the X axis of the stationary $X-Y-Z$ coordinate system(SCS) with a constant angular velocity ω.

For convenience, we further assumed that the axis $O-x$ of the rotating coordinate system(RCS) $x-y-z$ coincides with the $O-X$ axis of the (SCS) $X-Y-Z$. For the oil film of crankshaft main journal bearings is so thin, and the axial movement of the crankshaft system is constrained by the crankshaft thrust bearings.

For a joint i in the vibration system, we can define the displacement vectors $\{u\}_{si}$ and $\{u\}_{ri}$, respectively, for the SCS and the RCS as shown in Fig. 3.

$$\left.\begin{aligned}\{u\}_{si} &= [u_X, u_Y, u_Z, \theta_X, \theta_Y, \theta_Z]_i^T \\ \{u\}_{ri} &= [u_x, u_y, u_z, \theta_x, \theta_y, \theta_z]_i^T\end{aligned}\right\} \qquad (1)$$

Here, as shown in Fig. 3, u_X, u_Y, and u_Z are the translating displacements along the X, Y, and Z axes and θ_X, θ_Y, and θ_Z are angular displacements about X,

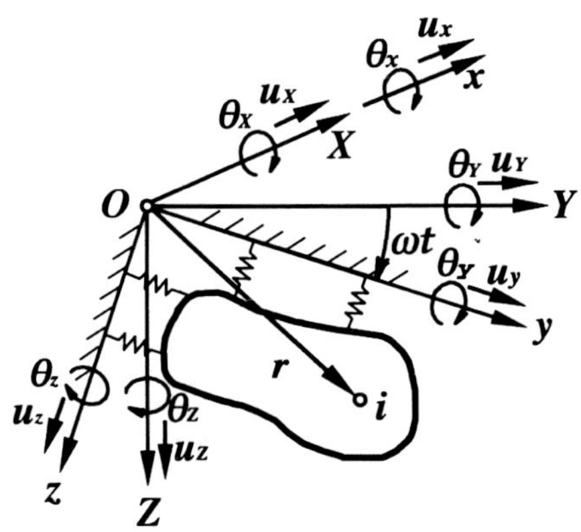

Fig. 3 Idealized rotating system

Y, and Z axes in the SCS. Similarly, u_x, u_y, u_z, θ_x, θ_y, and θ_z are the translating and angular displacements in the SCS. Here, $\{u\}_{si}$ designate the absolute displacements, but $\{u\}_{ri}$ designate the displacements measured in the $x-y-z$ coordinate system of the RCS.

From the assumption that the axes $O-x$ and $O-X$ are coaxial, the translating displacement u_X must coincide with u_x. However, since the angular displacement θ_X consists of the angular displacement θ_x about the x axis and the rotating angle ωt of the RCS with respect to the X axis, one can write as follows:

$$\theta_X = \theta_x + \omega t \qquad (2)$$

Nextly, one can define the locational vectors $\{r_0\}_{si}$ and $\{r_0\}_{ri}$ for the joint i without any displacements as follows:

$$\left.\begin{aligned} \{r_0\}_{si} &= [P_X, P_Y, P_Z,\ 0,\ 0,\ 0]_i^T \\ \{r_0\}_{ri} &= [P_x, P_y, P_z,\ 0,\ 0,\ 0]_i^T \end{aligned}\right\} \qquad (3)$$

Here $\{r_0\}_{si}$ is a time-variant vector; however, the $\{r_0\}_{ri}$ is a time-invariant, namely, a constant vector fixed to the RCS.

Then, the locational vectors $\{r\}_{si}$ and $\{r\}_{ri}$ which include joint displacements can be described as follows:

$$\left.\begin{aligned} \{r\}_{si} &= \{r_0\}_{si} + \{u\}_{si} \\ \{r\}_{ri} &= \{r_0\}_{ri} + \{u\}_{ri} \end{aligned}\right\} \qquad (4)$$

If the angular displacements θ_x, θ_y, and θ_z are very small, the relationship between the $\{r\}_{si}$ and $\{r\}_{ri}$ can be described as follows from the coordinate transformation, taking account of Eq. (2):

$$\{r\}_{si} = [L]_e\{r\}_{ri} + \omega t\{\alpha\}_e \qquad (5)$$

where

$$[L]_e = \begin{bmatrix} 1 & 0 & 0 & 0 & 0 & 0 \\ 0 & \cos\omega t & -\sin\omega t & 0 & 0 & 0 \\ 0 & \sin\omega t & \cos\omega t & 0 & 0 & 0 \\ 0 & 0 & 0 & 1 & 0 & 0 \\ 0 & 0 & 0 & 0 & \cos\omega t & -\sin\omega t \\ 0 & 0 & 0 & 0 & \sin\omega t & \cos\omega t \end{bmatrix}$$

$$\{\alpha\}_e = [0,0,0,1,0,0]^T$$

3.2 Vectors of velocity and acceleration

The vector of the absolute velocity for a joint i can be derived as $\{v\}_{si}$ in terms of the components for $X-Y-Z$ axes; similarly the velocity vector $\{v\}_{ri}$ can be derived in terms of the components for $x-y-z$ axes. Then the absolute acceleration vector $\{a\}_{si}$ for $X-Y-Z$ axes and $\{a\}_{ri}$ for $x-y-z$ axes can also be derived as follows.

The absolute vector $\{v\}_{si}$ and the absolute acceleration vector $\{a\}_{si}$ can be derived simply by differentiating the locational vector $\{r\}_{si}$ with respect to time t. Therefore, from Eq. (5), we can obtain

$$\{v\}_{si} = \{\dot{r}\}_{si}$$
$$= [\dot{L}]_e \{r\}_{ri} + [L]_e \{\dot{r}\}_{ri} + \omega\{\alpha\}_e \qquad (6)$$
$$\{a\}_{si} = \{\ddot{r}\}_{si}$$
$$= [\ddot{L}]_e \{r\}_{ri} + 2[\dot{L}]_e \{\dot{r}\}_{ri} + [L]_e \{\ddot{r}\}_{ri} \qquad (7)$$

From the coordinate transformation, the $\{v\}_{si}$ and $\{a\}_{si}$ can be described by $\{v\}_{ri}$ and $\{a\}_{ri}$ as follows:

$$\{v\}_{si} = [L]_e \{v\}_{ri} \qquad (8)$$

$$\{a\}_{si} = [L]_e \{a\}_{ri} \qquad (9)$$

From Equations (6),(7),(8), and (9) one can derive the following equations:

$$\{v\}_{ri} = [G]_e \{v\}_{si}$$
$$= [G]_e [\dot{L}]_e \{r\}_{ri} + \{\dot{r}\}_{ri} + \omega\{\alpha\}_e \qquad (10)$$
$$\{a\}_{ri} = [G]_e \{a\}_{si}$$
$$= [G]_e [\ddot{L}]_e \{r\}_{ri} + 2[G]_e [\dot{L}]_e \{\dot{r}\}_{ri} + \{\ddot{r}\}_{ri} \qquad (11)$$

where $[G]_e = [L]_e^{-1} (= [L]_e^T)$.

As $\{r_0\}_{ri}$ is constant and $\{\dot{r}\}_{ri} = \{\dot{u}\}_{ri}$, $\{\ddot{r}\}_{ri} = \{\ddot{u}\}_{ri}$, then,

$$\{v\}_{ri} = [G]_e [\dot{L}]_e \{\{r_0\}_{ri} + \{u\}_{ri}\}$$
$$+ \{\dot{u}\}_{ri} + \omega\{\alpha\}_e \qquad (12)$$

$$\{a\}_{ri} = [G]_e [\ddot{L}]_e \{\{r_0\}_{ri} + \{u\}_{ri}\}$$
$$+ 2[G]_e [\dot{L}]_e \{\dot{u}\}_{ri} + \{\ddot{u}\}_{ri} \qquad (13)$$

Here, the expressions for the vectors in complicated matrix form shown in Equations (6)–(12) might be understood easily from the following similarity. In classical dynamics, when a particle has a radius vector $\boldsymbol{r}_f$ in the fixed coordinate system (FCS) and a radius vector $\boldsymbol{r}_m$ in the moving coordinate system (MCS), while at the same time the MCS has a translational velocity $\dot{\boldsymbol{R}}$ and an angular velocity $\boldsymbol{\omega}$ with respect to the FCS, the particle has the following acceleration $\boldsymbol{a}_f$ in the FCS[8,9].

$$\boldsymbol{a}_f = \ddot{\boldsymbol{R}} + \boldsymbol{\omega} \times (\boldsymbol{\omega} \times \boldsymbol{r}_m) + \dot{\boldsymbol{\omega}} \times \boldsymbol{r}_m + \ddot{\boldsymbol{r}}_m + 2\boldsymbol{\omega} \times \dot{\boldsymbol{r}}_m \qquad (a)$$

If we have $\ddot{\boldsymbol{R}} = 0$ and $\dot{\boldsymbol{\omega}} = 0$, $\boldsymbol{a}_f$ can be expressed as follows:

$$\boldsymbol{a}_f = \boldsymbol{\omega} \times (\boldsymbol{\omega} \times \boldsymbol{r}_m) + 2\boldsymbol{\omega} \times \dot{\boldsymbol{r}}_m + \ddot{\boldsymbol{r}}_m \qquad (a')$$

Recall that the term $\boldsymbol{\omega} \times (\boldsymbol{\omega} \times \boldsymbol{r}_m)$ represents the centrifugal acceleration, the term $2\boldsymbol{\omega} \times \dot{\boldsymbol{r}}_m$ represents the Coriolis acceleration, and the term $\ddot{\boldsymbol{r}}_m$ represents the acceleration of the particle with respect to the MCS. By comparing Eq. (7) with Eq. (a'), we may notice that each of the terms in Eq. (7) represents the vector of the centrifugal acceleration, the Coriolis acceleration, and the relative acceleration, too.

3.3 Equation of motion and characteristic matrices

When the exciting forces $\{f\}_r$ for the RCS are applied on the vibration system, the equations of motion for the rotating system shown in Fig.3 can be described as follows:

$$[M]\{a\}_r + [C]\{v\}_r + [K]\{u\}_r = \{f\}_r \qquad (14)$$

Here $[M]$, $[C]$, and $[K]$ are the mass, damping, and stiffness matrices for the $x-y-z$ coordinate system. And $\{a\}_r$, $\{v\}_r$, and $\{u\}_r$ are the vectors for the total vibration system, in which the expressions of $\{a\}_{ri}$, $\{v\}_{ri}$, and $\{u\}_{ri}$ were expanded from a simple joint to a total engine system. Substituting Eqs. (12) and (13) into Eq. (14), one can obtain the next equations:

$$[M]\{\ddot{u}\}_r + \{[C]+2[M][G][\dot{L}]\}\{\dot{u}\}_r + \{[K]+[C][G][\dot{L}]+[M][G][\ddot{L}]\}\{u\}_r =$$
$$\{f\}_r - \omega[C]\{\alpha\} - \{[M][G][\ddot{L}]+[C][G][\dot{L}]\}\{r_0\}_r \qquad (15)$$

where

$$[L] = \begin{bmatrix} [L]_e & & 0 \\ & \ddots & \\ 0 & & [L]_e \end{bmatrix}$$

$$[G] = [L]^{-1} = [L]^T, \qquad \{\alpha\} = \left[\{\alpha\}_e^T, \cdots, \{\alpha\}_e^T\right]^T$$

If the rotating system is rotating with a constant angular velocity ω as we assumed previously, we can see that the next equation will hold:

$$[G][\dot{L}] = \omega[E] \qquad (16)$$

$$[G][\ddot{L}] = -\omega^2[I_0] \qquad (17)$$

where

$$[E] = \begin{bmatrix} [E]_e & & 0 \\ & \ddots & \\ 0 & & [E]_e \end{bmatrix}, \qquad [I_0] = \begin{bmatrix} [I_0]_e & & 0 \\ & \ddots & \\ 0 & & [I_0]_e \end{bmatrix}$$

$$[E]_e = \begin{bmatrix} 0 & 0 & 0 & 0 & 0 & 0 \\ 0 & 0 & -1 & 0 & 0 & 0 \\ 0 & 1 & 0 & 0 & 0 & 0 \\ 0 & 0 & 0 & 0 & 0 & 0 \\ 0 & 0 & 0 & 0 & 0 & -1 \\ 0 & 0 & 0 & 0 & 1 & 0 \end{bmatrix}, \quad [I_0]_e = \mathrm{diag}\{0,1,1,0,1,1\}$$

and these matrices are constants.

Therefore, we can derive the equation of motion by using the vectors for the RCS:

$$[M]_r\{\ddot{u}\}_r + [C]_r\{\dot{u}\}_r + [K]_r\{u\}_r = \{f\}_r - \omega[C]\{\alpha\} + \{\omega^2[M][I_0] - \omega[C][E]\}\{r_0\}_r \tag{18}$$

where

$$[M]_r = [M] \tag{19}$$

$$[C]_r = [C] + 2\omega[M][E] \tag{20}$$

$$[K]_r = [K] + \omega[C][E] - \omega^2[M][I_0] \tag{21}$$

We call $[M]_r$, $[C]_r$ and $[K]_r$ the characteristic matrices of the mass, damping, and stiffness matrix for the rotating system in the RCS. The characteristic matrices $[M]_r, [C]_r, [K]_r$ are constant matrices, because $[M], [C], [K]$ are constants and $[E], [I_0]$ are also constants. And the second and third terms on the left side of Eq. (18) are constant vectors because $\{\alpha\}$ and $\{r_0\}_r$ are constants.

If the equations of motion (18) and (20) were applied to a rotating rigid rotor, the gyroscopic terms of $J_p\omega\dot{\theta}_y$ and $J_p\omega\dot{\theta}_z$ would appear in the equations of motion.

In this research, the characteristic matrices $[M], [C], [K]$ for the idealized crankshaft system shown in Fig.2 were derived by the superposition of the DSM for each constituent member. Where the mass matrix and stiffness matrix were derived by the method in FEM for the beam and plate elements. The stiffness and damping matrices for the crank journal bearings were calculated from the Sommerfeld numbers for engine operating conditions.

4 Derivation of the DSM(dynamic stiffness matrix) for cylinder block

4.1 Reduction of the DSM of the cylinder block

As shown in Fig. 1 and 2, the crankshaft system is supported by the oil film filled between the crank journals and the main bearings. Since the main bearings are a part of the cylinder block, we must investigate the influence of the dynamic behavior of the cylinder block on the crankshaft vibrations. To simplify the analysis as much as possible, we drastically reduced the size of the DSM of the cylinder block.

For a structure of n DOF(degrees of freedom), one can partition the displacement vector, the force vector and the DSM in the following form, when only p of the forces: $f_1, f_2, \cdots, f_p$ are exerted to the structure; here $p < n$ [11].

$$\begin{bmatrix} [D] & [F] \\ [F]^T & [E] \end{bmatrix} \begin{Bmatrix} x_1 \\ \vdots \\ x_p \\ x_{p+1} \\ \vdots \\ x_n \end{Bmatrix} = \begin{Bmatrix} f_1 \\ \vdots \\ f_p \\ 0 \\ \vdots \\ 0 \end{Bmatrix} \tag{22}$$

By expanding the Eq. (22) and solving it for the $\{x_{p+1}, \cdots, x_n\}^T$, we obtain

$$\{x_{p+1}, \cdots, x_n\}^T = -[E]^{-1}[F]^T\{x_1, x_2, \cdots, x_p\}^T \tag{23}$$

Substituting Eq. (23) into Eq. (22), we obtain

$$\left([D]-[F][E]^{-1}[F]^T\right)\begin{Bmatrix} x_1 \\ \vdots \\ x_p \end{Bmatrix}=\begin{Bmatrix} f_1 \\ \vdots \\ f_p \end{Bmatrix} \tag{24}$$

Here, one can see that for a structure of n DOF, the DOF can be reduced from n to p if only p of the forces are exerted to it. Therefore, even for the cylinder block of complicated structure, since only a limited number of excitation forces would be exerted to the main bearing positions, due to the crankshaft system, one can drastically reduce the size of the DSM for it.

For the cylinder block shown in Fig. 1, since the excitation forces would be exerted on five of the main bearings, in the x, y and z directions, one can reduce the size of the DSM to only 15×15.

4.2 Derivation of the DSM by experimental modal analysis.

To derive the DSM for the cylinder block shown in Fig. 1, we measured the transfer functions between the main bearings, from a series of hammering tests. The compliance transfer functions $G_{ij}(\omega)$ were calculated from the experimental values of the modal parameters: Ω_r, m_r, k_r, c_r and ϕ_r for each mode.

For the r th mode in the frequency range ω_p to $\omega_{p'}$, we applied the following expressions:

$$G_{ij}(\omega)=\sum_{r=p}^{p'}\frac{\dfrac{\phi_{ri}\phi_{rj}}{k_r}}{1-\left(\dfrac{m_r}{k_r}\right)\omega^2+j\left(\dfrac{c_r}{k_r}\right)\omega}+\frac{C}{\omega^2}+D \tag{25}$$

where

$$\frac{C}{\omega^2}=-\frac{1}{\omega^2}\sum_{r=1}^{p-1}\frac{\Omega_r^{\,2}}{k_r/(\phi_{ri}\phi_{rj})} \tag{26}$$

$$D=\sum_{r=p'+1}^{n}\frac{1}{k_r/(\phi_{ri}\phi_{rj})} \tag{27}$$

C is the inertia restraint, and D is the residual compliance.

By assembling each term of $G_{ij}(\omega)$ together, one can establish the compliance matrix $[G(\omega)]$. Here, recalling that

$$\{X\}=[G(\omega)]\{F\},\quad \{F\}=[K(\omega)]\{X\} \tag{28}$$

the DSM $[K(\omega)]$ can be calculated as follows:

$$[K(\omega)]=[G(\omega)]^{-1} \tag{29}$$

5 Applications to forced vibration under firing conditions

By superposing the DSMs of the crankshaft system, the cylinder block, and the oil film, one can establish the DSM for an operating engine.

If the n th harmonic exciting force vector $\{F_n\}e^{jn\omega t}$ is applied to such vibration system, the n th order harmonic displacement vector $\{u_n\}$ induced in the system can be derived by solving the following equation for the steady-state condition:

$$[K_{dyne}(n\omega)]\{u_n\} = \{F_n\} \quad (30)$$

where $[K_{dyne}(n\omega)]$ is the dynamic stiffness matrix for the n th harmonic vibration of the system.

Similarly, in the RCS, the n th harmonic displacement vector for the crankshaft system which is rotating with a constant angular velocity ω and is under steady-state condition can be calculated by replacing the n th harmonic component of $\{f\}_r$ in Eq. (18) by the $\{F_n\}$ in Eq. (30), and by defining the dynamic stiffness matrix as follows:

$$[K_{dyne}(n\omega)] = ([K]_r - (n\omega)^2[M]_r) + jn\omega[C]_r \quad (31)$$

6 Experiments

In the experiments, the crankshaft vibration behaviors for a four-cylinder in-line turbocharged diesel engine (4-ϕ79×86) were measured when the masses and moments of inertia of the front pulley were changed step by step.

The layout of the experiment is shown in Fig. 4. We measured simultaneously:

- the noise level at 1 m distance from the engine's left side,
- the horizontal accelerations of cylinder block surfaces at the locations near each of the main bearings,
- the axial velocity (x - dir) of the crankshaft center on the front side.

The masses and moments of inertia were changed by attaching additional rings to the base pulley, as shown in Fig. 5. The masses and moments of inertia of the experimental front pulley are shown in Table 1. In this research, the experimental pulleys are designated Type 0 pulley to Type 4 pulley, from the lightest pulley to the heaviest pulley.

Furthermore, we measured the vibration behavior when a torsional damper was attached to the crankshaft.

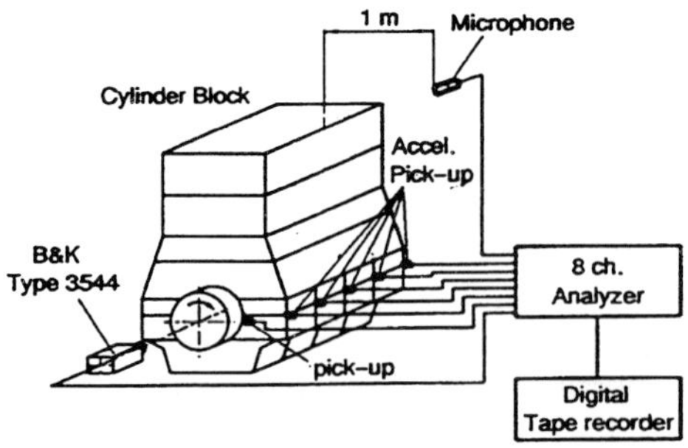

Fig. 4 Experiment layout

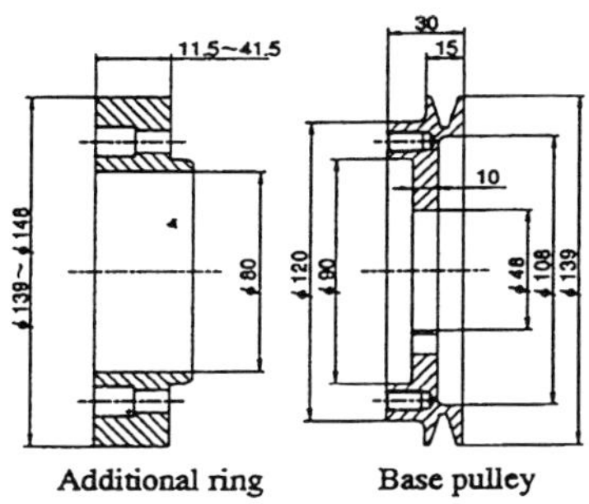

Fig. 5 Experiment pulley

Table 1 Masses and moments of inertia of experimental pulleys

	Mass [kg] [ratio]	Moment of Inertia [kg·m^2]		
		J_x [ratio]	J_y [ratio]	J_z [ratio]
Type 0 Pulley (Base Pulley)	1.59 [—]	4.347×10^{-3} [—]	2.231×10^{-3} [—]	2.231×10^{-3} [—]
Type 1 Pulley	2.59 [1.63]	7.448×10^{-3} [1.71]	3.792×10^{-3} [1.70]	3.792×10^{-3} [1.70]
Type 2 Pulley	3.59 [2.25]	1.065×10^{-2} [2.45]	5.600×10^{-3} [2.51]	5.600×10^{-3} [2.51]
Type 3 Pulley	4.64 [2.93]	1.500×10^{-2} [3.45]	7.794×10^{-3} [3.49]	7.794×10^{-3} [3.49]
Type 4 Pulley	5.65 [3.55]	1.855×10^{-2} [4.27]	9.904×10^{-3} [4.43]	9.904×10^{-3} [4.43]

7 Experimental results and calculated results

7.1 Horizontal vibration at the cylinder block surface

In Figs. 6(a),(b), and (c), the calculated results(for a solid cylinder block) of the 7th order components of the horizontal acceleration at the center of No.4 crank journal are shown, with the experimental results of the overall and 7th order components, for three cases: when Type 0 and Type 1 pulleys were attached, and when a torsional damper was attached.

From Figs. 6(a) and (b), one can see that 1) for the Type 1 pulley, the calculated value of the 7th order component was a little larger than the experimental result; however, there is good agreement for the Type 1 pulley, and 2) there is good agreement between the calculated and experimental engine speeds at which the peaks of 7th order components appeared.

Since the $(n+1)$th and $(n-1)$th order bending vibrations on the cylinder block surface can be attributed to the nth order crankshaft vibration in which torsion and "bending in the $x-y$ plane"(see Fig. 2) would be induced simultaneously[4,5], the dominant 7th order components in Fig. 6(a) and (b) must be induced by the 6th or 8th order crankshaft vibrations.

In Fig. 6(c), one can see drastic reduction of the 7th order component due to the reduction of the crankshaft torsional vibration by a torsional damper.

7.2 Axial vibration of crankshaft

The calculated 6th order harmonic components of the pulley center vibrations in axial(x or X) direction are shown in Fig. 7 with the measured overall and predominant 6th order harmonic components. Figs. 7(a) and (b) show the vibration behaviors of the crankshaft with Type 0 pulley and Type 1 pulley, respectively.

In Figs. 7(a) and (b), one can see good agreement between the calculated and the measured engine speeds at which the peaks of the 6th order components appeared, and the calculated and experimental amplitudes for the Type 1 pulley. However the calculated amplitude for the Type 0 pulley was greater than the measured amplitude. We think this might occur because we did not consider the dynamic behavior of the thrust bearings in our idealized vibration model of the crankshaft system. If more accuracy of the modeling and analysis were required, we should consider the dynamic behavior of the thrust bearings in the modeling of the crankshaft system.
The axial vibrations of the crankshaft pulley center are shown in Fig. 7(c) for the case when a damper pulley was attached. We can see that the axial vibration could also be reduced slightly when the torsional damper pulley was attached to reduce the crankshaft torsional vibration.

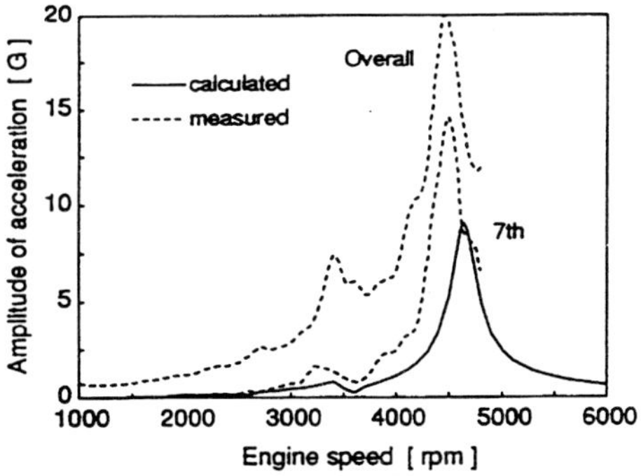

(a) with Type 0 pulley

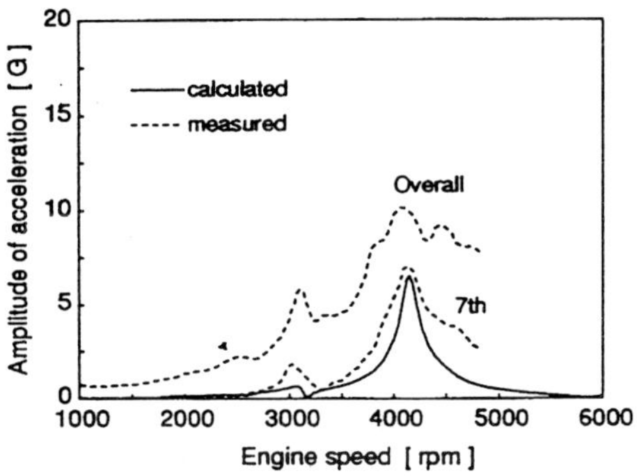

(b) with Type 1 pulley

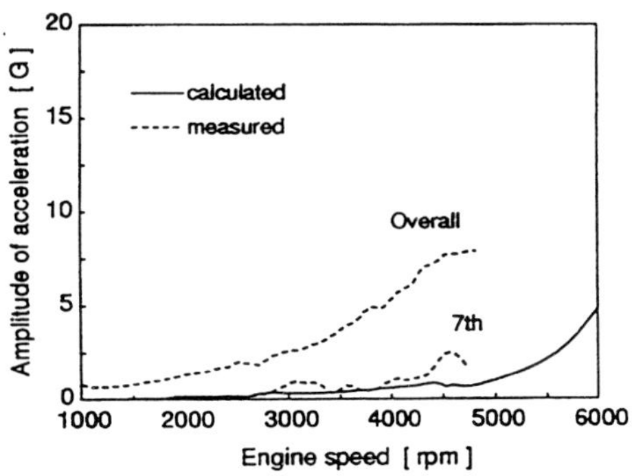

(c) with damper pulley

Fig. 6 Experimental results of the cylinder block surface vibration at No.4 main bearing position, and calculated results of the horizontal vibration at No.4 main journal center

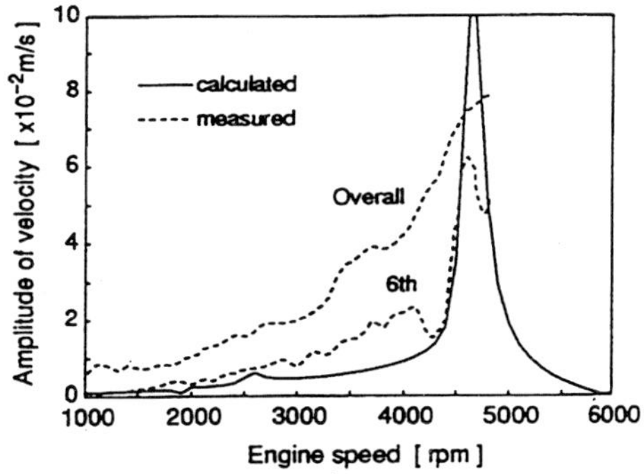

(a) with Type 0 pulley

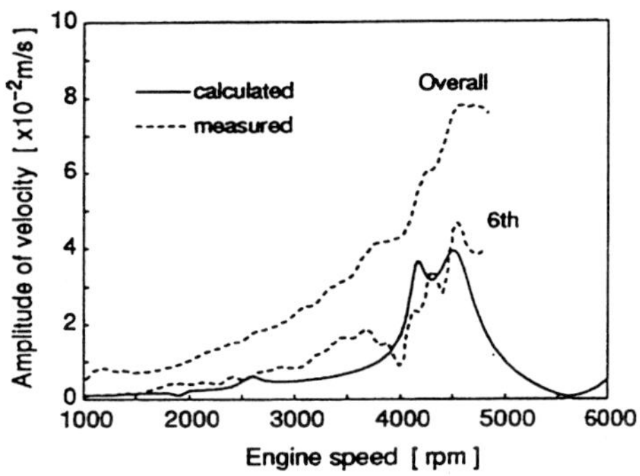

(b) with Type 1 pulley

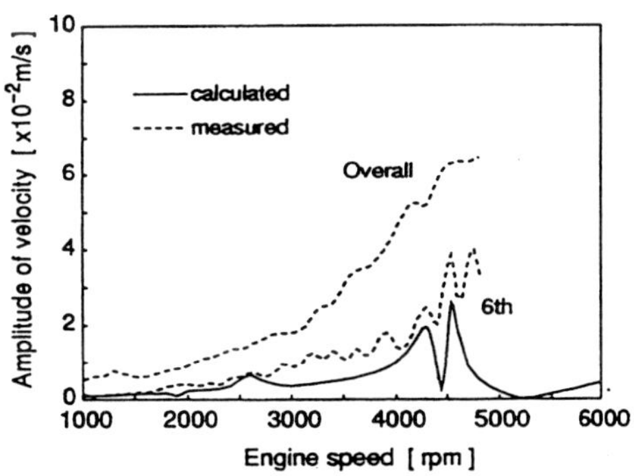

(c) with damper pulley

Fig. 7 Calculated and experimental results of the pulley center vibration in axial direction

7.3 Axial and torsional vibration

Figs. 8(a) and (b) show the calculated 6th order harmonic components of the axial and the torsional vibrations at the crankshaft front end, for different dimensions of the front pulley. In Fig. 8, the subscripts for the Type 0 pulley designate the percentage of the added or reduced masses and moments of inertia with respect to those of the Type 0 pulley, i.e., Type $0_{+25\%}$ designates a virtual pulley which has the masses and moments of inertia larger than those of the Type 0 pulley by 25%.

From Fig. 8(a), we can see the two resonance peaks for the axial vibration when the Type 1 pulley was attached. However, as the pulley was changed to the smaller ones step by step, each of the resonance engine speeds would approach the other one; then they will become only one peak, and finally the largest amplitude would be induced by the Type 0 pulley.

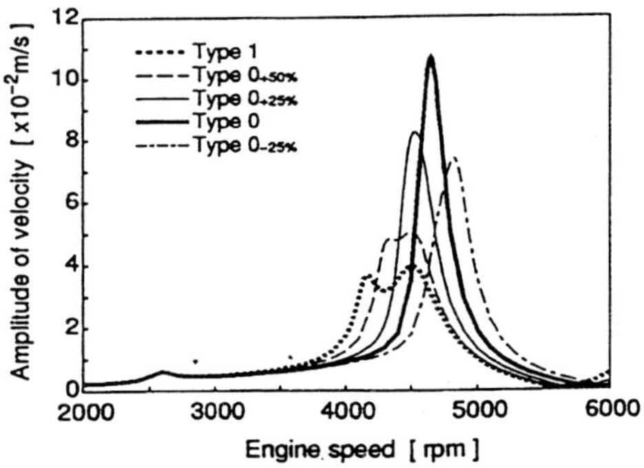

(a) 6th order harmonic components of axial vibration

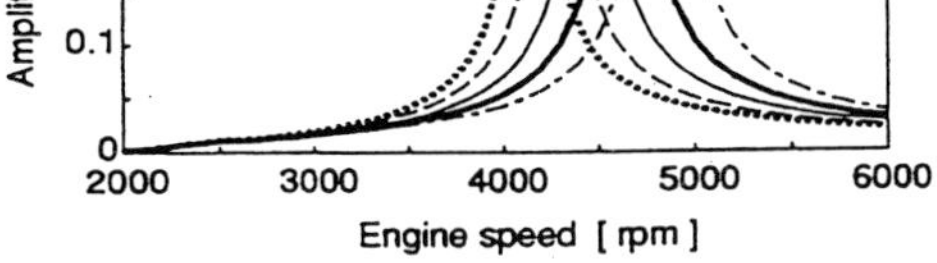

(b) 6th order harmonic components of torsional vibration

Fig. 8 Influences of the masses and moments of inertia on the front pulley

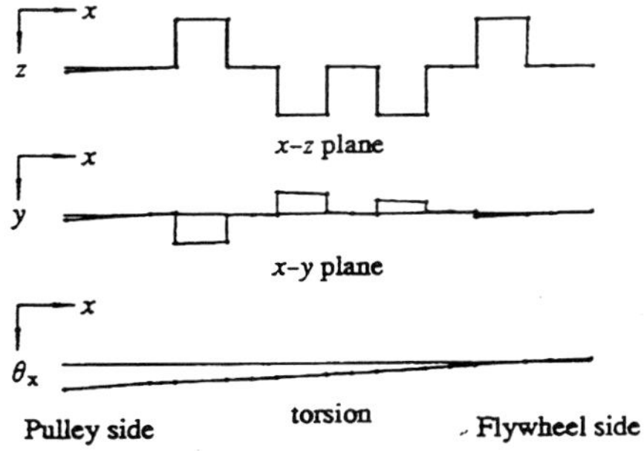

(a) at engine speed = 4160 rpm

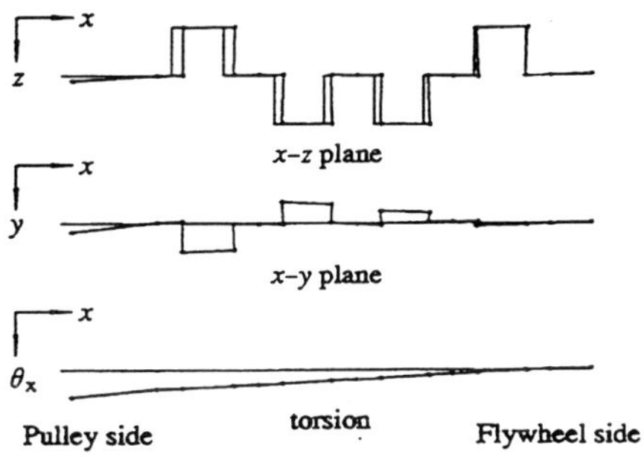

(b) at engine speed = 4520 rpm

Fig. 9 Resonance mode shapes of crankshaft system with Type 1 pulley, due to the 6th order harmonic forces

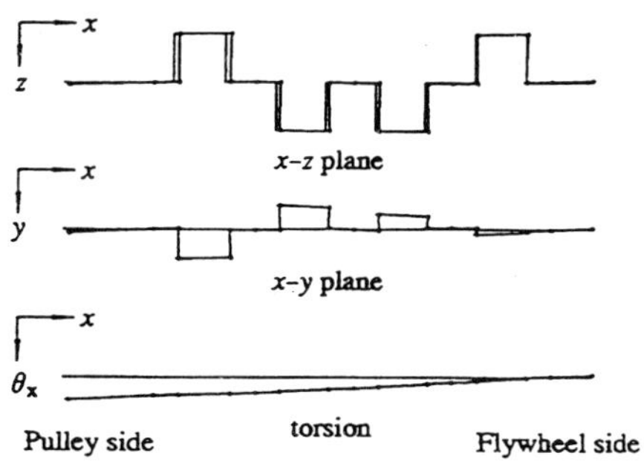

Fig. 10 Resonance mode shape of crankshaft system with Type 0 pulley at engine speed 4640 rpm, due to the 6th order harmonic forces

In Fig. 8(b), we can see that only one peak of the torsional vibrations would appear for each of the pulleys. In Figs. 8(a) and (b), we can also see that at engine speed 4160 rpm only one peak of the torsional vibrations would appear, simultaneously with the lower peak of the axial vibrations, when the Type 1 pulley was attached.

The mode shapes for each of the two peaks which appeared at engine speeds of 4160 rpm and 4520 rpm are shown in Figs. 9(a) and (b) for the Type 1 pulley. The mode shape which corresponds to the lower peak shown in Fig. 9(a) could be recognized as the typical torsional mode of a crankshaft, and the other one corresponding to the higher peak shown in Fig. 9(b) could be recognized as that of the longitudinal mode of a crankshaft.

In Fig. 10, the mode shape which corresponds to that of the engine speed 4640 rpm for the Type 0 pulley is shown. We can recognize that this mode shape consists of both the torsional and the longitudinal modes of a crankshaft.

From Figs. 8 to 10, we can draw the following conclusions for the crankshaft system we used for this research:

- In a crankshaft, axial vibration coupled with torsional vibration can occur simultaneously, although the two vibrations can also occur independently.
- When the resonances of the torsional mode and the longitudinal mode were induced simultaneously, the vibration induced in the axial direction of the rankshaft system is more significant than in the cases when the resonances were induced separately.
- The influences of the torsional vibrations on the axial vibrations were more predominant than those of the axial vibrations on the torsional vibrations.

The coupling behavior mentioned above can be recognized from the coupling terms included in Eqs. (18),(19),(20) and (21).

Here, we can explain the reason why the axial vibration could be reduced by the torsional damper pulley, as shown in Fig. 7(c). We think that it is because the axial vibrations coupled with torsional vibrations were slightly reduced by the attached damper pulley. However, since the axial vibration under the longitudinal vibration mode can not be reduced by the attached torsional damper pulley, some slight axial vibration remained there.

If predominant axial vibrations were induced in the crankshaft, the crankshaft would strike the bulk heads of cylinder block at the thrust bearings and engine structure noise would be induced. Here, if such axial vibrations were induced in coupling with the tortional vibration, the axial vibration could be reduced by the torsional damper, so as to reduce the torsional vibration only. However, if the axial vibration is induced in some other way, other kinds of vibration dampers must be attached to reduce it.

8 Conclusions

1. A new method was proposed to analyze the vibration behavior of the rotating crankshaft system under firing conditions in the RCS. In principle, this new analysis method in the RCS can be applied to the analysis of the vibration behaviors of a general rotating vibration system having non-isometric properties.

2. The experimental and calculated results showed good agreement. So the validity of the new analyzing method could be confirmed.

3. Under engine operating conditions, we found that *"in-plane bending and axial vibrations"* and *"out-of-plane bending and torsional vibrations"* could be coupled in *the planar-structure crankshaft*, as well as in the solid-structure crankshaft.

4. A numerical simulation showed that, if the resonance engine speed of the torsional vibration mode coincides with that of the longitudinal vibration mode, predominant axial vibration can be induced in the crankshaft simultaneously with the torsional vibration.

REFERENCES

[1] Martin, I. T., and Low, B., *Prediction of Crankshaft and Flywheel Dynamics*, I.Mech.E. Autotech 89, C382/046, November 1989.

[2] Priebsch, H. H., and Affenzeller, J., and Kuipers, G., *Structure Borne Noise Prediction Techniques,* SAE Paper 900019.

[3]Okamura, H., Shinno, A., Yamanaka, T., Suzuki, A., and Sogabe, K., *A Dynamic Stiffness Matrix Approach to the Analysis of Three-Dimensional Vibrations of Automobile Engine Crankshaft: Part 1 – Background and Application to Free Vibrations*, Proceedings of Vehicle Noise, ASME, NCA-Vol. 9, 1990.

[4]Morita, T., and Okamura, H., *A Dynamic Stiffness Matrix Approach to the Analysis of Three-Dimensional Vibrations of Automobile Engine Crankshaft: Part 2 – Application to Firing Conditions,* Proceedings of Vehicle Noise, ASME, NCA-Vol. 9, 1990.

[5]Okamura, H., and Morita, T., *Influence of Crankshaft-Pulley Dimensions on Crankshaft Vibrations and Engine-Structure Noise and Vibrations,* SAE Paper 931303.

[6]Den Hartog, J. P., *Mechanical Vibrations,* McGraw-Hill, 1956, pp.184-185.

[7]Japan Society of Mechanical Engineers, ed., *Data book on the static and the dynamic characteristics of the journal bearings,* Japan Industrial Publication, 1984. (in Japanese).

[8]Housner, G. W., and Hudson, D. E., *Applied Mechanics Dynamics 2nd Edition,* D. VAN NOSTRAND, 1959, pp38-47.

[9]Marion, J. B., *Classical Dynamics of Particles and Systems,* Academic Press, 1965, pp342-348.

[10] Morita, T., and Okamura, H., *Analysis of Crankshaft Three-Dimensional Vibrations in a Rotating Coordinate System,* SAE Paper 951292.

[11] Prentis, J. M., and Leckie, F. A., *Mechanical Vibrations: An Introduction to Matrix Methods,* Longmans,1963.

© With Author 1998

C553/020/98

The potential of rotors supported by floating bushing journal bearings in driving high-speed generators

C LEONTOPOULOS BEng, MSc, PhD, DIC, ACGI, AMIMechE, **D A ROBB** Btech, MSc, and **C B BESANT** BSc, PhD, DIC, FIMechE, FEng
Imperial College of Science, Technology, and Medicine, London, UK

Synopsis
The favourable features of floating/rotating bushing journal bearings are explored by experiments to examine their high-speed stabilising characteristics despite the limited knowledge of their behaviour in the non-laminar flow regime. It is argued that these bearings can possibly be used to support rotors which drive mechanical loads at very high speeds. In particular, existing high speed turbomachines - turbochargers with such bearing systems, could be used to directly drive generators at high speed, i.e., without the need of a gearbox. Thus, a new market can be exploited with the recent advent of the new, small-size, high-speed generators. An attempt is made here to explain the experimental findings from monitoring the vibration of a turbo-compressor engine as a driver unit, using simplified models and concepts studied by various researchers.

Nomenclature:

$C_{1,2}$	radial clearances
W	mass
C_1x	small displacement of the rotor
C_1y	small displacement of the journal
$C_1\eta_1$	small displacement of the floating bush
k_s	the bending stiffness of the shaft
σ	mass ratio
β	mean radial clearance ratio
ϕ	shaft flexibility index
δ	outer to inner diameter ratio of floating bush
λ_1	inner bearing number
t	time
μ	oil viscosity
p_m	mean pressure
r	radius
l	length
D	diameter

© IMechE 1998 C553/020

1. INTRODUCTION

The knowledge of the dynamic characteristics of journal bearings, acquired by analysis and by experiment, has been of great value in improving the understanding of the various types of vibration observed in turbomachinery, and thereby facilitating the work of the turbine engineer in dealing with such vibration problems. The numerical assessment of linear dynamic characteristics has been of much assistance in critical speed calculations, in spite of the gaps in the knowledge of characteristics of bearings in the non-laminar flow regime. Information about bearing dynamic characteristics from analysis of test rigs has, however, been of limited service to turbine engineers in designing bearings for resistance to particular types of instability, see (**1**), (**2**). Information as to how bearing design and operating conditions affect rotor stability has come mainly from observation and empirical modification of sets in service.

For a long time, bearing designers have known that a oil-lubricated plain cylindrical bearing has inherent stability problems. Improvements however, have been achieved by modifying bearing geometry. Elliptical bearings, bearings with grooves and lobes and bearings with mobile parts such as tilting pad bearings are generally more stable than plain cylindrical bearings. These improved features are associated with changes in circumferential lubricant flow which in all these improved designs has a lower rate, due to boundary conditions, than in plain cylindrical bearings. Thus, it is claimed that by decreasing the circumferential flow of the oil wedge(s), the stability threshold can be increased. One of these features is the existence of a floating or rotating bushing in the journal bearing. The technology is not uncommon to automotive turbo-chargers which operate at very high speeds of the order of 100,000 rpm and above and use extensively this design or similar variations.

This experimental investigation is based on the vibrational behaviour of a typical small size high-speed gas-turbine engine initially designed to provide compressed air. The rotor of the engine is supported on two journal bearings with rotating bushing. It is claimed that the construction of this bearing type leads to an increase of the stability threshold and also to improved cooling of the working bushing by flowing oil. The investigation has a two-fold aim: Primarily, to assess the validity of basic rotordynamics theory and models found in the literature for the understanding of the machine's dynamic behaviour and second to study the vibrational behaviour when the rotor is modified to drive a mechanical load, such as a generator at high speed. The tests are performed on real turbo-machinery rather than on an experimental rig and are planned to reveal the potential of the floating bushing for use in applications where a mechanical load is applied, or is driven directly at very high speeds.

The journal bearing with the rotating bushing or floating ring has been previously studied by several researchers from 1912 including Campbell. Since then a small number of studies have been made throughout the years, both of theoretical and experimental nature, by Tondl (**3**), Dworski (**4**), Lund (**5**) and others. Most recently, Tanaka and Hori (**6**) and Leung, et al (**7**) studied the behaviour of the floating ring bearing together with the characteristics of a cylindrical-spherical floating ring bearing. However, this subject is still under investigation as the requirements for high power-to-weight ratio turbocompressors-turbogenerators are continuously increasing.

C553/020 © IMechE 1998

It is generally the case that the rotors of high speed machines with floating bushing run at super-critical speeds and, in general, are not used to drive mechanical loads (such as gearboxes, generators), rather to provide compressed air or gas (e.g. turbochargers), mainly due to the demands of high speed and the relevant risk and reliability factors involved. However, with an understanding of the bearing system parameters and after eigen-value analysis, it can be shown that under certain conditions, the system's dynamic behaviour can be predictable and stable with low vibration. This is important if one considers that with the advent of the high-speed, high power-to-weight generators, a new market can be exploited, whereby existing high-speed turbomachines with such bearing systems can be used to direct drive such generators, removing the need for step-down gearboxes and reducing the overall volume and weight of the machine.

2. THE GAS-TURBINE ENGINE WITH THE FLOATING BUSHING BEARING SYSTEM

The engine under test was a 50 kW single stage radial compressor-radial turbine with a pressure ratio of 3:1. The bearings were cylindrical journal fluid film bearings, with a loose rotating bushing lying between the rotating shaft and the stationary bearing babbit. The design speed of the engine under test is 60,000 rpm but variations of speed can be achieved due to the nature of the engine's fuel system control. The oil pressure is regulated via a small oil pump and is adjusted to 3.5 bar during normal operation. The oil viscosity varies according to temperature and is 25 mm^2/s at 40 degrees C.

The engine rotor consists of a shaft and two disks (radial compressor, radial turbine) installed at each end of the shaft and the two journal bearings supporting the shaft between the two disks. The total length is about 27 cm. The rotor has been suitably modified with a coupling adaptor installed at the nose of the compressor (see **figure 1**), in order to drive a potential mechanical load such as a high-speed generator.

3. MATHEMATICAL MODEL

3.1. The dynamic characteristics of a fluid-film bearing

An oil-film bearing can be broadly described as a spring damper element, with the special property that the displacement and the force in general do not occur in the same direction. As a result of the peculiar force relationships, the possibility of instability arises, and large number of theoretical and experimental works appear in the literature, see Bently and Muszynska (**8**). The oil film has a spring action. The inner fluid film supports the external load acting on the journal while the outer fluid film is required to support both the external load and the weight of the floating ring. The steady-state solution of the floating ring bearing is mainly based on two equilibrium conditions; primarily, balance of forces i.e., the sum of the hydrodynamic force of the inner film plus the weight of the floating ring should be equal to the hydrodynamic force of the outer film. Secondly, balance of frictional moments on the rings; i.e., under steady state operation, the viscous moment acting on the inside surface of the ring should be equal to the viscous moment acting on the outside surface of the ring. The equations of motion have been solved by Tanaka and Hori (**6**), Leung (**7**), Lund (**5**), and others, in an attempt to match laboratory experimental data to theoretical data. In **figure 2**, there is an orthogonal coordinate system on a plane normal to the shaft. O, O_1, O_2 and S are centres of the fixed bearing, the floating bush, the journal and the rotor mass respectively. The equations of motion for such system, according to Tanaka (**9**) are:

$$\frac{2W_1}{g}C_1\ddot{x} + k_s(C_1x - C_1\xi_1) = 0 \qquad (1)$$

$$\frac{2W_1}{g}C_1\ddot{y} + k_s(C_1y - C_1\eta_1) = 0 \qquad (2)$$

$$k_s(C_1x - C_1\xi_1) = f_{\xi_1} \qquad (3)$$

$$k_s(C_1y - C_1\eta_1) = f_{\eta_1} \qquad (4)$$

$$\frac{W_2}{g}C_2\xi_2 + f_{\xi_2} - f_{\xi_1} = 0 \qquad (5)$$

$$\frac{W_2}{g}C_2\eta_2 + f_{\eta_2} - f_{\eta_1} = 0 \qquad (6)$$

where the dots represent differentiation with respect to time t, $2k_s$ is the bending stiffness of the shaft, (C_1x, C_1y), $(C_1x_1, C_1\eta_1)$ and $(C_2x_2, C_2\eta_2)$ represent small displacements of the rotor, the journal and the floating bush, respectively, from their equilibrium points. However, the fact that the oil film is represented by linear springs is merely an assumption and therefore the approximate nature of the equations (1)-(6) must be acknowledged. Furthermore, the elastic members of the system such as the floating ring -being thin-walled- are subject to deformation at high pressures. However, in this somewhat simplified model the flexible body motions have been also ignored.

The stability of the system depends on the following factors:

a. $\sigma = \frac{W_2}{W_1}$ = mass ratio (7)

b. $\beta = \frac{C_2}{C_1}$ = mean radial clearance ratio (8)

c. $\phi = \frac{W_1}{\frac{k_s}{C_1}}$ = shaft flexibility index (9)

d. $\delta = \frac{D_3}{D_2}$ = outer to inner diameter ratio of floating bush (10)

e. $\lambda_1 = \frac{\mu}{p_{m1}}\left(\frac{r_1}{C_1}\right)^2\left(\frac{l_1}{2r_1}\right)^2\frac{1}{2\pi}\sqrt{\frac{g}{C_1}}$ = inner bearing number (11)

According to Tanaka (**9**) theoretical predictions, a typical such system with low clearance ratio becomes unstable as soon as it starts rotating, then it becomes stable and remains stable until it reaches its design threshold of stability. Predictive stability maps for different sizes and variations of floating bushing bearings have been published, see also Kalogerakos, (**10**) and Dewhirst, et al (**11**).

C553/020 © IMechE 1998

Furthermore, it has been shown by Bently and Muszynska (**8**), (**15**), that the threshold of stability (for isotropic rotor supported in one rigid and one plain cylindrical oil-film bearing) is inversely proportional to the fluid circumferential average velocity ratio λ and to the square root of the rotor modal mass i.e.,

$$\Omega_{st} \approx \frac{1}{\lambda}\left(\sqrt{\frac{2k_s}{M}}\right) \quad (12)$$

assuming that the rotor stiffness, k_S , is much higher than the fluid film radial stiffness, k_B, although both factors, i.e. λ and k_B play an important role on rotor stability. Bently and Muszynska (**8**), (**15**), have calculated the variation of λ and k_B with the eccentricity ratio, ε. However, according to them, (**8**), (**15**), no laboratory test has been performed yet for observation of λ, although a few indirect measurements have been accomplished on plain cylindrical bearings. Their predicted variation with eccentricity ratio can be seen in **figure 3**. Hence, in a simplified physical interpretation, the higher eccentricities increase the bearing stiffness k_B nearly exponentially, which in turn results in the increase of the overall system stiffness thus shifting the system's natural frequencies higher. This will delay the start of whirl which effectively takes place at frequencies within the range of variation of the first mode of the system. In a similar way the higher eccentricity will reduce the λ, and thus the system's damping will increase or the "quadrature dynamic stiffness" will decrease by dissipating vibrational energy on the fluid. Therefore, according to their theory, the attachment of a rotating load on the rotor of the turbomachine, should provide a "friendly misalignment" which will exert a steady radial force on the shaft, increasing the eccentricity ratio and thus improving the machine's stability characteristics.

For the current case of a rotor supported on journal bearings with rotating bushing, the stiffness of the bearing can be approximated by a simple model with two springs in series. The individual spring stiffness K_{01}, K_{02} can be approximated by the Sommerfield Number method while the stiffness of the rotor can be determined with the aid of a finite element model confirmed by modal tests. Calculations have shown that the bearing stiffness for this particular bearing geometry for a range of L/D=0.5 to L/D=1 for the inner and outer fluid films, is much more sensitive to the clearance values C1, C2, than to the speed of rotation. The ratio C2/C1 i.e. outer over inner clearance, determines the relationship between the inner and the outer film eccentricity ratio which in turn determines the ratio between the speed of the bushing and the speed of the rotating shaft. It can also be shown that the ratio of the frictional moments between the inner and outer surfaces of the bushing is proportional to the fourth power of the ratio between the inner and outer radius, see Leung et al (**7**) i.e.,

$$\frac{M_1}{M_2} \approx \frac{R_2^4}{R_1^3}\frac{C_1}{C_2} \quad (13)$$

where M_1, M_2 are the frictional moments acting at inside and outside surfaces of the bushing and R represents the radii and C the clearances.

3.2. Modal analysis.

According to preliminary modal tests of the engine rotor in a free-free condition, conical and bending modes influenced by the bearing stiffness were expected to appear well within the operating range below the running speed. The results from modal tests at zero speed were then checked against those from a finite element model of the rotor **(12)**, **(13)**, **(14)**, **(16)**. Normally, the theoretical mass and stiffness matrices were adjusted in order to correlate the results with those from the modal tests under static conditions (i.e. at ω=0 rpm). Also, due to the nature of the rotor with two disks overhanging at the end of the shaft and the bearings situated between the disks, it was expected that the natural frequencies would change substantially due to gyroscopic effects, see Dimarogonas **(16)**, Childs **(17)**, Bently et al **(8)**, Lalanne **(18)** and Leontopoulos **(19)**.

The operation of a floating bushing journal bearing when passing through multi-mode whirl-whip involves simultaneous variations of bearing stiffness with speed of rotation. This in turn causes variations of speed of the rotating bushing and of the eccentricity ratio and thus controls damping and stability thresholds. Therefore, as will be shown later, the bearing stiffness and damping can not be monotonic functions (i.e. increasing or decreasing) of speed of rotation as assumed in the FE model. The accurate prediction of the critical speeds and stability and in particular their variation over the speed range becomes quite a complicated task.

4. VIBRATION MONITORING

As has been shown in **figure 1** there are two orthogonal displacement probes in the test engine on the compressor side, pointing to the coupling adaptor. In real turbomachinery it is often difficult to install the appropriate instrumentation at suitable positions. Although it would have been more desirable to install displacement probes next to the bearings, constraints due to high temperature and lubrication requirements as well as the small size and the thickness of the bearing housing made their installation impossible. An accelerometer was also installed on the bearing housing in the radial direction in order to measure the acceleration response of the casing, and the overall vibrational levels.

The dynamic behaviour of the rotor of the machine, as measured by the displacement probes (eddy current probes) and the accelerometer (piezoelectric transducer), was recorded during a speed change and at steady-state conditions by a four channel data acquisition and analysis system developed at Imperial College. The output of the voltage of the probes was proportional to the gap between the probe and the material surface (coupling adaptor surface). Their sensitivity is approximately 8mV/μm.

A cascade spectrum or a spectrogram was then produced as frequency spectra stacked next to each other. This has been regarded as a valuable tool for the visualisation of the rotor-dynamic. The data in the spectrogram are presented in a 2D chart of frequency versus speed of rotation which uses colour contours for the vibration amplitude information. Also, with the aid of two orthogonal probes, phase information can be extracted and the forward-backward whirl separation can be plotted in "full spectrogram" format separating the forward and backward components of vibration in the frequency spectrum, see Bucher et al **(20)** and Leontopoulos **(19)**, and **(21)**. For ease of interpretation, the value of the running speed frequency component will be given in Hz and rpm, while other frequency components will be given in Hz only.

C553/020 © IMechE 1998

4.1. Case A: Engine only, at no-load condition

Figure 4 shows the spectrogram from the run-up of the turbine engine from 0 to 60,000 rpm as seen by one of the displacement probes pointing at the coupling adaptor at the end of the shaft. The brightest line is the running speed frequency component and is the one having the highest displacement amplitude throughout the test. Firstly, it is possible to observe another small-in-amplitude, low frequency component which varies from approximately below 100 to 210 Hz.

As the speed approaches 60,000 rpm another two low frequency components around the major one become more intense but with amplitudes always much lower than that of the running speed. The second major observation involves the initial phase of the run-up, experiencing a temporary instability which appears as whirl. However, as speed increases the whirl disappears and a new state of equilibrium is reached while the overall vibration level decreases. It is worth noting that a drop in the stiffness of the system must be experienced soon after the whirl disappears as the eccentricity ratio decreases and the system re-stabilises. This is evident from the sub-synchronous frequency component (**figure 4**) which decreases and then increases again as speed increases.

The initial phase of the run-up procedure and the first 10s of the acquired data are further processed and shown in **figure 5**.

A graph at the bottom of the spectrogram shows the corresponding time domain and the display of any individual "instantaneous" spectrum out of the cascade spectra, by moving the cursor on the spectrogram (blue line) to desirable positions (graph at the right of the spectrogram). From the spectrogram it can be seen that the subsynchronous whirl which is always about 0.5 times the running speed, starts at 110 Hz (6600 rpm) and tends to split into two frequency components. An unusual sudden step can be observed at 250 Hz running speed after which the amplitudes of both the 1X component and that of the whirl, drop until the whirl (0.5X) vanishes completely at 543 Hz (32580 rpm).

Although the floating bushing is supposed to reduce λ to values less than 0.5 (which is a typical value for the case of simple cylindrical bearings), this is not so and it seems that the floating bushing plays no special role for the initial phase of acceleration. That is, due to the rotor weight being dominant at low speeds, the floating ring does not rotate fast enough to produce a second oil film, and therefore the model of a simple cylindrical bearing can be assumed. This phenomenon has also been reported by Tondl, (**3**) and Kalogerakos (**10**), and has been experimentally investigated by special optical probes installed directly in the bearing support which identified possible rotation of the floating bushing.

During the run-up of the engine it was observed from the shaft orbit displayed on the oscilloscope that a sudden lurch of the trace sideways occurred around 410 Hz (24,600 rpm) which corresponds to the observed "step" of the whirl frequency component. This indicates the start of a faster rotation of the rotating bushing and the proper formation of the inner and outer films. At about 300 Hz (18,000 rpm) the increasing dynamic eccentricity ratio due to the increasing 1X vibration amplitude in the area of the natural frequency (conical mode), increases the direct stiffness of the system and thus raises the threshold of stability to a higher frequency. At the same time the speed of the floating bushing increases reducing the total equivalent λ and a new equilibrium state of forces is established. So the self-excited vibrations start to decay until the corresponding frequency component vanishes.

Figure 6 focuses on the area of the maximum speed displaying again the spectrogram, the time domain and an instantaneous spectrum. Three subsynchronous components can be observed at 214, 323 and 128 Hz, with amplitude always smaller than that of the running speed. The most dominant of the three is the 214 Hz component whose amplitude relative to the running speed is between 16 and 19 times smaller. The other two components are even smaller as can be seen from the instantaneous spectrum of **figure 6**, while the 323 Hz one is of pulsating nature. These components have been predicted from the modal analysis, to be the conical rigid body modes of the engine rotor.

The overall vibrational velocity level recorded on the bearing housing did not exceed 5 mm/s which is regarded as satisfactory by established standards. Hence, the system is stable under the normal operating conditions.

4.2. Case B: The Turbo-Generator assembly.

A prototype high speed generator, designed and developed at Imperial College, was directly coupled to the engine shaft via a flexible coupling, see **figure 7**. The coupling used was of ball-type arrangement with a cylindrical tube-spacer fitted in between two coupling adaptors at the end of each shaft. This type of coupling has been patented by Imperial College. It was designed so that it is capable of transmitting the driving torque and allowing for some lateral misalignment and axial displacement of the order of 0.1 mm at the area of the coupling adaptor. The axial movement can be adjusted to take into account possible thermal shaft expansion or other types of thrust load from either of the rotors. The operating speed was set at 42,000 rpm and the engine's fuel system and idling speed were adjusted accordingly.

The rotor of the generator was supported on rolling element bearings and it behaved as a rigid body throughout the operating range of speeds. Modal analysis at zero speed together with FE analysis of the whole assembly revealed the possible critical speeds both lateral and torsional.

Figure 8a shows the spectrogram from the run-up of the rotor of the engine. It can be seen that there is an initial whirl, a sub-synchronous frequency component which locks onto a frequency that coincides with the predicted rigid body modes of the turbine engine's rotor coupled now to the rotor of the generator. In this case, the whirl (0.5X) quickly disappears at (344 Hz) 20,640 rpm. As was expected, the "friendly misalignment" due to the radial preload caused by the attachment of the generator, improved the engine's vibrational characteristics giving an overall vibration level of 2.5mm/s at the operating speed. **Figure 8b** compares the differences in the start and finish of whirl for the cases of with and without the generator attached to the engine.

5. CONCLUDING REMARKS

Basic rotordynamics theory, modal testing and FE analysis have been used to obtain an understanding of the rotor system's dynamic behaviour. The experimental analysis was accomplished with the aid of spectrograms which display joint time-frequency analysis and allow detailed interpretation of the vibration at any operating condition. Also, the installation of two orthogonal probes provided separation of the forward and backward vibration components ("full spectrogram").

The role of the floating bushing is to increase damping by reducing the average oil circumferential velocity ratio, λ, and hence increase the stability threshold. From the vibration monitoring of the engine at low speeds it is evident that due to the weight of the rotor and the ring, the rotating bushing does not initially form a second oil film. As speed increases, the eccentricity ratio increases due to the unbalance force, and the stiffness of the system increases. At the same time, the increase in eccentricity ratio generates a high frictional power, causing the rotating bushing to rotate faster, in order to form two oil-films and decrease the overall λ. The splitting and vanishing of the initial whirl frequency (0.5X) indicates the presence and the formation of two bearing oil films.

A light misalignment of the engine rotor due to its attachment to a mechanical load (coupling, generator) improves the vibrational and stability characteristics of the system as a whole. This has been has been justified theoretically and confirmed experimentally by directly coupling a high speed prototype generator onto the shaft of a test engine, dispensing with the need of a gearbox.

ACKNOWLEDGMENTS

The authors would like to thank Prof. Ewins's research group in the Dynamics Section of Imperial College and in particular Goetz Von Groll and Dr I. Bucher for their kind help in the data acquisition and processing.

REFERENCES

1. **Smith, D.M.**, 1969, Journal bearings in turbomachinery, Chapman & Hall, London,
2. **Bently D.E., Muszynska A., Olas A.**, 1986, "Identification of the modal parameters by perturbation testing of a rotor with strong gyroscopic effect", Proceedings of the Inernational Conference on Rotordynamics, 87/19297 Tokyo, pp 189-194
3. **Tondl A.**, 1965, Some Problems on Rotor Dynamics, Chapman and Hall
4. **Dworski J.**, 1958, "High speed rotor suspension formed by fully floating hydrodynamic radial and thrust bearings, ASME paper, Vol. 80, pp1756-1764
5. **Lund J.W.**, 1988, "Rotor bearing dynamics design technology, part VII: The three lobe bearing and floating ring bearing", Mechanical Technology Inc.
6. **Tanaka M.**, and **Hori Y.**, 1972, "Stability characteristics of floating bush journal bearings", Trans. ASME, Journal of Lubrication Technology, Vol.94, No.3, pp.248-259Campbell C., 1987, "Floating bush bearings" Chartered Mechanical Engineers, pp14
7. **Leung P.S., Craighead, I.A., Wilkinson T.S.**, 1988, "An analysis of the steady state and dynamic characteristics of a cylindrical-spherical floating ring bearing", Proceeding of Institution of Mechanical Engineers, C242/88, UK
8. **Bently, D.E.**, and **Muszynska, A.**, 1988, "Role of circumferential flow in the stability of fluid-handling machine rotors, Texas A&M 5th Workshop on Rotordynamics Instability Problems in High Performance Turbomachinery, College Station,
9. **Tanaka M.**, 1996, "A theoretical analysis of stability characteristics of high speed floating bush bearings", IMechE, C500/087
10. **Kalogerakos E.**, 1996, "The design of journal bearings with floating bushing", Interim Report IV, IMASCO, Athens, pp5-56
11. **Dewhirst K., McEwen J.A., Marsh K.W.**, and **Firth MR**, 1990,"Design and development of the Holset HX series of turbochargers", SAE, 900356, pp57-74

12. **ANSYS,** 1995, The ANSYS users manual for revision 5.1, Swanson Analysis Systems, Houston
13. **ICATS,** 1995, ICATS reference manual Ver 4.1, In-house toolkit, Imperial College of Science, Technology and Medicine, London
14. **ROTGUI,** 1995, "Rotating machinery FE analysis software", in house toolkit, MATLAB Ver 4.2c, Imperial College of Science, Technology and Medicine, London
15. **Muszynska A.** and **D.E. Bently,** 1989,"Fluid generated instabilities of rotors, Orbit, April
16. **Dimarogonas A.D.** and **Paipetis S.A.,** 1983, Analytical Methods in Rotor Dynamics, Applied Science Publisher
17. **Childs D.,** 1993, Turbomachinery Rotordynamics, John Wiley
18. **Lalanne M.** and **Ferraris G.,**1990, Rotordynamics Prediction in Engineering, John Wiley
19. **Leontopoulos C.,** 1996, Vibration analysis for the design of a turbo-generator based powertrain for hybrid vehicles, PhD Thesis, Imperial College, London, UK
20. **Bucher I., Ewins D.J., Robb D.A.,** and **Schmiechen P.,** 1995, "Multi-Dimensional, Directional Campbell (ZMOD) Diagrams for Rotating Machinery", ASME, Design Engineering Technical Conferences, Vol.3
21. **LABVIEW**, 1995, The LABVIEW users manuals for revision 3.1, National Instruments Corporation, 6504 Austin, 78730-5039, Texas

 C553/020 © IMechE 1998

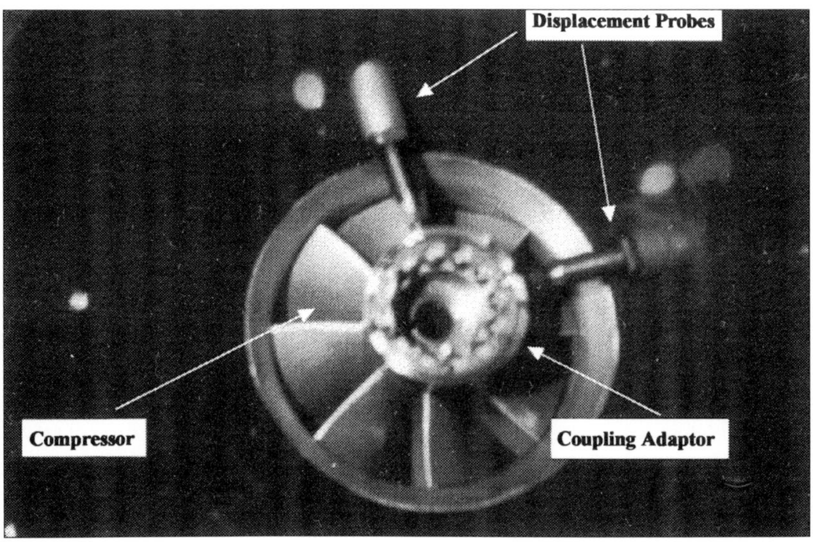

Figure 1: Photos of the actual 50 kW gas-turbine engine from the compressor side

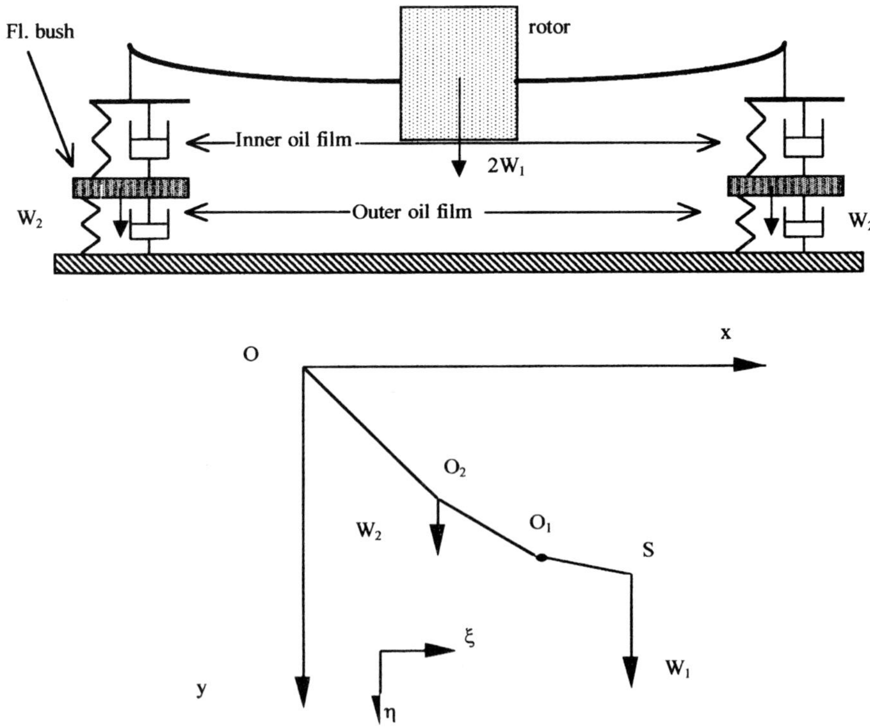

Figure 2: Schematic model of the rotor bearing arrangement and co-ordinate system

C553/020 © IMechE 1998

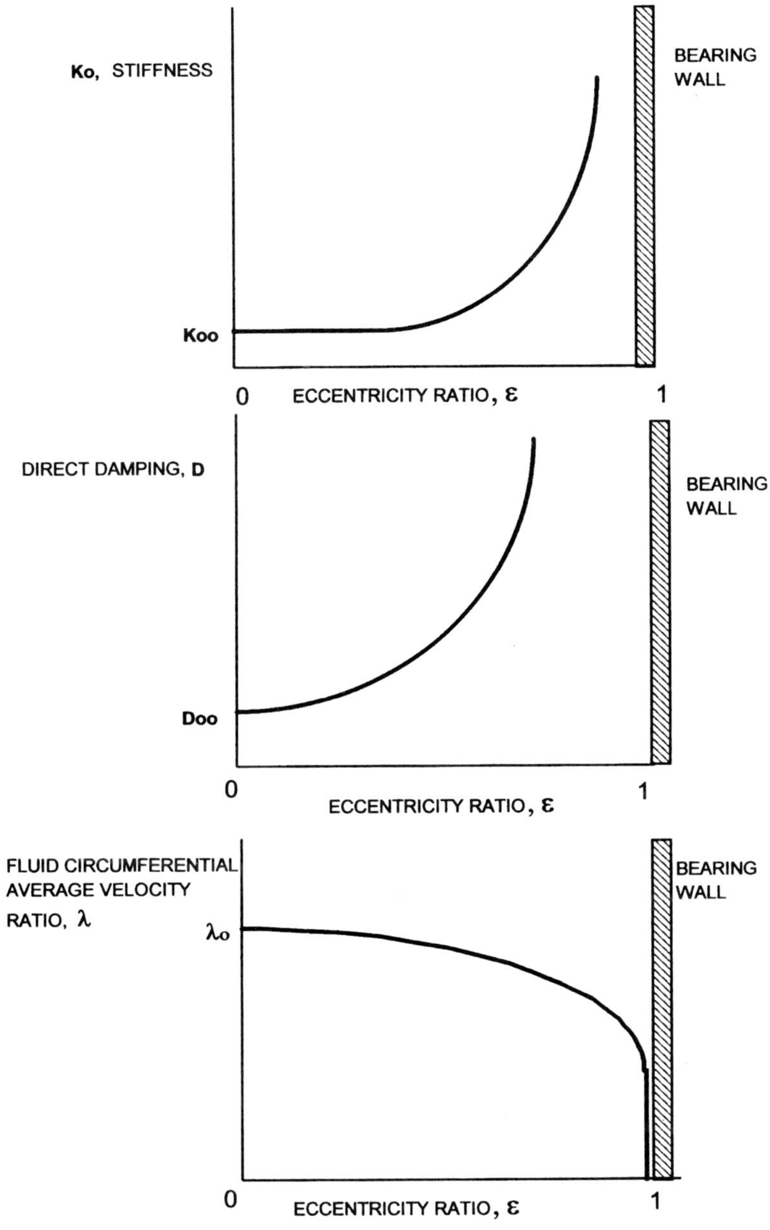

Figure 3: Variation of rotor-system parameters with eccentricity ratio according to Bently and Muszynska (1988)

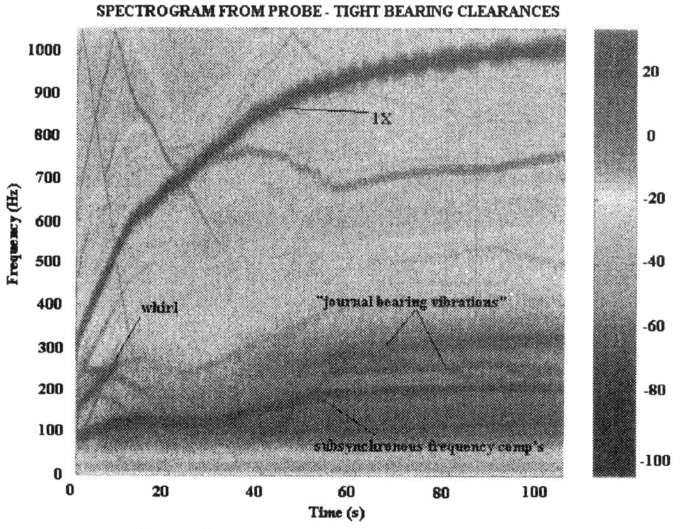

Figure 4: Spectrogram showing the run-up of the engine

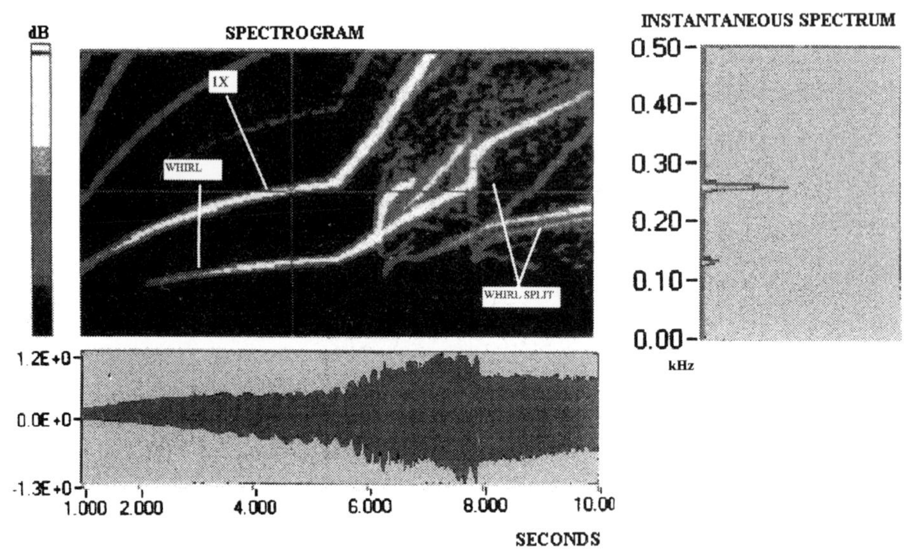

Figure 5: Data from part of the run-up phase of the turbine rotor showing the subsynchronous whirl

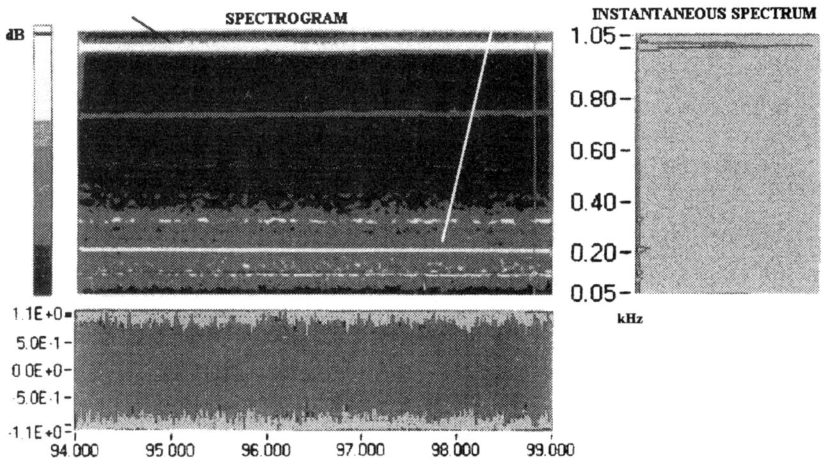

Figure 6: Data captured from the maximum speed of the turbine engine (61,000 rpm) showing the sub-synchronous frequency components

Figure 7: Photo of the prototype turbo-generator assembly tested at Imperial College, Leontopoulos (1996)

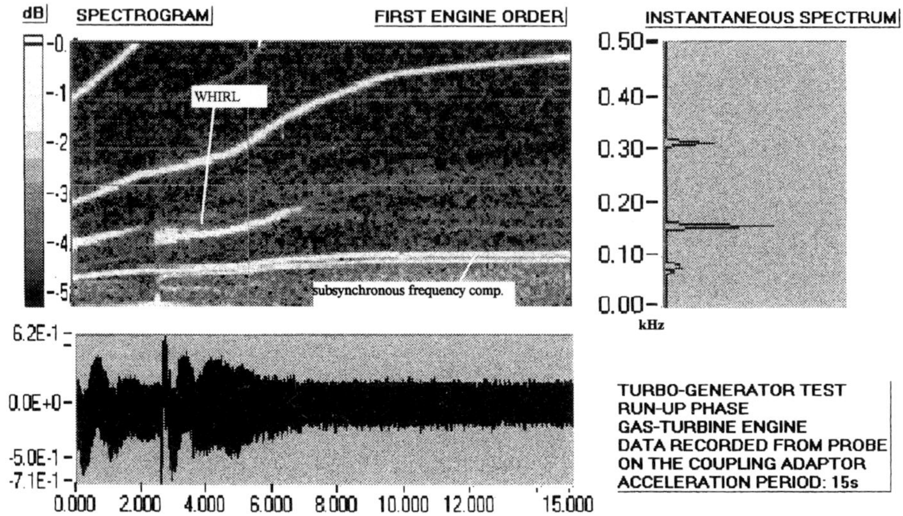

Figure 8a: Data from part of the run-up phase of the turbogenerator assembly

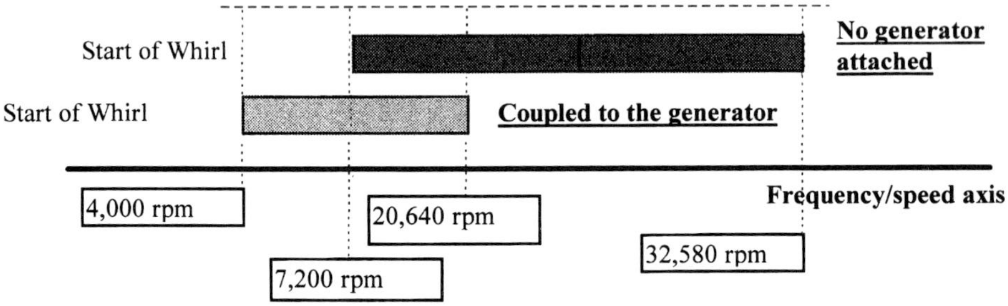

Figure 8b: The start and finish of whirl for the cases with and without the generator attached to the engine

C553/020 © IMechE 1998

C553/044/98

Vibrations associated with ball bearings

R GOHAR PhD, DIC, MIMechE
Imperial College of Science, Technology, and Medicine, London, UK
N AKTURK
Department of Mechanical Engineering, Gazi University, Ankara, Turkey

ABSTRACT

The vibration properties of ball bearings supporting mechanical systems such as small coupled high speed turbo alternators or machine tool spindles and their cutting tools, can have an important influence on their performance.

Even if the bearings are geometrically perfect, they can still be a source of untoward vibrations in a dynamical system. Clearly, these vibrations are also a contributary factor in the life of the bearings themselves and hence, indirectly, of the complete system.

This paper discusses the dynamical behaviour of a simple model comprising a uniform rotor supported by two angular contact ball bearings. The significance of certain vibrations associated with these bearings is discussed, firstly when they are geometrically perfect and then when the bearing rings are misaligned.

NOTATION

a	distance between the misalignment centre and bearing centre
or,	point of application of external loads (Fig. 4)
a_l	position of LHS bearing from the C. of G. (Fig.4)
b_l	position of RHS bearing from the C. of G. (Fig4)
A	distance between centres of curvature of inner and outer race grooves(m)
B	A/d_b
d	diameter (m)
d_b	ball diameter (m)
E	Young's Modulus (Nm^{-2})
$F(\rho)$	curvature difference
f	frequency (s^{-1})
g	acceleration due to gravity (ms^{-2})
I	moment of inertia ($Kg\ m^2$)
k	stiffness ($N\ m^{-1}$)
M	mass of rotor (Kg)
m	number of balls
P_{ra}	preload (N)

r radius (m)
t time (s)
W restoring force (N)
x,y,z coordinates attached to the rotor and displacements of rotor from X,Y,Z
X,Y,Z fixed coordinate set at initial position of rotor
x_c,y_c,z_c coordinate set moving with the cage
α contact angle (rad.)
β offset angle of cage from x axis (rad.)
ϑ angle defining position of arbitrary fixed axis from x axis (rad.)
δ contact deflection (m)
$\delta^{\bullet}$ dimensionless contact deflection
δ_c maximum displacement due to misalignment (m)
υ Poisson's ratio
ϕ rock angle of rotor about y axis (rad.)
γ angle between adjacent balls (rad.)
ρ curvature (equation (2) (m^{-1}))
or, misalignment angle (equation (26) (rad.))
θ angular position of a ball (rad.)
ω angular velocity (rad./s)
ω_c cage angular velocity (rad./s)
ψ rock angle of rotor about y axis (rad.)
R_i radius of locii of groove centres for inner ring (m)
R_o radius of locii of groove centres for outer ring (m)

Subscripts
a relating to axial
b relating to balls
c relating to cage
g relating to groove
i relating to ith ball
or, relating to inner race
n relating to natural frequency
m relating to pitch diameter
o relating to outer race
p relating to preloaded
r relating to radial

Abbreviations
BPF Ball Passage Frequency
BPV Ball Passage Vibrations

1 INTRODUCTION

This paper discusses certain aspects of vibrations associated with ball bearings. The vibrational behaviour of rotors can be influenced by this type of support because of its non linear and time varying stiffness.

The way the load distribution on the bearings varies, as the rolling element set rotates round the rings, causes the ball bearing itself to become a vibration generator. Such behaviour can even come from what is nominally a geometrically perfect bearing as well as one which possesses imperfections from manufacture, installation, lubrication, ambient running conditions or one experiencing changing properties due to wear and fatigue [1-6]. Thus, it is important to carry out correct design procedures as well as detecting incipient bearing deterioration at an early stage in order to minimise more extensive damage at a later stage.

When quiet running is desirable, for example for domestic purposes, the bearing vibration, is often termed *noise* [2]. Another occurence of bearing noise is in submarines where the noise footprint, as it travels through the water, may be detected and thus help to identify the type and whereabouts of its source [2].

Ball bearings which support a machine tool spindle are a major feature of its running accuracy and vibrational behaviour. The quality of the spindle's supporting bearings, their disposition, assembley and lubrication, determine its running accuracy, especially when operating near its design speed and under a machining load [7, 8, 9].

This paper will discuss some of these features, concentrating mainly on the authors' research experience of the topic [4, 10, 11, 12].

2 MODELING THE SYSTEM

To investigate the vibration characteristics of ball bearings, the rotor and its bearings can be considered as a spring- mass system, where the rotor is the mass and the bearing rings and balls form a series of non linear rotating contact springs. The resultant reaction force caused by these springs has to be determined. The local Hertzian contact force-deflection relationship between two elastic bodies is [13]:

$$W_i = K\delta_i^{3/2} \tag{1}$$

where the stiffness coefficient K for the same material properties of the two contacting bodies is:

$$K = \frac{2\sqrt{2}}{3}\frac{\left(\frac{E}{1-\nu^2}\right)}{\left(\sum\rho\right)^{1/2}}\left(\frac{1}{\delta^*}\right)^{3/2} \tag{2}$$

where δ^* is the dimensionless deflection factor given in graphical form by Harris [13] as a function of the curvature difference *F(ρ)* and expressed for computational purposes [5] as:

$$\delta^* = -329 + 1883(F(\rho)) - 3798(F(\rho))^2 + 3270(F(\rho))^3 - 1027(F(\rho))^4 \tag{3}$$

For the *i*th ball there are two contact surfaces: outer ring / ball and inner ring / ball, as in Fig.1. The equivalent stiffness coefficient, K is given by Harris [13] as :

$$K = \left[\frac{1}{\left(\frac{1}{K_i}\right)^{2/3} + \left(\frac{1}{K_o}\right)^{2/3}}\right]^{3/2} \tag{4}$$

In the case of angular contact ball bearings, K in Eq.(4), varies with the contact angle [12].

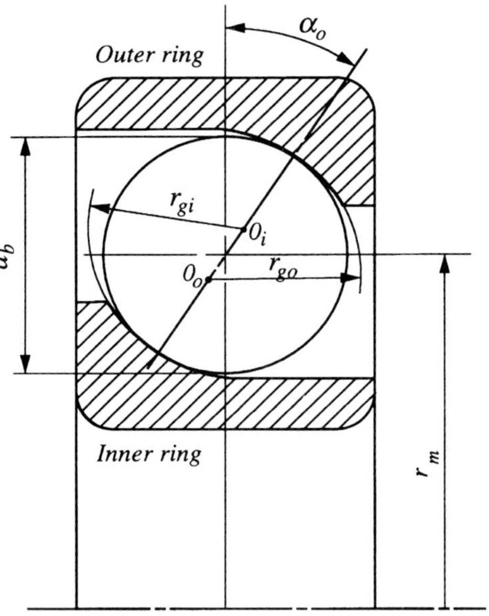

Fig. 1 Geometry of the *i*th Ball/Ring Contact

2.1 Deflections at the Contacts

When the bearing rotates, each ball of the set is continually in contact with different points on the raceways. In order to find the total reaction force, the deflection at the *i*th ball must first be calculated. Generally, the bearing is subjected to combined radial and axial loads. Thus, referring to Fig.1, which is a section through the *i*th ball, initially in touching contact with the races, points 0_i and 0_o are respectively the centres of curvature of the inner and outer race grooves and move with them. Assuming the outer race is fixed, take 0_o as a fixed origin. Now take an auxiliary plane containing an axis parallel to the rotor spin axis (z) at 0_o and a radial axis defining the pitch angle of the *i*th ball centre with respect to the *x* axis. Also contained in this plane is the line $0_i0_o = A$, the initial distance between the centres of curvature of the inner and outer races (Fig. 2).

Thus, starting at 0_i (zero load) defined by (A, α_0), 0_i will move through z_p under the preload to $(0_i)_1$, creating a new contact angle α_p and preload deflection δ_p (same for all balls). Under external axial and radial load components, $(0_i)_1$ will move through z to $(0_i)_2$ and δ_r to its final position $(0_i)_3$. The total radial approach of the races at the ith ball is given by

$$\delta_r = x\cos\theta_i + y\sin\theta_i \tag{5}$$

where θ_i defines the position of the *i* th ball with respect to the vertical x axis.

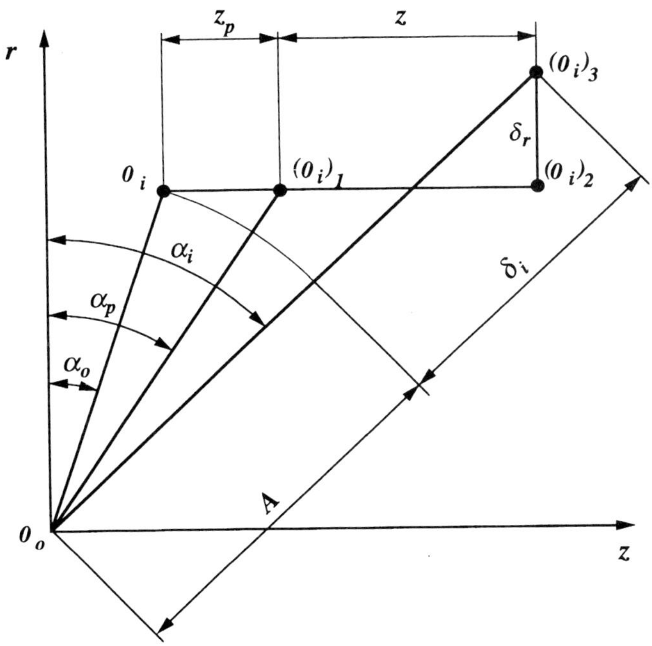

Fig 2 Deflections at the *i*th ball/race Contact

For a five degrees of freedom system, the total relative approach at the races at the ith ball will also depend on total movement in the *z* direction from preload and an additional *z* displacement, as well as the pitching and yawing movements of the rotor [8].

For a simple three degrees of freedom system, with no pitching or yawing of the rotor axis, referring again to Fig. 2, the total relative approach is given by:

$$\delta_i = \left[\left(A\sin(\alpha_0) + z_p + z\right)^2 + \left(\delta_r + A\cos(\alpha_0)\right)^2\right]^{1/2} - A \tag{6}$$

and the final contact angles can be found from:

$$\tan\alpha_i = \frac{A\sin(\alpha_0) + z_p + z}{\delta_r + A\cos(\alpha_0)} \tag{7}$$

where

$$A = r_{go} + r_{gi} - d_b \tag{8}$$

and

$$z_p = Bd_b \frac{\sin(\alpha_p - \alpha_0)}{\cos(\alpha_p)} \tag{9}$$

where α_p is the preloaded contact angle and can be calculated from:

$$P_{ra} = mK_i\left[Bd_b\left(\frac{\cos(\alpha_i)}{\cos(\alpha_p)} - 1\right)\right]^{3/2}\sin(\alpha_p) \quad \textbf{(10)}$$

where P_{ra} is the applied preload in the axial direction and α_p is solved by iteration. If the preload equals zero, $\alpha_p = \alpha_0$. From the above equations, the normal force W_{r_i} on the ith ball can easily be found. Hence the force in the X direction is:

$$W_{X_i} = W_{r_i}\cos\theta_i \quad \textbf{(11)}$$

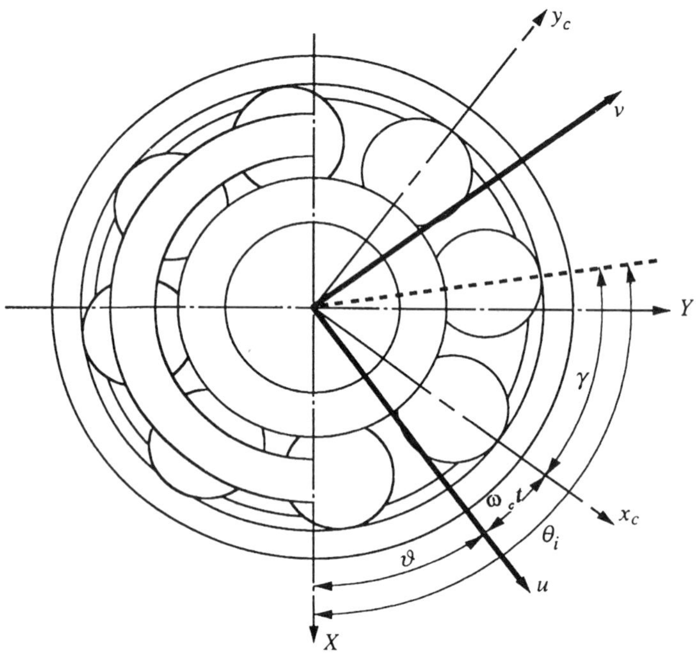

Fig. 3 The Reference Axes

where θ_i is the angle between the X axis and the axis of the ith ball. This angle is a combination of different angles as shown in Fig. 3. Axes (X, Y and Z) are fixed in space with the X axis in the gravitational direction. Axes x, y and z (not shown in Fig.3), move with the rotor but do not spin with it. For a three degrees of freedom system, the x axis will always be parallel to the X axis. Axes u, v, and w are fixed in space with the u axis offset at angle ϑ anticlockwise from the x axis. Normally, ϑ is zero. The last set of axes $(x_c, y_c$ and $z_c)$ is

C553/044 © IMechE 1998

rotating with the ball set at ω_c . Thus, at time t it is at angle $\omega_c t$ from the x axis and the angle between adjacent balls is defined as :

$$\gamma = \frac{2\pi}{m} \tag{12}$$

The ith ball is now easily located since its position is fixed with respect to the (x_c, y_c, z_c) set is:

$$\theta_i = \vartheta + \omega_c t + i\gamma \tag{13}$$

where the cage speed ω_c is [5]:

$$\omega_c = \frac{1}{2}\omega_i\left[1 - \frac{d_b}{d_m}\cos\alpha\right] + \frac{1}{2}\omega_o\left[1 + \frac{d_b}{d_m}\cos\alpha\right] \tag{14}$$

2.2 Equations of motion.

A real example of a rotor supported by ball bearings is difficult to model because of its complication and myriad defects. Thus, some simplifications for this paper are made:

- **(1)** A three degrees of freedom system is assumed.
- **(2)** Balls are massless.
- **(3)** Races are flexurally rigid and only undergo local deformation at the contacts
- **(4)** All contact deformations follow Hertzian theory.
- **(5)** Balls are equipitched with no interaction between them.
- **(6)** The effects of elastohydrodynamic conditions at the contacts are ignored.
- **(7)** All moving elements rotate at constant speed.
- **(8)** Both bearings are positioned symmetrically with respect to the uniform rotor (except for one example where one of the inner rings is misaligned). The balls are assumed always to be in phase.
- **(9)** For a five degrees of freedom system, the yaw and pitch angles are assumed to be small, so second order terms are ignored in the derivation of the equations of motion, given below. A diagram of the forces acting on the rotor from one of the bearings, is shown in Fig. 4. Thus, the general equations of motion for a five-degrees -of -freedom system are as follows:

$$\begin{aligned} &M\ddot{x} + \sum_{i=1}^{m}\left[(W_i)_L \cos(\alpha_i)_L \cos(\theta_i)_L\right] \\ &+\sum_{i=1}^{m}\left[(W_i)_R \cos(\alpha_i)_R \cos(\theta_i)_R\right] + Q_x - Mg = 0 \end{aligned} \tag{15}$$

$$\begin{aligned} &M\ddot{y} + \sum_{i=1}^{m}\left[(W_i)_L \cos(\alpha_i)_L \sin(\theta_i)_L\right] \\ &+\sum_{i=1}^{m}\left[(W_i)_R \cos(\alpha_i)_R \sin(\theta_i)_R\right] + Q_y = 0 \end{aligned} \tag{16}$$

$$M\ddot{z} - \sum_{i=1}^{m}\left[(W_i)_L \sin(\alpha_i)_L\right] + \sum_{i=1}^{m}\left[(W_i)_R \sin(\alpha_i)_R\right] + Q_z = 0 \tag{17}$$

$$I_{xx}\ddot{\psi} + I_{zz}\dot{\phi}\omega - \sum_{i=1}^{m}\left[(W_i)_L \cos(\alpha_i)_L \sin(\theta_i)_L\right]a_1$$
$$+\sum_{i=1}^{m}\left[(W_i)_R \cos(\alpha_i)_R \sin(\theta_i)_R\right]b_1 - \sum_{i=1}^{m}\left[(W_i)_L \sin(\alpha_i)_L \sin(\theta_i)_L\right]R$$
$$+\sum_{i=1}^{m}\left[(W_i)_R \sin(\alpha_i)_R \sin(\theta_i)_R\right]R - Q_y(a_1 + a) = 0 \tag{18}$$

$$I_{yy}\ddot{\phi} - I_{zzz}\dot{\psi}\omega + \sum_{i=1}^{m}\left[(W_i)_L \cos(\alpha_i)_L \cos(\theta_i)_L\right]a_1$$
$$-\sum_{i=1}^{m}\left[(W_i)_R \cos(\alpha_i)_R \cos(\theta_i)_R\right]b_1 + \sum_{i=1}^{m}\left[(W_i)_L \sin(\alpha_i)_L \cos(\theta_i)_L\right]R$$
$$-\sum_{i=1}^{m}\left[(W_i)_R \sin(\alpha_i)_R \cos(\theta_i)_R\right]R + Q_x(a_1 + a) = 0 \tag{19}$$

Here, R is taken as the pitch circle radius of the bearing.

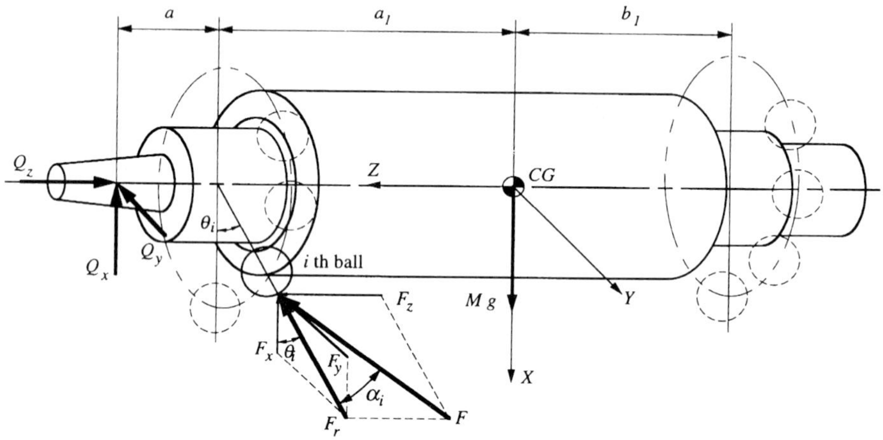

Fig. 4 Forces on the Rotor

In all the examples quoted below, because the model is assumed to be a horizontal, symmetrical two bearing system supporting a uniform rotor, having equal preloads on both bearings and under gravity load only, equations (15)-(19) are simplified, with $a_1 = b_1$ and a $= Q_x = Q_y = Q_z = 0$.

The solution of the simplified versions of equations (15)-(19) were obtained by using the Runge-Kutta Iteritive method [5]. The initial conditions were: $x(0) = 1\mu m$, $y(0) = 0.1\mu m$, $z(0) = 0.01\mu m$, and all initial velocities were zero.

C553/044 © IMechE 1998

3 BALL PASSAGE FREQUENCY (BPF) AND NATURAL FREQUENCY

A ball bearing can be considered to be a series of equipitched weightless balls, acting as non linear contact springs which roll round the inner and outer bearing races. As mentioned in the introduction, the nature of its design causes it to possess vibration generating properties. This is because the balls are discrete elements whose angular positions change cyclically with time relative to the load line as the bearing rotates.

Eqs. (15-19) are of the Non Linear Mathieu type [14] and have to be solved numerically because of their complexity.

To describe the behaviour of this system, let Fig. 5 be an example of a geometrically perfect radial ball bearing with m = 6 balls, zero radial clearance with a unidirectional radial load. Under these conditions, the applied load is only supported by a few balls restricted to a narrow load region. The balls will deform as they enter the loaded zone, where mutual convergence of the races takes place and then they rebound as they move into the unloaded zone. The reason for this behaviour is the instantaneous change in nett stiffness of the ball set with respect to the load line. The left hand view in Fig. 1 shows the situation with Ball 1 directly under the load and three balls are deformed. The middle view is 1/12 of a revolution of the cage later. Now, the load is supported by only 2 balls of the set instead of three as

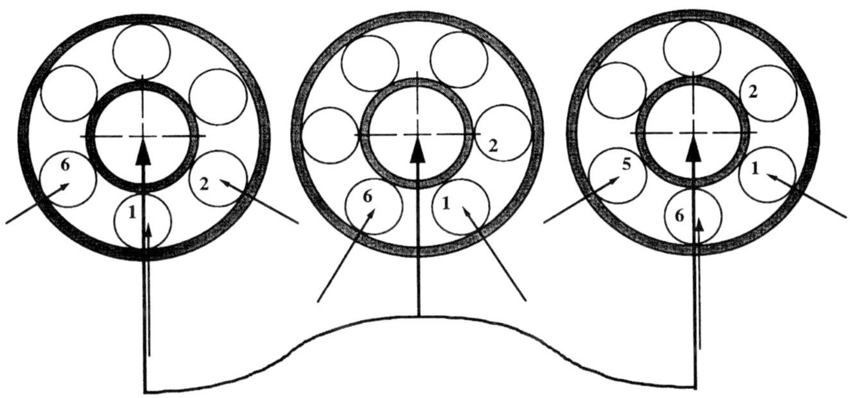

Fig. 5 Variation Of Total Spring Force

previously. The rotor therefore moves downwards in response to the decreased contact stiffness force (which is non linear). Thus, the radial contact force changes cyclically as shown by the radial force vector in Fig. 4. In the right hand view, the first position has been regained. The time taken in the cycle is therefore:

$$t = \frac{\text{time for a complete revolution of the cage}}{m} \qquad \textbf{(20)}$$

As the time for one cage rotation is $2\pi/\omega_c$, the rotor will be excited at a frequency of:

$$\omega_{bpo} = m\omega_c \tag{21}$$

where ω_c is found from Eq. (14). This frequency is called the *Ball Passage Frequency* (BPF) relative to the outer race.

Because the inner race is rotating, the BPF relative to the inner race is given by:

$$\omega_{bpi} = m(\omega_i - \omega_c) \tag{22}$$

The vibrations associated with the BPF are called the *Ball Passage Vibrations* (BPV) or Elastic Compliance Vibrations [2]. The fact that the BPV are associated with a time varying stiffness, which is itself non linear, suggests that there will exist, in addition to the BPF itself, both superharmonics and perhaps subharmonics, in addition to the vibrations being non sinusoidal. In addition to the BPF, a rotor supported by bearings having the above properties will possess natural frequencies along principle axes.

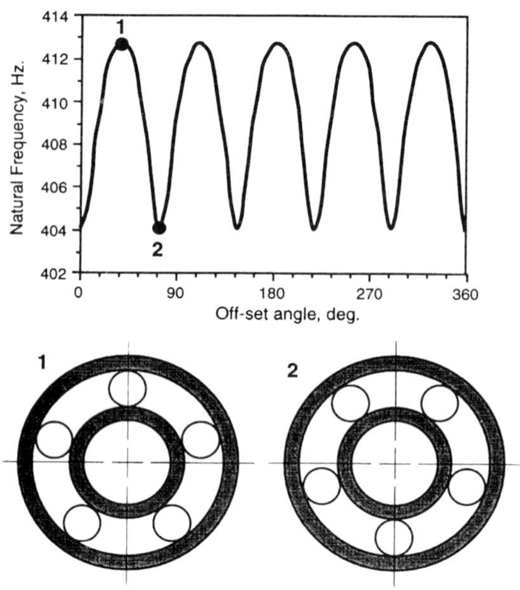

Fig. 6 Effect Of Ball Position On Natural Frequency (m=5, P_{ra}= 5N)

As a demonstration of this property, the simplified model discussed above was employed (specification in Appendix 1). First, a five ball set was employed with its balls in various offset positions relative to the vertical, as in positions1 and 2 in Fig. 6. With the rotor speed set to zero, its centre of gravity was released from an arbitrary initial position. For each position of the ball set, the system natural frequency in the x direction was calculated by a method described by Akturk [15], which is taken from Rao [16]. The method is based on plotting the phase- plane of the x direction vibrations. As the system is not rotating here, the

C553/044 © IMechE 1998

phase plane oscillations will describe an ellipse (Fig. 6 of Ref. 5). The Natural frequency is given by: $\omega_n = b/_{2\pi a}$ where b = amplitude of $\dot{x}$ oscillations and a = amplitude of x oscillations. Fig. 6 shows a cyclic variation in natural frequency which amounts to no more than a ripple, changing five times over one revolution of the cage. Increasing the number of balls to eight gave a cyclic change of eight over one revolution, though with a smaller amplitude. This suggests that the natural frequency increases with the number of balls because the system becomes stiffer. Fig. 7 shows this variation, with the rate of increase of natural frequency diminishing and the ripple height around the average decreasing as the number of balls increases.

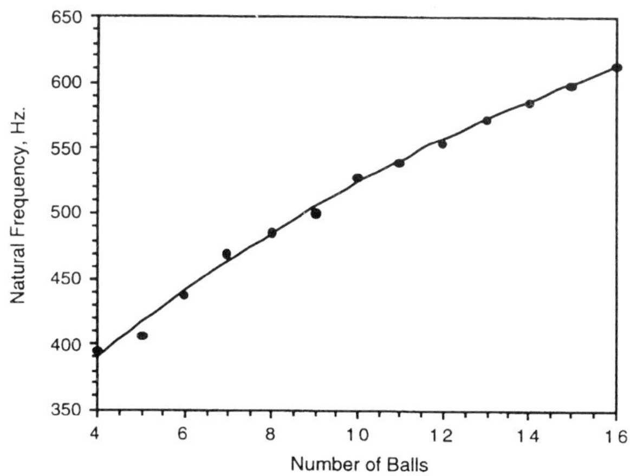

Fig. 7 Effect of the Number of Balls on the Natural Frequency (m= 8, P_{ra}= 5N)

Likewise, changing the axial preload will stiffen the bearing and increase the natural frequency. Fig. 8 shows this change. For lower preloads, the rate of change is fast but diminishes as the balls become stiffer under increasing load.

An approximate expression for the natural frequency of the system, which demonstrates the significance of these variables, can be writen as (see Appendix 2):

$$\omega_n = \left(\frac{3S}{M}\right)^{1/2}\left[\left(\frac{P_{ra}}{mK\sin\alpha_p}\right)^{1/6} + \frac{1}{4}\left(\frac{\Delta\delta}{\left(\frac{P_{ra}}{mK\sin\alpha_p}\right)^{1/2}}\right)\right] \quad \textbf{(23)}$$

where

$$S = \sum_1^m K\left\{\cos(\theta_i - \beta)\cos^2\alpha_p + \sin^2\alpha_p\right\}^{3/2}\cos\alpha_p\sin\theta_i \quad \textbf{(24)}$$

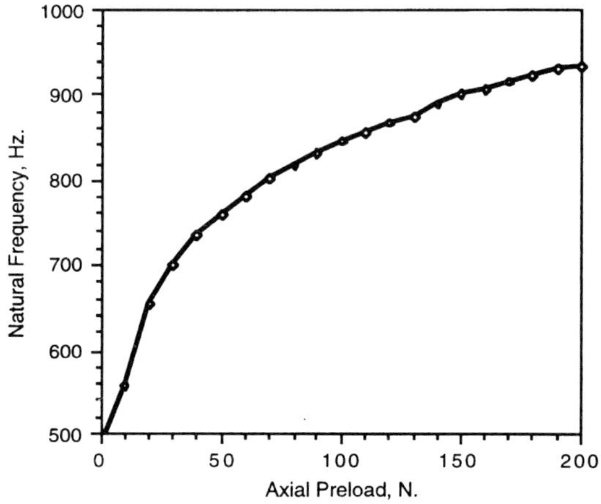

Fig. 8 Effect Of The Axial Preload On The Natural Frequency (P_{ra}= 5N)

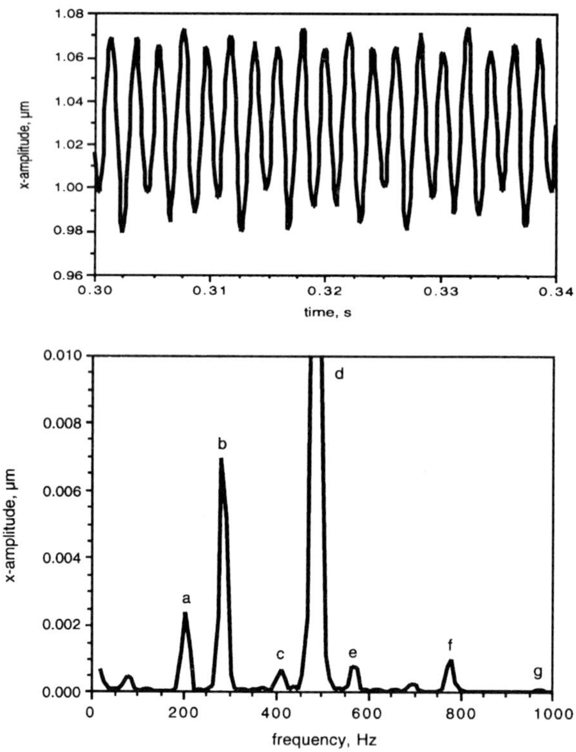

Fig. 9 Vibrations Along the x axis in the Time Domain and its FFT
(m = 8, P_{ra} = 5N, n = 5000 rev/min, ω_{bp} = 280 Hz, ω_n = 500 Hz)

C553/044 © IMechE 1998

In Eq. (23), $\Delta\delta$ is the amplitude of the deflection of the rotor centre vibrations after preload has been applied. It shows that the amplitude can be an important factor in non linear vibrations and would be even more significant, compared with the first term, if there was a low preload, P_{Ra} and/or a high value of m. The value of preload angle, α_p, can be obtained from Eq.(10). It varies little with P_{Ra}. However, in the examples quoted, $\Delta\delta$ is small and the preloads are relatively high. Thus the second term in Eq. (23) can be neglected, leaving only the 'static' natural frequency. Fig. 9 shows vibrations along the x axis in the time and frequency domains (FFT). For this speed, the BPF is 280 Hz relative to the outer race and the *natural frequency* is:

510 *Hz.* these two frequencies are present in the FFT as peaks (b) and (d). The effect of the Natural Frequency along the z axis (FFT not shown) appeared at 200 Hz. Along the x axis it is peak (a). Peaks (c) and (f) are the second and third superharmonics of the z axis natural frequency. Peak (e) is the first superharmonic of the BPF and peak (g) is the second. All these peaks also appear on the y axis FFT. Thus, the base frequency of the BPF and its superharmonics, are present to excite the natural frequency at certain rotor speeds. This situation can be demonstrated by varying the rotor speed, introducing arbitrary damping into the system on all three axes and plotting the steady state frequency response function. Fig. 10 shows that, as the number of balls increases, the system will stiffen, causing the *natural frequncy* to increase.

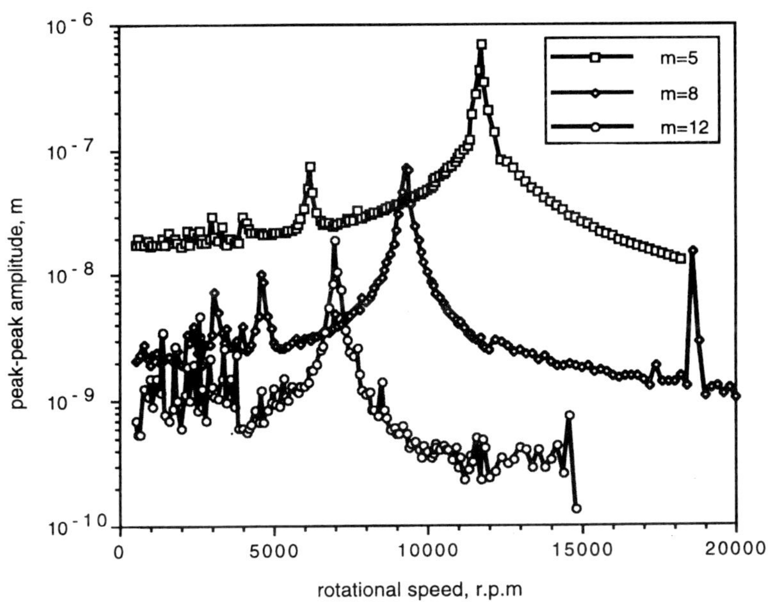

Fig. 10 Effect of Varying the Number of Balls on the Frequency Response Function ($P_{ra} = 10$ N, $c = 300$ Nsm^{-1})

4 THE MODEL WITH MISALIGNED RACES

Defects in ball bearings may be due to machining inaccuracies, such as waviness on the race and ball surfaces [6, 17], or ball size variation as a result of poor selection [4, 18]. Another defect, which sometimes occurs, is misalignment of the bearing races caused by inaccurate machining of the rotor or bearing housing or by poor assembly. The vibrations caused by these factors can have the same character as some of the faults arising from poor quality bearings. In early studies, misalignment was considered only because it imposed additional stresses on the bearings [13]. Brandlein [19] and Ellis[20], discussed the effect of misalignment on *ball speed variation* which they found was caused by contact angle variation as each ball tries to rotate about the skewed bearing axis at a different speed, with the sequence being repeated each cage revolution. The situation is exacerbated under combined loading. Other theoretical studies of the effects of misalignment were studied by Meyer et al. [1] and Gad et al [3].

4.1 Deflection and Contact Angle Variation due to Misalignment

Misalignment of a bearing is generally defined about an axis at right angles to its spin axis. It is expressed as a specified angular displacement, about the bearing rotation centre, together with a displacement. In the general case the maximum displacement due to misalignment will occur at an angle θ_c to the vertical x axis, depending on which way it is misaligned. The maximum deflection can be found from Fig. 11:

$$\delta_c = \begin{bmatrix} \left((R_i + a)\cos(\rho) - (R_0 + a\ )\right)^2 + \\ \left((R_i + a)\sin(\rho) + Bd_b \sin(\alpha_0)\right)^2 \end{bmatrix}^{1/2} - Bd_b \tag{25}$$

where ρ is the misalignment angle and

$$R_i = \frac{d_m}{2} + \left(r_{gi} - \frac{d_b}{2}\right)\cos\alpha_0, \text{ and}$$

$$R_0 = R_i = Bd_b \cos\alpha_0$$

The resulting deflection, due to misalignment only, on the ith ball is

$$\delta_{ic} = \delta_c \cos(\theta_i - \theta_c) + \delta_c \sin(\theta_i - \theta_c)\sin\theta_i \tag{26}$$

where θ_c defines the ball experiencing the maximum displacement due to misalignment.

The *total* deflection at each ball, δ_i, is found by adding vectorially to δ_{ic}, the additional deflections due to preload and displacement (Eq. 6).

Using Eq. (7),the total contact angles for each ball, are estimated in a similar way [15].

 C553/044 © IMechE 1998

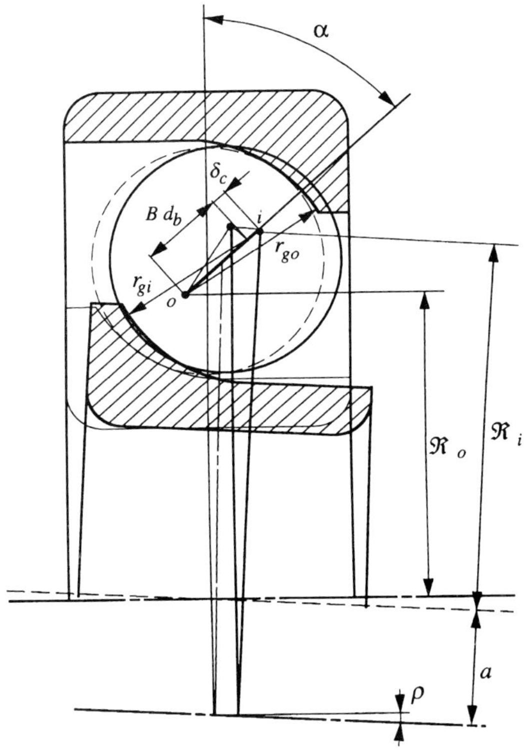

Fig. 11 Misalignment of Angular Contact Ball Bearings

4.2 Bearing Outer Rings Misaligned

Two examples of misalignment will now be given. Let the LHS bearing outer ring be misaligned, with the RHS bearing outer ring having an equal and opposite misalignment. This simplifies the problem by eliminating z direction oscillations. In addition, make $a = \theta_c = 0$, so that ρ has its centre on the bearing axis of symmetry.

With ρ = 0.02 rad., P_{ra} = 10N and rotor speed = 3000 rev/min., Fig. 12 shows the FFT of the x direction oscillations. There are vibration responses at two principle frequencies: the cage speed (21Hz) and the system Natural Frequency (820Hz). A third response is also apparent at the BPF (168Hz). The reason for the one at cage speed is because the outer ring misalignment produces a situation where the same position of the balls is obtained once per cage revolution. The response at Natural Frequency is probably caused by a superharmonic of the BPF or the cage speed. The misaligned system is relatively more non linear due to misalignment because, while the balls in the lower half of the bearing become more preloaded, those in the upper half are relieved. Half a cage rotation later, the situation reverses. Note also, for the chosen misalignment angle, the amplitudes of vibration are very high compared with the previous examples of perfect bearings.

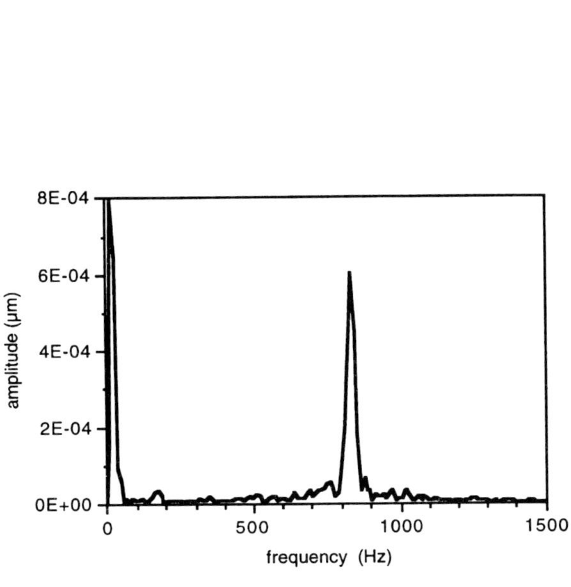

Fig. 12 FFT of x Direction Vibrations due to Outer Ring Misalignment
(m = 8, ρ = 0.02 rad. P_{ra} = 10 N, c = 300 Nsm^{-1}, n = 3000 rev/min, ω_c = 21 Hz, ω_n = 820 Hz)

4.3 Inner Ring Misalignment

Let the LHS inner ring be misaligned by 0.02 rad. and the RHS bearing have an equal and opposite misalignment. Thus, again there will be no vibrations in the z direction. Fig. 13 shows the effect of this inner ring misalignment on system vibrations. Three rotor speeds were tested. In each case the dominant vibration occurs at rotor speed because one ball in the set will be loaded most once per revolution of the rotor.

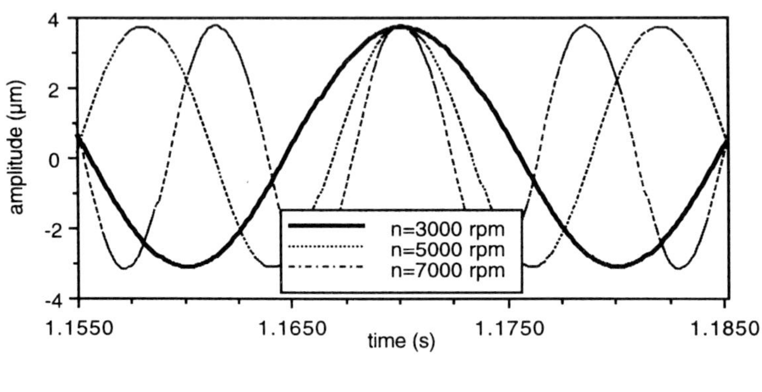

Fig. 13 Time Domain Vibrations due to Inner Ring Misalignment
(m = 8, ρ = 0.02 rad. P_{ra} = 10 N, c = 300 Nsm^{-1})

C553/044 © IMechE 1998

When only the LHS bearing inner ring is misaligned by 0.02 rad. and the RHS bearing has no misalignment, with the rotor orientation assumed to be constant, Fig. 14 shows that violent vibrations are produced about the *x*, *y*, and *z* axes. For a rotor speed of 5000 rev/min.(83Hz), the 9th superharmonic of the shaft speed coincides with the natural frequency in the x direction and excites it. For the same reason, a smaller peak occurs in the *y* direction although its natural frequency is slightly different because of the misalignment. A large peak occurs in the z direction spectrum at twice the rotor speed and others at rotor speed harmonics as the inner ring is excited twice in the z direction for one revolution of rotor.

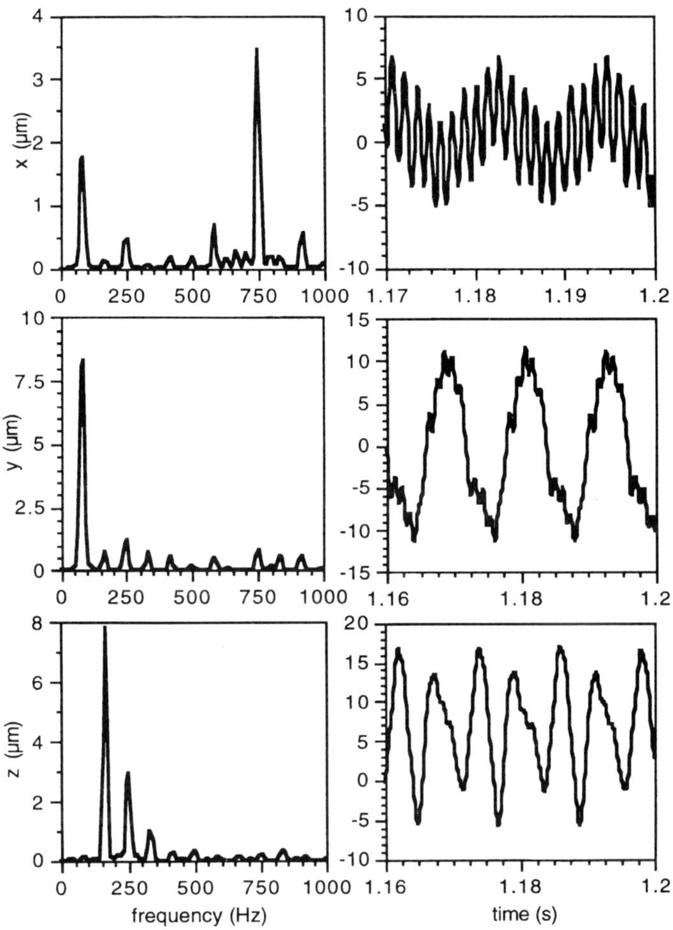

Fig. 14 Vibrations due to Misalignment on the LHS Bearing Only
(m = 8, ρ = 0.02 rad. P_{ra} = 10 N, c = 300 Nsm^{-1}, n = 5000 rev/min, ω = 83 Hz ω_c = 35 Hz, ω_n= 750 Hz)

© IMechE 1998 C553/044

5 CONCLUSIONS

Some examples of vibrations associated with ball bearings have been given, firstly for a geometrically perfect bearing and then for one with misalignment defects.
the main conclusions are :

(a) Natural Frequency

- The Natural Frequency varies cyclically as the ball set changes its position.
- The number of variations of Natural Frequency, per cage revolution, equals the number of balls in the set.
- Increasing the number of balls and / or the preload, increases the Natural Frequency and reduces the amplitude of this variation.

(b) Ball Passage Frequency

The BPF is caused by the cyclic variation in non linear stiffness of the ball bearing. It becomes important when it, or one of its harmonics, coincides with the Natural Frequency. Furthermore, as the BPF depends on the number of balls in the cage, resonances may occur at quite low rotor speeds.

(c) Bearing Outer Ring Misalignment

Equal and opposite outer ring misalignments on the bearings increase the maximum preloads and hence raise the system natural frequency. Significant frequencies are at the cage speed and Natural Frequency.

(d) Bearing Inner Ring Misalignment

When the two inner rings are misaligned in opposite senses, resulting significant vibrations are at the rotor speed. When only one inner ring is misaligned, there are severe vibrations of high amplitude at a superharmonic of the rotor speed, coinciding with the Natural Frequency as well as at the rotor speed itself.

6 REFERENCES

1 Meyer, L. D.; Ahlgren, F.F. and Weichbrodt, B. "An Analytical Model for Ball Bearing Vibrations." Trans. ASME J.of Mechanical Design **102**, 1980, 205-210.

2 Wardle, F. P. and Poon S. Y., "Rolling Bearing Noise Cause and Cure" Chart. Mech. Engineer, July/August, 1983.

3 Gad, E.H.; Fukata, S. and Tamura, H., "Computer simulation of Rotor Axial and Radial Vibrations Based on Ball Bearings." Memoirs of Faculty of Engineering, Kyushu University, **44**, (2), 1984b,169-189.

4 Franco, J.; Akturk, N. and Gohar R., "Vibrations of a Rigid Shaft Supported by Ball Bearings with Several Defects.", ASME Paper No. 92-GT-164, Int. Gas Turbine and Aero Engine Congress and Exhibition, Cologne, Germany,1992, 1-4.

5 Akturk, N.; Uneeb, M. and Gohar, R. "The Effects of the Number of balls and Preload on Vibrations Associated with Ball Bearings." Trans ASME J. Tribology,1997, **119**, 747-763.

6 Akturk, N. and Gohar, R. "The Effect of Ball Size Variation on Vibrations Associated with Ball Bearings", I.Mech.E., Proc. Part J, J. of Engineering Tribology, 1998, **212**, 101-110.

7 Lacey, S. J., Wardle, F.P. and Poon, S.Y., "High Speed Bearings for CNC Machine Tool Spindles", Chart. Mech. Engr., Dec. 1983, 51-56.

8 Aini, R., Rahnejat, H. and Gohar, R. "A Five degrees of freedom analysis of Vibration in Precision Spindles" Int. Journ. of Mach. Tools and Manufact., 1990, **30**(1),1-18.

9 Aini, R., Rahnejat, H. and Gohar, R. "Vibration Modelling of Precision Spindles", Proc. Inst. Mech. Engrs. Part C: Journ. of Mech.Engr. Sci. 1995, **209**, 107-114.

10 Rahnejat, H. and Gohar, R. "Vibrations of Radial Ball Bearings" Proc. Inst. Mech. Engrs. Part C,1985, **199**(C3), 181-193.

11 Akturk, N. and Gohar, R. "A Technique for Estimating the Dynamic Stiffness and Loss Coefficients of Elastomeric Materials from their Material Properties" Machine Vibration, 1993, **2**,168-176.

12 Akturk, N. and Gohar, R. "Damping the Vibrations of a Rigid Shaft Supported by Ball Bearings by Means of External Elastomeric O-Ring Dampers" Proc. Inst. Mech. Engrs., J.of Eng.Trib.,1994, **208**, 183-190.

13 Harris, T. A., *Rolling Element Bearing Analysis*, 1991, (John Wiley, New York).

14 Hayashi, C., *Non Linear Oscillations in Physical Systems*, McGraw-Hill, 1964

15 Akturk, N., Dynamics of a Rigid Shaft Supported by Angular Contact Ball Bearings, PhD thesis, Imperial College, London, 1993.

16 Rao, S. S., *Mechanical Vibrations,* Addison Wesley, 1986, No.C3.

17 Wardle, F. P. "Vibration Forces Produced by Waviness of the Rolling Surfaces of Thrust Loaded Ball Bearings" Part 1: Theory and Part 2: Experiment, Proc. I. Mech. E. **202**, C5, 1988, 303-319.

18 Akturk, N., "The Effects of Waviness on Vibrations Associated with Ball Bearings." Trans ASME J. Tribology, 1998, accepted for publication.

19 Brandlein, J., "Properties of Rolling Bearings Supporting Machine Tool Spindles" Ball and Roller Bearing Engineering, 1985, 44 -54.

20 Ellis, E. E., "Misaligned Ball Bearings" Tribology, Feb.1970.

APPENDIX 1

The rotor is assumed to be uniform and rigid, supported by two angular contact ball bearings at each end, having the following specification:

Inner ring bore	(D_i)	0.04 m
Inner ring diameter	(d_i)	0.046 m
Outer ring diameter	(d_o)	0.062 m
Outer ring outside diameter	(D_o)	0.068 m
Inner ring groove radius	(r_{gi})	0.00408 m
Outer ring groove radius	(r_{go})	0.00461 m
Ball diameter	(d_b)	0.00794 m
Number of balls	(m)	8 or less
Unloaded contact angle	(α_0)	15 deg.
Pitch diameter of the ball set	(d_m)	0.054 m
Mass of rotor	(M)	5.5 kg

APPENDIX 2

Assume that Fig. A2.1 is an angular contact bearing, initially under a preload causing deflection δ_0 at each ball. Let the inner ring now be given a deflection

$$\Delta\delta = \sqrt{x^2 + y^2 + z^2} \quad \textbf{(A2.1)}$$

where the offset angle is given by

$$\beta = \tan^{-1}\frac{y}{x} \quad \textbf{(A2.2)}$$

Assume that, because of the relatively low external loads employed here, the contact angle is constant at its preload value α_p.

The radial deflection at the ith ball due to this offset and the preload combined is

$$\delta_{ir} = \delta\left[\cos(\theta_i - \beta)\cos^2\alpha_p + \sin^2\alpha_p\right] \quad \textbf{(A2.3)}$$

where $\delta = \delta_0 + \Delta\delta$

Thus, the total forces in the x, y and z directions are:

$$W_x = \delta^{3/2}\sum_{i=1}^{m} KA_i\cos\alpha_p\cos\theta_i \quad \textbf{(A2.4)}$$

$$W_y = \delta^{3/2}\sum_{i=1}^{m} K\ A_i\cos\alpha_p\sin\theta_i \quad \textbf{(A2.5)}$$

$$W_z = \delta^{3/2}\sum_{i=1}^{m} K_i A_i\sin\alpha_p \quad \textbf{(A2.6)}$$

where $A_i = \left[\cos(\theta_i - \beta)\cos^2\alpha_p + \sin^2\alpha_p\right]^{3/2}$

C553/044 © IMechE 1998

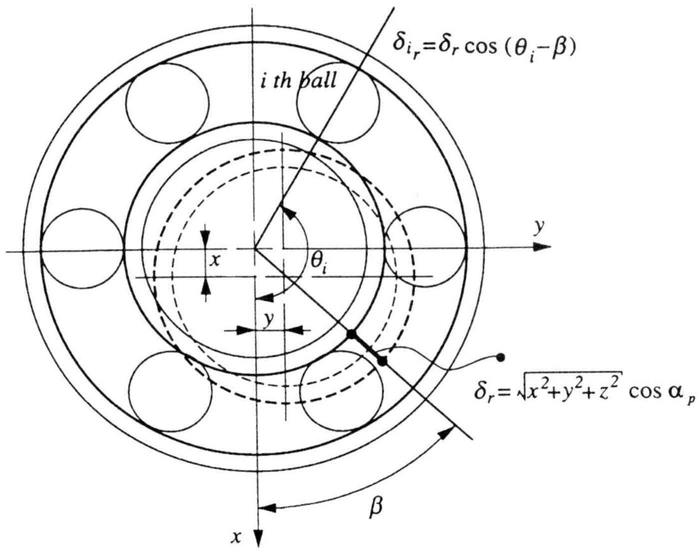

Fig. A2.1 Force on the *i*th Ball due to Rotor Centre Deflection

From these forces, the stiffness coefficients in the *x*, *y* and *z* directions can be derived. Choosing only the *x* direction, the force can be written as:

$$W_x = S\,\delta^{3/2} \tag{A2.7}$$

where

$$S = \sum_{i=1}^{m} KA_i \cos\alpha_p \cos\theta_i \tag{A2.8}$$

Eq. (A2.7) suggests that, at a given instant of time t_1, S is constant. Therefore the derivation of Eq. (A2.7) with respect to deflection will give an instantaneous stiffness for one of the bearings around the value of δ_0. Therefore:

$$(k_0)_x = \frac{\partial W_x}{\partial x} = \frac{3}{2} S(\delta)^{1/2}_{\delta=\delta_0} \tag{A2.9}$$

Thus, the approximate *natural frequency* of the model in the *x* direction will be

$$(\omega_n)_{\delta=\delta_0} = \left(\frac{3S}{M}\right)^{1/2} \left(\delta^{1/4}\right)_{\delta=\delta_0} \tag{A2.10}$$

As $\delta = \delta_0 + \Delta\delta$ and we assume that $\Delta\delta$ is small compared with δ_0, :

$$(\omega_n)_{\delta=\delta_0} = \left(\frac{3S}{M}\right)^{1/2} \delta_0^{1/4} \left(1 + \frac{1}{4}\frac{\Delta\delta}{\delta_0}\right) \tag{A2.11}$$

From Eq. (A2.6):

$$P_{ra} = mK(\delta_0)^{3/2} \sin\alpha_p$$

Therefore, Eq. (A11) will become:

$$(\omega_n)_{\delta=\delta_0} = \left(\frac{3S}{M}\right)^{1/2} \left[\left(\frac{P_{ra}}{mK \sin\alpha_p}\right)^{1/6} + \frac{1}{4} \left\{ \frac{\Delta\delta}{\left(\frac{P_{ra}}{mK \sin\alpha_p}\right)^{1/2}} \right\} \right] \tag{A2.12}$$

C553/044 © IMechE 1998

C553/040/98

New design, analysis, and performance techniques in hydrodynamic fluid film bearings

J K MARTIN MSc, PhD, CEng, MSTLE, MSAE, MIMgt, FIMechE
The Open University, Milton Keynes, UK

SYNOPSIS

New developments in fluid film bearings include the ability to adjust hydrodynamic performance characteristics during operation. Versions have been devised achieving this continuously by proactive means, irrespective of the loading conditions on the bearing. Such bearings offer potential future benefits in a wide range of applications. Three versions of such an adjustable hydrodynamic bearing are briefly described along with a specially written computer analysis model and a practical test rig.

Practical results verified performance predicted by the theoretical model which was based on firmly established Newtonian principles, in particular the equilibrium and non–equilibrium of forces, and dynamic viscosity.

NOTATION

A	Substitution parameter
a_{xx}, a_{yy}	Oil film displacement coefficients
B_L	Bearing axial length
D	Diameter of central shaft with no adjustments
e	Eccentricity of shaft and rotor axes
h	Oil film thickness

© IMechE 1998 C553/040

H	Non-dimensionalised oil film thickness
i, j	Bearing surface general node location
m	Finite difference mesh, segment circumferential end node
n	Finite difference mesh, segment axial end node
P	Lubricant hydrodynamic pressure
$\overline{P}$	Non-dimensionalised hydrodynamic pressure
P^*	Substitution pressure parameter
t	Time
$\bar{t}$	Non–dimensionalised time
T	Lubricant temperature
u_0	Inner rotor surface tangential speed
x, y, z	Directions circumferential, radial, axial
η	Lubricant dynamic viscosity
$\overline{\eta}$	Non-dimensionalised lubricant dynamic viscosity
η_1	η for inlet supply conditions
ω	Rotor angular velocity

1. INTRODUCTION

Hydrodynamic fluid film bearings are widely used in many areas of mechanical engineering where rotating parts are supported. Generating their own lubricant film pressures they can carry large loads, provide significant stiffness, and damping, while preventing physical contact between the bearing surfaces. They can be of radially and axially supporting forms. The radial type most commonly comprises a shaft or journal rotating within a stationary bearing housing. Less common is the inverse arrangement of a rotor on a stationary shaft, spindle or journal. The axial form is the thrust bearing where the rotating member bears on thrust faces reacting axial loads. The principles of hydrodynamic lubrication apply equally to all forms.

Hydrodynamic lubrication is a self sustaining fluid film separation of the two bearing surfaces. Loads are carried by pressures generated independently of supply pressure, the latter merely ensuring adequate supply of lubricant. The lubricant, almost always oil, must have significant viscosity, there must also be considerable relative motion of the surfaces, and the oil film shape must at some part be convergent. The phenomenon was put on a mathematical footing by Reynolds in 1886 and most hydrodynamic lubrication theory involves some form of Reynolds' pressure field equation. This equation relates velocity and pressure induced fluid flows based on the assumptions of continuity of volume flow rate and equilibrium of fluid film forces. It predicts that substantial pressures are generated which maintain full film lubrication under large loads.

The continuing desire for better performance in terms of higher speeds, lower energy losses, vibration suppression, accuracy of location, and noise signatures has led to increased attention focusing on bearing behaviour. In addition some hitherto satisfactory bearings are failing drastically under increasingly severe performance demands.

A novel form of adjustable fluid film bearing has been devised whereby the hydrodynamic conditions can be changed in an externally and continuously controlled manner during operation. The principle can be applied to conventionally orientated journal bearings; to inverse orientations; and to thrust bearings. Investigations have been carried out on theoretical and practical models for journal bearings of both orientations. The novel bearing has demonstrated significant improvements over conventional bearing designs in terms of stiffness, damping, rotational accuracy, power losses and temperature rise.

Performance characteristics predicted by the computer model have been demonstrated in practice.

2. ADJUSTABLE BEARING SCHEMES

Figure 1 shows an adjustable hydrodynamic bearing of conventional orientation supporting a rotating shaft or journal. A number of pads (only one is shown) form the bearing support segments, each producing independent hydrodynamic pressure fields in the lubricant film between the pad and the journal. A continuously variable mounting device supports the pad providing a radially stiff load path between the loaded journal, lubricant film and the casing. The term continuous can refer to both the infinitely small variation of adjustment and the time sense.

Figure 2 shows an inverse orientation where a rotor is supported on a stationary shaft, spindle or journal. A number of adjustable segments, in this case 4, are formed within the shaft. Each produces independent hydrodynamic pressure fields in the lubricant film between the segment and the rotor. Again a continuously variable mounting device supports each segment providing a radially stiff load path between the loaded rotor, lubricant film and stationary shaft.

Figure 3 shows 2 forms of thrust bearing. For each form only one of a number of pairs of support pads is shown. Independent hydrodynamic pressure fields are formed in the lubricant film between the rotating member and pad. Continuously variable mounting devices support each pad providing axially stiff load paths between the rotating member and stationary pads via the lubricant film. One pair of thrust pads shown has 2 adjustable supports for each pad. In principle the pads for both orientations of journal bearing could also have 2 independently adjustable supports instead of just one. It should be noted that in Figs.1, 2 and 3 the displacements and adjustments shown are greatly exaggerated for clarity. In practice pad adjustments of the order 20 μm producing changes in converging film shape gradient of about 0.1 ° are sufficient to greatly alter the hydrodynamic conditions.

3. TEST RIGS

There are some difficult practical issues to be resolved when designing test rigs for fluid film bearings. One is that as the quantities to be measured are usually small in relation to the magnitudes of characteristic data ultimately yielded there is great potential for significant errors to be introduced and magnified. For example, fluid film thicknesses are very small in relation to other dimensions such as journal diameter and changes in film thickness and eccentricities smaller still. Loads however tend to be quite large, so properties such as stiffness in N mm^{-1} comprise a mixture of both.

To avoid unwanted motions and disturbances therefore supporting structures need to be massive and stiff. It is still not possible however to provide perfectly rigid support systems so it is very difficult to produce reliable data on absolute measurements such as the precise location of the 'fixed' bearing centre. This is particularly true when the structure is subject to thermal influences when the bearing is warming up or otherwise changing its operating temperatures. There is also a need to provide static and dynamic loads from a firm base but without unduly influencing the bearing properties.

Positional measurements tend to be made with inductive displacement transducers. These have the advantages of being comparatively cheap (compared with lasers, for example) and, importantly, are not affected by the variable presence of any oil mist (and other liquids) in the gap between probe tip and bearing member. They can however be prone to drift errors over time, and with temperature changes, requiring frequent re–calibration if small absolute displacements are to be accurately assessed. It has been found however that linear characteristic gradients arc usually consistent despite these drift errors.

A number of authors have addressed these issues in various ways. Swanson and Kirk (1) have surveyed and summarised these in tabular form, and make recommendations on some weaknesses which should be addressed in future experimental works. Flack et al (2) have described a test rig with dynamic measurement capabilities. This has load vibrators set at ± 45° from the vertical above the test bearing, requiring a degree of rigid support well above floor level but permitting non influencing links to transmit forces. Parkins (3) describes experimental techniques using imposed vibration orbits arising from two mutually perpendicular external oscillating forces of variable relative magnitude and phase. Comparisons were later made of results arising from two different techniques using the same basic test rig (4). In both cases, electromagnetic vibrators applied dynamic forces through connectors which transmitted both tension and compression, but exerted no lateral constraint over the limited bearing housing displacement.

Figure 4 shows the overall view of the main test rig assembly used for testing a practical version of the concept as applied to an inverse orientation form of the bearing, documented in full by Martin (5) The shaft and rotor system are shown in Figs. 5 and 6. For this orientation the central shaft remained stationary at all times and the rotor was supported on it by full film lubrication, using mineral oil to ISO VG 32. The rotor was belt driven from a variable speed DC motor. This bearing and rotor system simulated a proposed design of machine tool spindle. There were two hydrodynamic bearings of the novel adjustable configuration widely spaced along the shaft and a single 4 pocket hydrostatic bearing in between. The purpose of the hydrostatic bearing was to provide full film lubrication between the rotor and shaft for zero speed and all other conditions when the hydrodynamic bearings were not operating.

The steel shaft was 70 mm nominal diameter and provided the hydrostatic bearing pockets, their feeds and restrictors, the hydrodynamic bearings, their oil feeds, adjustment devices, and a combined scavenge oil return. Oil supplies entered via end blocks containing galleries and 'O' ring seals. This manifold supply arrangement (not shown) minimised external pipework allowing efficient access to the adjuster controls and other items. Also shown in Fig. 5 is one of eight adjuster pins (4 at each end) which form part of the hydrodynamic bearings and bear beneath the eight adjustable segments. Each adjustable segment was of a specially shaped cantilevered form supported by the adjuster pin. There were 4 adjustable segments associated with each hydrodynamic bearing.

Each adjuster pin comprised a small location diameter, a tapered support length, engagement thread and square end for locating a key. The tapered portion supported the under surface of the adjustable segment providing a radially stiff load path. Turning the adjuster pin caused the taper to move longitudinally creating a wedging action. The effect was to lift (or lower) the segment surface slightly in a controlled and continuous manner, thereby changing the hydrodynamic conditions. All adjuster pins engaged to produce lift in a smooth and repeatable manner for about $1\frac{1}{2}$ complete turns, which equated to a bearing surface lift greater than the available radial clearance in situ in the bearing.

Figure 7 shows the main rotor loading scheme. The rotor was steel and of composite construction designed to interact with 2 large specially made electro-magnets. The bore was coated with a tin rich white metal. Each electro-magnet was mounted in a cradle and supported by linkages and self-aligning ball bearing assemblies located on 2 circular mountings set concentrically with the central stationary shaft. By this means each magnet line of force could be set in a series of angular positions from horizontal to vertical. Setting both magnets at the same angle at the lowest position above horizontal (22.5°) however, allowed a load equivalent to the rotor weight to be quickly vectored in virtually any radial direction by suitable adjustment of the magnet forces, without the need for repositioning.

Inductive displacement transducers provided rotor radial displacement data which were processed based on techniques recently developed by Parkins (6). In these data relating to changes in rotor position could yield information on rotor centre of rotation, eccentricities and displacement coefficients. These relied on using the transducer calibrated characteristic gradients which were found to be thermally stable throughout testing. Figure 8 shows a typical set of observed position traces for the rotor centre as moved in a repeatable manner by the adjusters. The load, speed, oil inlet temperature etc. were held constant and the effects of

C553/040 © IMechE 1998

each adjuster observed in turn, a few seconds apart. The movement of the rotor centre of rotation in 4 orthogonal directions is clearly evident. Perhaps just as important if the load magnitude and or direction were changed it was a simple matter to re–position the rotor centre to a given location, including the zero eccentricity position, by selective adjustments. Thus a given centre of rotation position could be maintained irrespective of load. This adjustable feature has obvious benefits in many applications. Further experimental results are highlighted in Section 5.

4. THEORETICAL MODEL

The governing Reynolds' equation for the fluid film pressure field acting on 1 adjustable segment was derived with reference to the fluid element depicted in Fig. 9. Pressure, film thickness and viscosity were set to vary non-uniformly in two dimensions; circumferentially around and axially along the segment surface. The derivation was based on a number of other assumptions most of which are normally invoked as follows

- Constant fluid density (i.e. not affected by pressure or temperature).
- Fluid acceleration forces are considered negligible compared with viscous shear forces.
- Weight of fluid is negligible.
- Film thickness is small compared with bearing surface radii of curvature.
- Pressure remains constant radially through the film thickness at each point.
- Viscosity constant radially through film thickness at each point.
- Viscosity a variable independent of pressure (initially).
- Relative inclination of bearing surfaces is small.
- No fluid slip takes place at fluid-bearing surfaces.
- No loss of fluid occurs due to leakage.
- A constant volume flow rate is assumed.
- Fluid forces are assumed in equilibrium.
- Fluid is considered to be Newtonian.

In its usual form the pressure field Reynolds' equation as formulated, including squeeze film velocity is given by:

$$\frac{\partial}{\partial x}\left(h^3\,\eta^{-1}\,\frac{\partial P}{\partial x}\right)+\frac{\partial}{\partial z}\left(h^3\,\eta^{-1}\,\frac{\partial P}{\partial z}\right)=6\,u_0\,\frac{\partial h}{\partial x}+12\frac{\partial h}{\partial t} \qquad [1]$$

The variables are non-dimensionalised as:

$$\bar{x}=\frac{x}{\pi D},\ \bar{z}=\frac{z}{B_L},\ H=\frac{h}{C_{31}},\ \bar{\eta}=\frac{\eta}{\eta_1},\ \bar{P}=\frac{P}{\omega\eta_1\left(\frac{R}{C_{31}}\right)^2},\ \bar{t}=t\omega$$

The Reynolds' equation becomes:

$$\frac{\partial}{\partial\bar{x}}\left(\frac{H^3}{\bar{\eta}}\frac{\partial\bar{P}}{\partial\bar{x}}\right)+\left(\frac{\pi D}{B_L}\right)^2\frac{\partial}{\partial\bar{z}}\left(\frac{H^3}{\bar{\eta}}\frac{\partial\bar{P}}{\partial\bar{z}}\right)=12\pi\frac{\partial H}{\partial\bar{x}}+48\pi^2\frac{\partial H}{\partial t} \qquad [2]$$

The substitution parameter $P^*=\bar{P}\,H^{3/2}$ is used to smooth the pressure derivatives resulting in the Reynolds' equation:

$$\left(\frac{H^3}{\bar{\eta}}\,\frac{\partial}{\partial\bar{x}}\left(P^*H^{-3/2}\right)\right)+\left(\frac{\pi D}{B_L}\right)^2\frac{\partial}{\partial\bar{z}}\left(\frac{H^3}{\bar{\eta}}\,\frac{\partial}{\partial\bar{z}}\left(P^*H^{-3/2}\right)\right)=12\pi\frac{\partial H}{\partial\bar{x}}+48\pi^2\frac{\partial H}{\partial t} \qquad [3]$$

To expand this equation for the conventional type of journal bearing analysis it is frequently assumed that the surfaces are perfectly cylindrical, and their longitudinal axes are parallel, in which case both derivatives of H by $\bar{z}$ are zero. Furthermore, derivatives of H by $\bar{x}$ can usually be expressed by differentiation of a relatively simple expression relating H to the angular position. In this work none of these are assumed, owing to the non-uniform variation of one side of the oil film (the adjustable segment surface). Thus, expansion of equation [3] required full development of the differential products of P^*, H and $\bar{\eta}$, resulting in an extended Reynolds' equation:

$$
\begin{aligned}
& H^{3/2}\,\bar{\eta}^{-1}\frac{\partial^2 P*}{\partial \bar{x}^2} - H^{3/2}\frac{\partial P*}{\partial \bar{x}}\,\bar{\eta}^{-2}\frac{\partial \bar{\eta}}{\partial \bar{x}} + \bar{\eta}^{-1}\frac{\partial P*}{\partial \bar{x}}\,\frac{3}{2}\,H^{1/2}\frac{\partial H}{\partial \bar{x}} \\
& -\frac{3}{2}P*H^{1/2}\,\bar{\eta}^{-1}\frac{\partial^2 H}{\partial \bar{x}^2} + \frac{3}{2}P*H^{1/2}\frac{\partial H}{\partial \bar{x}}\,\bar{\eta}^{-2}\frac{\partial \bar{h}}{\partial \bar{x}} - \frac{3}{4}\,\bar{\eta}^{-1}\left(\frac{\partial H}{\partial \bar{x}}\right)^2 P*H^{-1/2} \\
& -\frac{3}{2}\,\bar{\eta}^{-1}\frac{\partial H}{\partial \bar{x}}\,H^{1/2}\frac{\partial P*}{\partial \bar{x}} + AH^{3/2}\,\bar{\eta}^{-1}\frac{\partial^2 P*}{\partial \bar{z}^2} - AH^{3/2}\frac{\partial P*}{\partial \bar{z}}\,\bar{\eta}^{-2}\frac{\partial \bar{\eta}}{\partial \bar{z}} + A\,\bar{\eta}^{-1} \\
& \frac{\partial P*}{\partial \bar{z}}\frac{3}{2}H^{1/2}\frac{\partial H}{\partial \bar{z}} - \frac{3}{2}AP*H^{1/2}\,\bar{\eta}^{-1}\frac{\partial^2 H}{\partial \bar{z}^2} + \frac{3}{2}\,AP*H^{1/2}\frac{\partial H}{\partial \bar{z}}\,\bar{\eta}^{-2}\frac{\partial \bar{\eta}}{\partial \bar{z}} - \frac{3}{4}\,A\,\bar{\eta}^{-1} \\
& \left(\frac{\partial H}{\partial \bar{z}}\right)^2 P*H^{1/2} - \frac{3}{2}\,A\,\bar{\eta}^{-1}\frac{\partial H}{\partial \bar{z}}\,H^{1/2}\frac{\partial P*}{\partial \bar{z}} = 12\pi\frac{\partial H}{\partial \bar{x}} + 48\pi^2\frac{\partial H}{\partial t}
\end{aligned}
\qquad [4]
$$

where: $A = \left(\frac{\pi D^2}{B_L}\right)^2$

Finite difference approximations are made for derivatives of pressure, film thickness and viscosity in x and z, and for temperature in the x direction, which are then used to solve the expanded Reynolds' equation for one adjuster segment. A uniform mesh is assumed comprising 143 finite difference nodes distribution in the circumferential and axial directions by m and n respectively over the fluid film surface of one adjuster segment. The general node integer locations is i, j. Full details of this and the temperature and viscosity models are reported by Martin (5). In brief, a suite of specially written computer programmes produces finite difference nodal values of lubricant temperature, viscosity and pressure using the Gauss–Seidel iteration technique, provided information about the film thickness profile is known.

A 3D finite element solid model of the adjustable segment is created and meshed in such a way that the finite difference nodal pressures would each be applied directly to a finite element face centroid, (see Fig. 10.) In this way a pressure field can be applied directly as a load case and the effect of such pressures on the adjustable segment shape be determined by a subsequent finite element analysis. A specially written computer programme interrogates the finite element analysis results and produces a film thickness profile, allowing for input adjustments and rotor eccentricities. A map of squeeze film velocities is also produced.

A computational processing technique combines repeated finite element analyses of each adjustable segment with updating finite difference computations in between. This process continues automatically until for all 4 adjustable segments the fields of film thickness, temperature, viscosity and pressure all mutually and simultaneously converge to a satisfactory solution. One pressure field is shown in Fig.11. Using this process a comprehensive parametric study was carried out providing useful information for the practical design. Results were also compared with subsequent test measurements.

5. DISCUSSION OF RESULTS.

The main feature – the ability to continuously adjust the position of the centre of rotation of the rotating member, on demand, with different load magnitudes and directions applied – has

C553/040 © IMechE 1998

been frequently and repeatedly demonstrated. This characteristic represents a significant step in the design of fluid film bearings with potential application in a number of fields. Despite extensive searches no reference to such designs or behaviour have been found, prompting British Technology Group to apply for a number of patent applications internationally with a view to licensing the technology (7). The bearing also demonstrated potentially beneficial performance characteristics in a number of other operating modes, divided broadly into conditions of zero eccentricity or of non-zero eccentricity.

For non-zero eccentricity conditions and with adjusters set at zero the bearing behaved like a conventional cylindrical journal bearing. When a load was applied the centre of rotation was displaced from the zero position to a new eccentric equilibrium position in which the hydrodynamic fluid film pressures provided a bearing force to balance the load. If, however, the adjusters were set equally to a given adjustment then upon application of the same load this eccentricity was reduced. The amount of reduction in eccentricity was influenced by the degree of prior adjustment, the more adjustment the less the eccentricity. The effects were similar whether the load was directed on to an adjustable segment (on–segment) or between adjustable segments (off–segment). This is shown graphically in Fig. 12. This property was manifest as an ability to increase the stiffness of the bearing. This effect was measured practically and predicted by the theoretical model in the form of displacement coefficients at zero load, Fig. 13. Similar effects were observed regarding other displacement coefficients, and damping as predicted by the theoretical model, but measurement of damping was not attempted.

Zero eccentricity could be maintained for a variety of loads by differential segment adjustments. It was observed both practically and theoretically that loads in any direction, and with a range of magnitudes, could be balanced while maintaining zero eccentricity. Significant stiffness and damping were maintained for all conditions studied. Moreover significant stiffness and damping were still present when loads were removed leaving the selective adjustments unchanged, producing an unloaded non-zero eccentricity position.

Experiments with a conventionally orientated journal bearing have been carried out using a non-continuous method of adjustment (8). These have produced many similar results to those reported above, in addition to demonstrating significant reductions in power absorbed.

6. CONCLUSIONS.

A novel bearing design has been devised comprising a non-linear multi-body system with interfaces between solid and fluid and solid to solid boundaries. A small degree of structural elastohydrodynamic action was also present.

The theoretical analysis and practical measurement of such a bearing have both used recently devised techniques combining finite element and finite difference models and specially written computer processing software. All such techniques were based on firmly established principles, however. In particular Newtonian models of solid and fluid mechanics were invoked and Reynolds' pressure field equation was derived and extended to cover the new application.

The theoretical model was used comprehensively in parametric studies and to optimise design details relevant to the manufacture of practical versions of the bearing. Test rigs were specially designed and built to demonstrate the adjustable bearing and use recently developed measurement techniques. Performance characteristics predicted by the theoretical model have been observed in practice.

The characteristics of the bearing give rise to properties of advantage over current bearing designs, and potential future benefit in a variety of applications in fields such as vehicles, plant and machinery.

ACKNOWLEDGEMENTS

The author is pleased to acknowledge the support of the Open and Cranfield Universities, British Technology Group Ltd., and Stress Elements Ltd.

REFERENCES

1 **Swanson, E.E., Kirk, R.G.,** 'Survey of Experimental Data for Fixed Geometry Hydrodynamic Journal Bearings'. *Trans. ASME. Jnl. Trib.* Vol. 119, Oct 1997.

2 **Flack, R.D., Kostrzewsky, G.J., Taylor, D.V.,** 'A Hydrodynamic Journal Bearing Test Rig with Dynamic Measurement Capabilities'. *STLE. Trib. Trans.* Vol. 36, No. 4, 1993.

3. **Parkins, D.W.,** 'Theoretical and Experimental Determination of the Dynamic Characteristics of a Hydrodynamic Journal Bearing'. *Trans. ASME.Jnl. Lub. Tech.* Vol. 101, April 1979.

4 **Parkins, D.W.,** 'Measured Characteristics of a Journal Bearing Oil Film'. *Trans. ASME.Jnl. Lub. Tech.* Vol. 103, Jan. 1981.

5 **Martin, J.K.,** 'Investigations into an Adjustable Fluid Film Bearing', *Cranfield University PhD Thesis,* March 1997.

6 **Parkins, D.W.,** 'Measurement of Oil Film Journal Bearing Damping Coefficients - an Extension of the Selected Orbit Technique'. *Trans. ASME. Jnl. Trib.* Vol. 117, Oct. 1995.

7 **Martin, J.K., Parkins, D.W.,** 'Fluid Film Bearings', *International* (PCT) *Patent Application.* WO-95/29346, November 1995.

8 **Muhsin, I.A.,** 'Design and Evaluation of a Novel Film Journal Bearing', *Cranfield University PhD Thesis*, March 1991.

 C553/040 © IMechE 1998

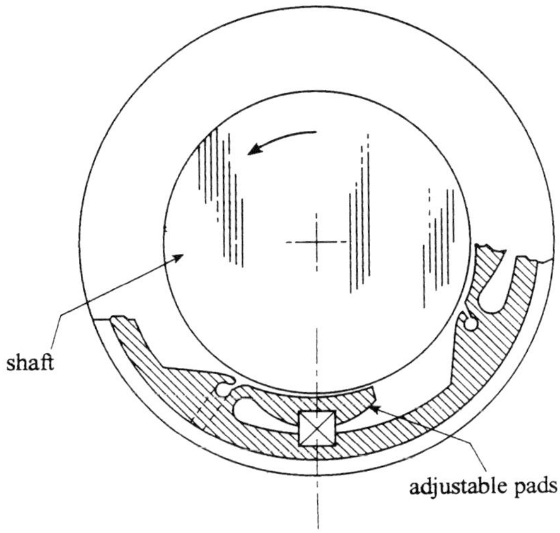

Fig. 1 Adjustable hydrodynamic bearing, journal type

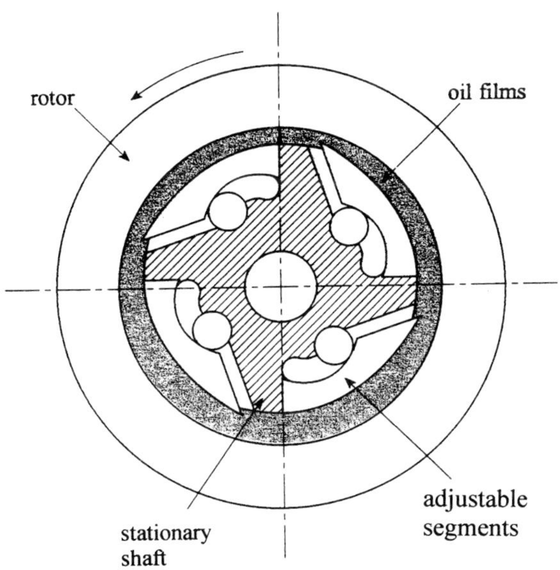

Fig. 2 Adjustable hydrodynamic bearing, rotor type

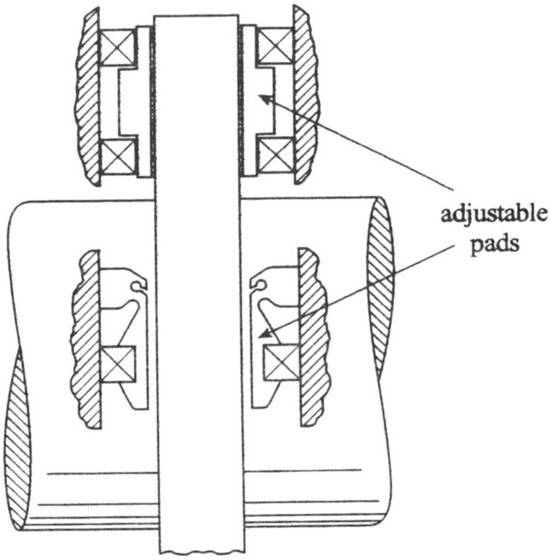

Fig. 3 Adjustable hydrodynamic bearing, thrust types

Fig. 4 Main test rig assembly

C553/040 © IMechE 1998

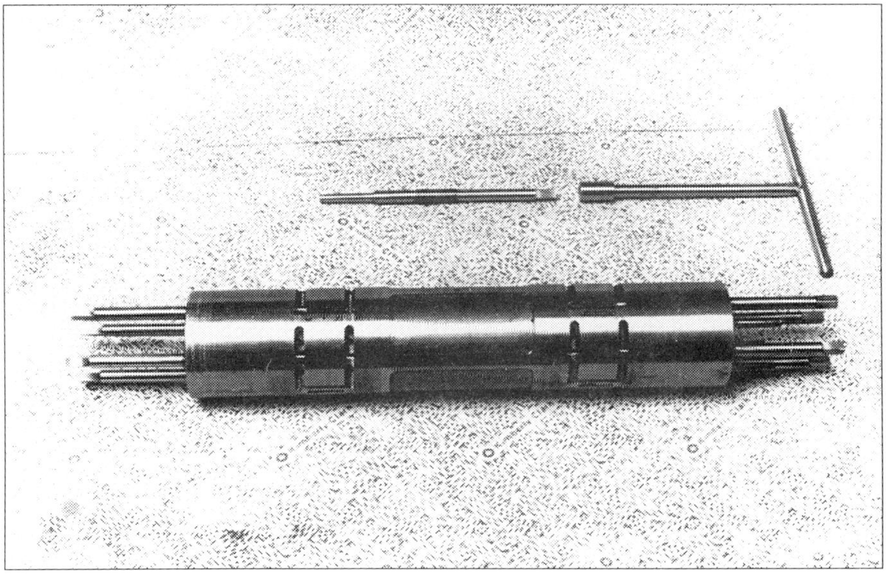

Fig. 5 Central shaft assembly

Fig. 6 Rotor assembly

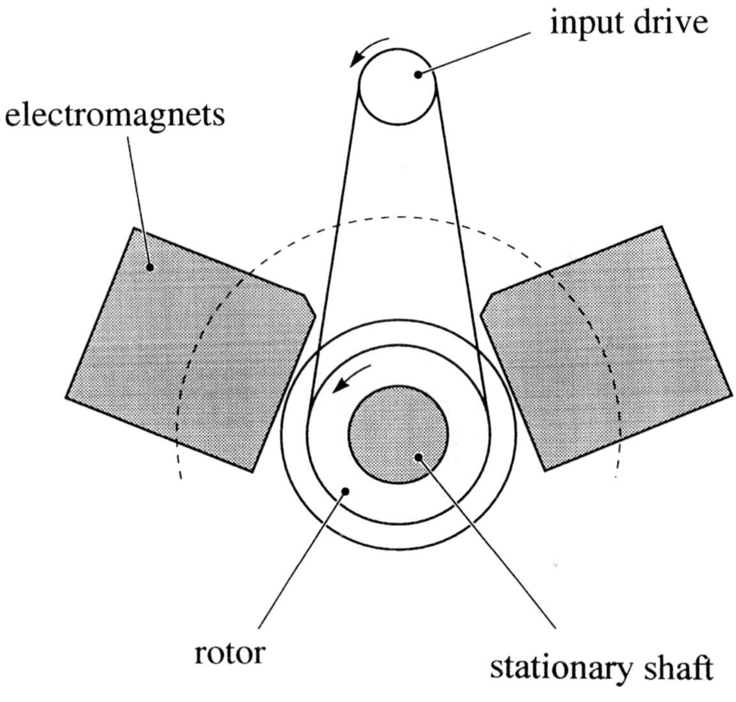

Fig. 7 Electro-magnet loading scheme

C553/040 © IMechE 1998

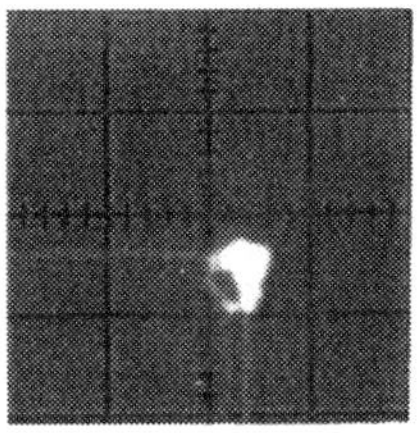

Rotor 3500 rpm
Segment 1 adjusted

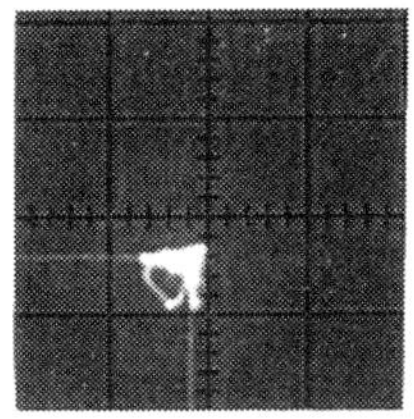

Rotor 3500 rpm
Segment 2 adjusted

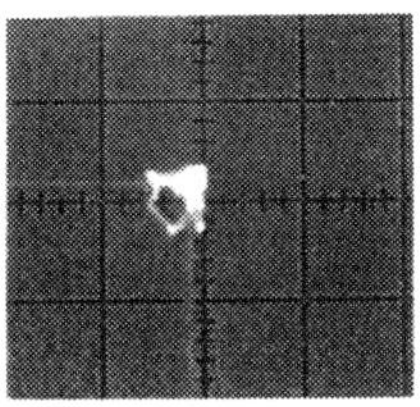

Rotor 3500 rpm
Segment 3 adjusted

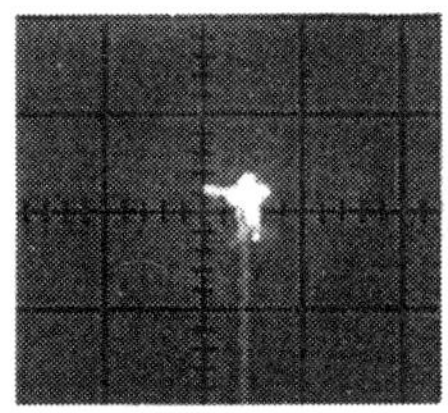

Rotor 3500 rpm
Segment 4 adjusted

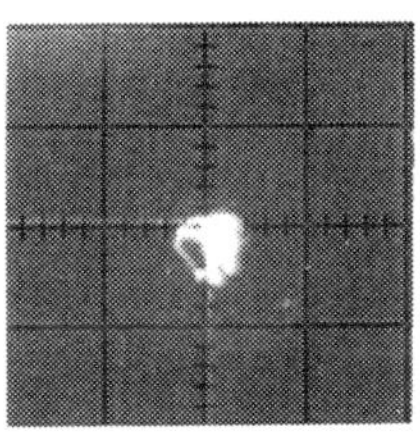

Rotor 3500 rpm
Zero adjustments

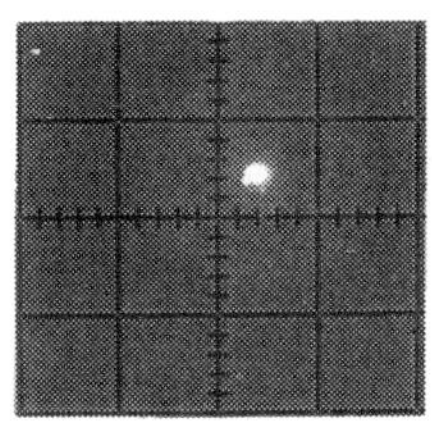

Rotor stationary
Zero adjustments

Fig. 8 Rotor position traces

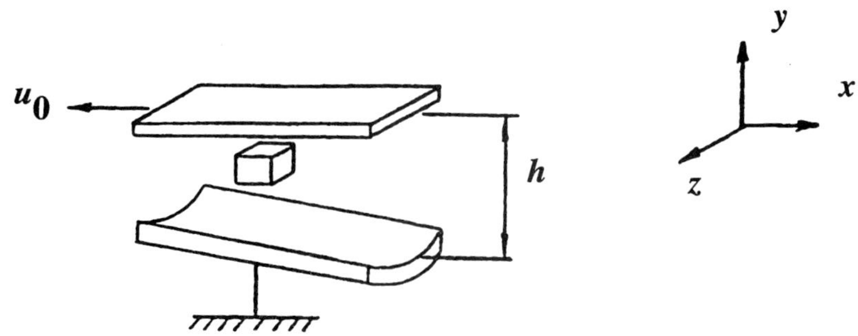

Fig. 9 Hydrodynamic lubrication fluid element

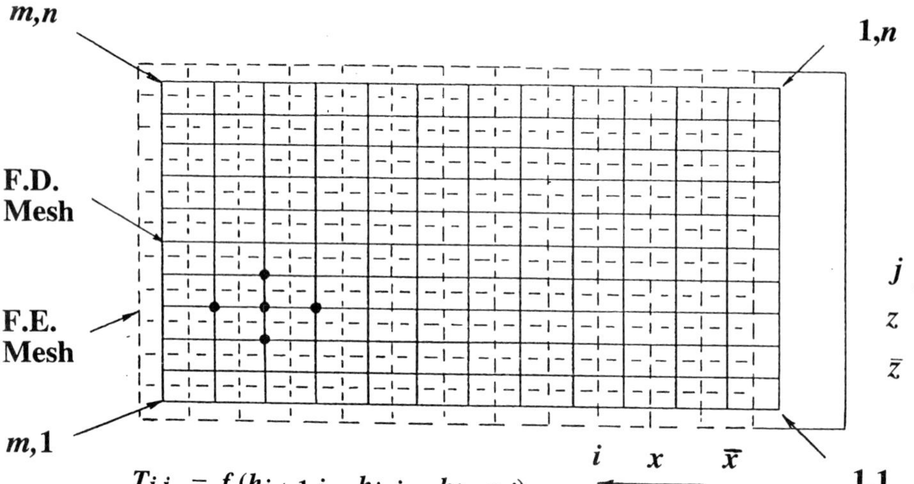

$$T_{i,j} = f(h_{i+1,j},\ h_{i,j},\ h_{i-1,j})$$

$$\eta_{i,j} = f(T_{i,j},\ P_{i,j})$$

$$P_{i,j} = f(h_{i+1,j},\ h_{i,j},\ h_{i-1,j},\ h_{i,j+1},\ h_{i,j-1},\ \eta_{i,j},\ v_p\ i,j)$$

Fig. 10 Segment oil film surface meshes

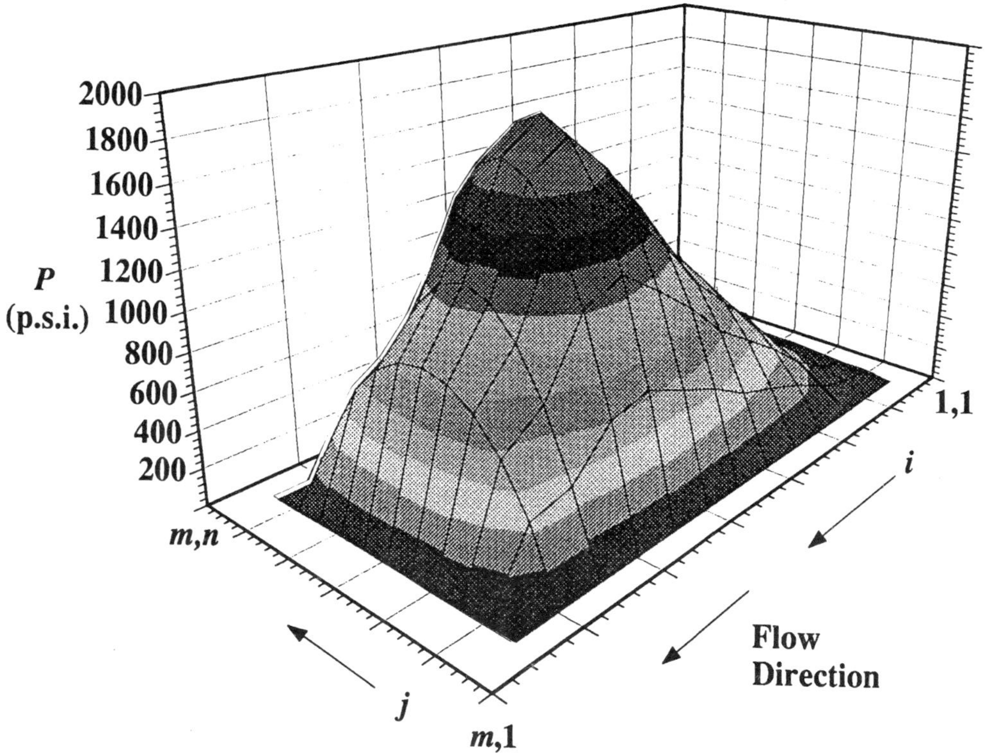

Fig. 11 Typical computed pressure field for 1 segment

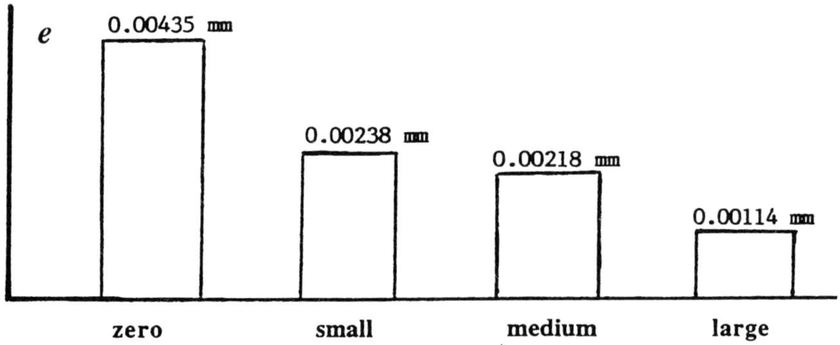

Segment adjustments, load on-segment

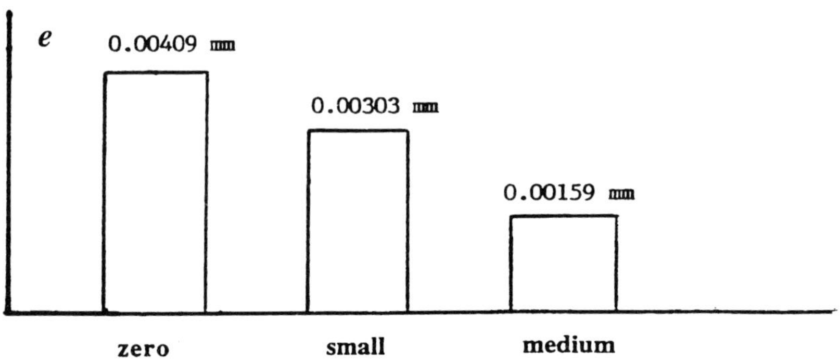

Segment adjustments, load off-segment

Fig. 12 Maximum eccentricities for segment adjustments

C553/040 © IMechE 1998

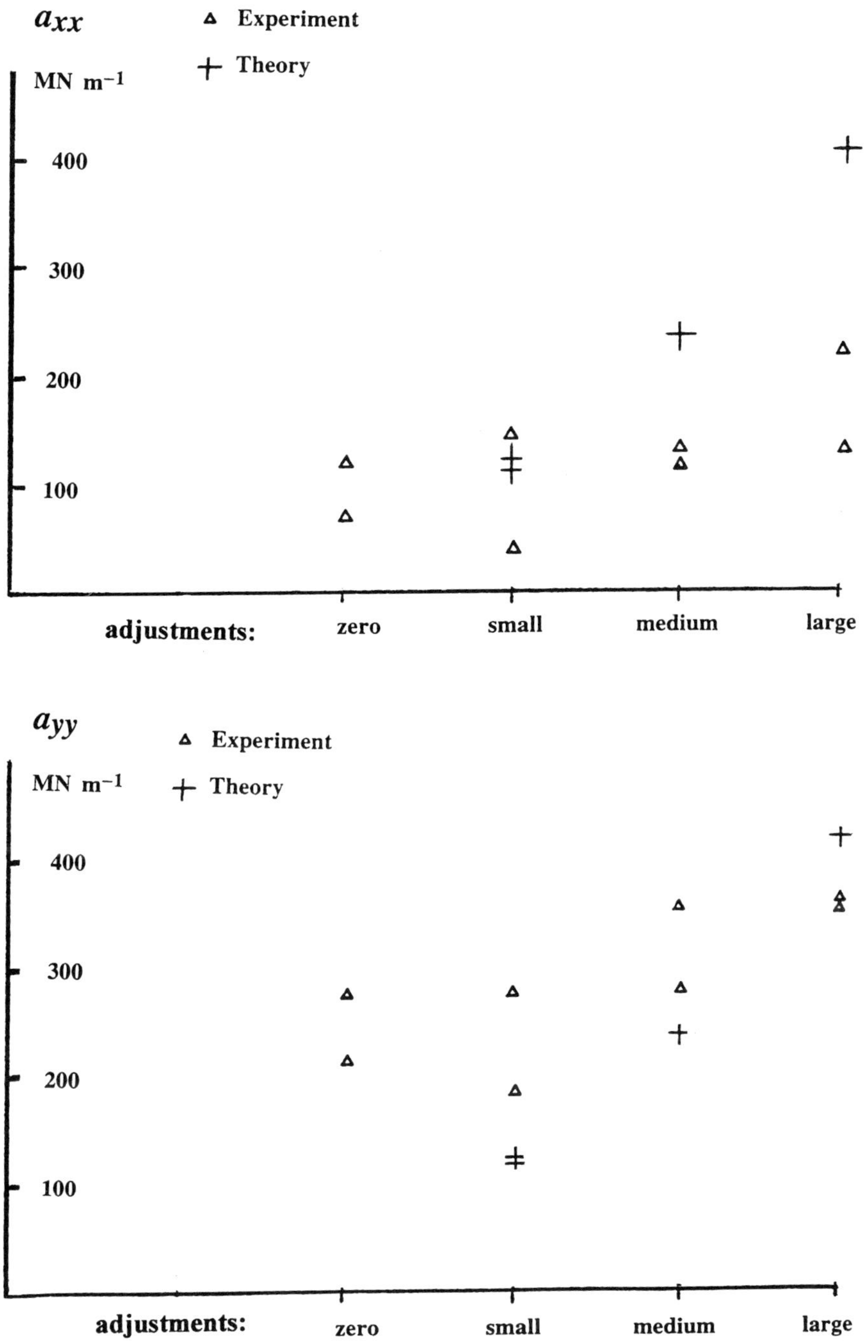

Fig. 13 Displacement coefficients, effect of adjustments

C553/041/98

Tribo-multibody dynamic analysis of valve-train systems

M KUSHWAHA BEng, **H RAHNEJAT** BSc, MSc, PhD, DIC, CEng, MIMechE, FCybS, and **Z M JIN** BEng, PhD, CEng, MIMechE, MESB
Department of Mechanical Engineering, University of Bradford, UK
P M JOHNS BSc, MSc, PhD, DIC
First Dynamics Limited, Ilkley, UK

SYNOPSIS

This paper presents a generic multi-body model of a vehicular valve train for cam and flat follower arrangement. The model incorporates the inertial members, the constraint functions, and the sources of compliance in the valve train. The sources of compliance include the valve train spring, as well as the contact compliance between the cam and the flat follower. The contact domain is treated as a counterformal concentrated lubricated region, subjected to elastohydrodynamic regime of lubrication (EHL). The prevailing contact geometry is one of finite line contact.

1. INTRODUCTION

The contact between cam and the follower is responsible for a sizeable proportion of the frictional losses experienced in the internal combustion engines (1,2). A progressive demand for increasing engine speeds in modern motor vehicles, and in particular in racing vehicles has resulted in the valve train performance becoming a critical factor. Faster valve accelerations are, therefore, desired which can lead to increases in inertial imbalances, thus rendering high levels of Noise, Vibration and Harshness (NVH) (3,4).

Difficulties also exist in the formation of a coherent lubricant film in the contact conjunction between the cam and follower (5,6). Indeed under steady state conditions theoretical analyses point to the loss of a lubricant film in the vicinity and immediately prior to the cam nose to follower contact (7,8). This problem is further exacerbated at cold start and at low speeds of rotation due to high frictional torques and/or thin film thicknesses. In fact it

© IMechE 1998 C553/041

is claimed, through theoretical investigations, that in the aforementioned regions the predominant regime of lubrication is due to boundary films. However, experimental evidence obtained by Hamilton (5) does not concur with the theoretical findings. Until recently the presence of a lubricant film measured in these regions could not be explained using the theory of elastohydrodynamic lubrication (EHL). Solutions obtained for transient EHL conditions point to a combined entraining and squeeze film action (7,9,10). These solutions indicate the critical role that the squeeze film action plays in the retention of a lubricant film, where the contribution due to rolling is insignificant (either because of low entraining speeds or a reversal in the position of inlet boundary near the cam nose). The effect of the squeeze film action is particularly significant at the start up and low speeds of entraining motion. Furthermore, its contribution is enhanced by the presence of trapped lubricant in the surface undulations formed by asperities. Therefore, the absence of wear is much owed to the combined effects of squeeze film motion and local trapped films.

The failure of the rolling surfaces is not only caused by the long process of wear, but also by fatigue spalling due to the generation of unacceptable sub-surface stresses caused by high lubricant pressures in the elastohydrodynamic conjunction (11,12). Therefore, the mechanical properties of materials for cam and followers, as well as the lubricant pressures and dynamic loads are important in the design of cam/follower pairs.

It is clear that design of new valve trains is critically affected by the cam and follower contact conditions. The trend for the future development of internal combustion engines is to increase valve accelerations, whilst minimising NVH effects. This points to a reduction in the mass of the inertial members (i.e. using lighter cams and followers and their attachments), thereby reducing the valve spring rate. However, lighter materials point to reduced fatigue strength of contacting members. A compromise approach is, therefore, the key for future developments to come. An integrated approach incorporating both the NVH aspects of the valve train and cam-follower lubricated contact dynamics is seen as essential.

This paper highlights a detailed methodology for such an integrated study.

2. The Multi-body Model

The multi-body dynamic model of the system is a lumped parameter model (see figure 1) that consists of the inertial components of the system referred to as "parts". The parts are constrained to move with respect to one another in a specified manner, thus rendering the ideal functions that are desired in the valve train system. The aim of the multi-body model is to incorporate the "parts" and the system constraints, as well as including the sources of system compliance. The overall model, therefore, can be used for a kinematic or a non-linear dynamic analysis of the system in order to investigate its NVH performance.

2.1 Inertial Parts

There are 3 parts in the model: ground (or the fixed datum), the cam and the follower. Mass and inertial properties of the latter two are provided in table 1.

C553/041 © IMechE 1998

Part Name	Mass (Kg)	Inertial Values (Kg mm^2)		
		Ixx	Iyy	Izz
Cam (inc. proportion of camshaft)	0.468	78.4	78.4	190.1
Valve (inc. retainer)	0.202	200.7	33.4	200.7

Table 1: Mass and inertial properties

2.2 Constraints

Parts in the multi-body model as described above, are constrained to one another by rigid constraints. These constraints are applied at given geometric locations between parts. Table 2 lists the joints in the model. Each joint includes a number of constraints that are represented by scalar algebraic constraint functions. These functions are non-linear algebraic equations. The number of constraints imposed by each formulated joint type is given in the key to table 2.

Part I	Part J	Joint/Constraint
Cam	Ground	Cylindrical
Cam	Valve	Curve-to-Curve
Valve	Ground	Translational

Key: No. of constraints: Cylindrical = 4 , Curve-Curve = 2 , Translational = 5

Table2: Contraints in the model

A curve-to-curve constraint function is employed between the flat follower and the cam radial profile. Such a constraint introduces 2 scalar algebraic functions that inhibit the separation of two bodies. Therefore, a point on the cam surface remains in contact with the tappet at all times. Contact separation is only allowed through elastic deformation in the contact due to the generation of elastohydrodynamic pressures. This small gap is responsible for the formation of a lubricant film due to the combined effect of entraining and squeeze film action. No other separation (as a result of camshaft local flexion) is allowed, nor the effect of tappet spin is included. However, in the real situation such motions can take place under dynamic conditions.

2.3 Sources of Compliance

The sources of compliance in a real valve train are manifold. These include the following:

- valve spring stiffness and damping (the latter ignored in the model)
- the contact compliance in the lubricated conjunction between the cam and follower described above
- the contact compliance between the valve stem and the valve seat (also ignored in the model due to its complexity- this requires a separate detailed analysis)
- tangential traction in the contact domain which is also omitted in the current analysis

The valve spring stiffness is considered to be constant at 38 N/mm. In automotive valve trains the valve spring coils are irregularly spaced, resulting in non-linear characteristics. Such springs can be modelled by lumped parameter formulation.

2.4 Cam-Tappet Concentrated Contact

The cam to flat follower contact at a particular cam angle can be regarded to be the same as the contact that a roller of the same radius as the cam's instantaneous radius makes with an elastic half-space. Such a dry contact is now well understood and follows the Hertzian stress analysis of a roller indenting a semi-infinite elastic half-space. The Hertzian theory assumes that the surfaces of the roller and the plane are perfectly frictionless and hence the generated reaction forces are normal to the plane of contact. When the two bodies are elastic, the contact length is assumed to be slightly shorter than the roller length and the strains that are applied remain within the elastic limit. The contact width is necessarily small in comparison to the principal radii of curvature of the bodies, in this case the instantaneous cam radius.

Using the Hertzian theory, the deflections under the induced contact pressure at any point in the contact region can be obtained as (13):

$$\delta(x,y) = [\frac{(1-\nu_1^2)}{\pi E_1} + \frac{(1-\nu_2^2)}{\pi E_2}] \int_{x_1}\int_{y_1} \frac{p(x,y)}{\sqrt{(x-x_1)^2+(y-y_1)^2}} dx_1 dy_1 \qquad [1]$$

The pressure distribution is assumed to be elliptical in the transverse direction for dry contacts following the Hertzian theory. This of course is not the case for lubricated contacts under the EHL condition. However, the EHL pressure distribution follows closely the Hertzian pressure profile closely, except for the inlet trail and the exit secondary pressure peak. The conformity to the Hertzian conditions is further emphasised with thin films and starved inlets, a feature of cam/follower contact. Therefore, for all intent and purposes the contact deformation can be obtained from the Hertzian pressure distribution. This makes the evaluation of the central deflection quite simple by analytic means. Hence, such an analytic expression can be included in a multi-body model in order to determine the contact deformation under load and the corresponding contact width, eliminating the need to solve the elasticity problem in each time step of the transient contact dynamics by a finite difference formulation. The EHL of the contact is described below and is based upon the extrapolated oil film thickness equation reported by Rahnejat (9). Hence the transient tribo-dynamic problem becomes more manageable in each step of time.

Assuming the lateral Hertzian elliptical pressure profile and that the pressure and the footprint half-width at any section are proportional to their respective values at the contact centre (13), then:

$$\frac{b_1}{p_1} = \frac{b_0}{p_0} = \frac{D}{4E} \qquad [2]$$

The central contact half-width and the corresponding pressure can be obtained from the classical Hertzian theory as:

C553/041 © IMechE 1998

$$b_o^2 = \frac{2}{\pi}\left[\frac{(1-\nu_1^2)}{E_1} + \frac{(1-\nu_2^2)}{E_2}\right]\frac{WD}{L} \qquad [3]$$

$$p_o = \frac{2W}{\pi b_o L} \qquad [4]$$

The lateral elliptical pressure profile in any cross-section of the overall pressure distribution is given by:

$$p(x_1, y_1) = p_1\left[1 - \left(\frac{x_1}{b_1}\right)^2\right]^{1/2} \qquad [5]$$

Substituting from equation [5] into equation [1], and integrating over the contact domain, the deflection at any point x,y can be obtained. This procedure can be carried out for all points within the contact domain. For the central contact point (i.e. x=y=0):

$$\delta(0,0) = b_0 p_0\left[\frac{1-\nu_1^2}{E_1} + \frac{1-\nu_2^2}{E_2}\right]\left(\ln\frac{4b}{b_0} + 0.5\right) \qquad [6]$$

The load deflection characteristics of the contact in terms of the central deflection can be obtained by replacing in equation [6] from equations [3] and [4]. This would indicate the non-linearity of the contact compliance. It should also be observed that the contact compliance alters with the instantaneous contact radius of the cam. Such variation of the contact stiffness itself is responsible for high frequency contents in the valve train spectrum, a fact that is largely ignored by most of the reported models.

The above analysis, whilst useful in the analytic evaluation of the contact deformation and dimensions, is based upon the assumption of an axial uniform pressure distribution. This assumption yields a rectangular footprint which deviates from reality, where the footprint shape has been shown to approximate to a "dog-bone" shape (14). To obtain the film thickness and the correct pressure distribution a solution to finite line EHL is required.

3. Formulation of Elastohydrodynamic Conjunction

The lubricated contact pressure distribution occurs axially along the depth of cam, and laterally along the direction of entraining motion, with the contact half-widths in all cross-sections being much smaller than the contact length. This gives rise to a line contact condition of finite length. Therefore, a finite line EHL solution is sought. For this purpose the simultaneous solution of the Reynolds' hydrodynamic equation and the elastic film shape is necessary. Solutions for finite line contact EHL problem have been attempted to a much lesser extent than those for elliptical point contact conditions. The main solutions have been observed by Mostofi (15), Mostofi and Gohar (10), and Rahnejat (9). The work in reference [9] includes the effect of squeeze film action. The other solution is reported by Dowson, Taylor and Zhu (7) for the case of cam-follower contact. This paper follows the same general finite line solution methodology reported in reference (9) for the case of combined entraining and squeeze film action.

3.1 Reynolds' Equation

For the velocity induced flow in the transverse contact direction y, subjected to normal approach of mating rolling members, the Reynolds' equation is:

$$h^{*3/2}[\frac{\partial}{\partial \bar{x}}(\bar{\rho}\frac{\partial \phi}{\partial \bar{x}})+\frac{\partial}{\partial \bar{y}}(\bar{\rho}\frac{\partial q \phi}{\partial \bar{y}})]-\frac{3}{2}\phi[\frac{\partial}{\partial \bar{x}}(\bar{\rho}h^{*1/2}\frac{\partial h^*}{\partial \bar{x}})+\frac{\partial}{\partial \bar{y}}(\bar{\rho}h^{*1/2}\frac{\partial h^*}{\partial \bar{y}})]= \qquad [7]$$

$$12\{u^*\frac{\partial(\bar{\rho}h^*)}{\partial \bar{y}}+\bar{\rho}w_s^*\}$$

Where: $\phi = q^* h^{*3/2}$ and:

$$q^* = \frac{1}{G^*}(1-e^{-G^*P^*}) \qquad [8]$$

The variation of density with lubricant pressure is obtained from:

$$\bar{\rho} = \frac{1+0.6P*}{1+1.7P*} \qquad [9]$$

3.2 Elastic Film Shape

The EHL film thickness can be written in dimensionless form as (9,10,14):

$$h^*(\bar{x},\bar{y}) = \{\bar{p}_r(\bar{x},\bar{y}) - [\bar{\delta}(0,0) - \bar{\delta}(\bar{x},\bar{y})]\} + h_0^* \qquad [10]$$

The term $\bar{p}_r(\bar{x},\bar{y})$ is the separation of the bodies in their undeformed state (i.e. the cam profile in this case). The term in the square brackets is referred to as the elastic film shape, and h_0^* is the dimensionless central oil film thickness.

It should be noted that the profile of the cam varies along the direction of entraining motion, $\bar{y}$ and is flat in the transverse direction, $\bar{x}$. Therefore:

$$\bar{p}_r(\bar{x},\bar{y}) = \frac{\bar{y}^2}{2} \qquad [11]$$

The elastic film shape components in equation [10] are obtained from equation [1] and put in the dimensionless form as:

$$\bar{\delta}(\bar{x},\bar{y}) = \frac{\delta(x,y)}{R} \qquad [12]$$

C553/041 © IMechE 1998

3.3 The Extrapolated Film Thickness Equation

A simultaneous quasi-static solution of the EHL problem with multi-body kinematic model renders an unacceptable computation time. The solution for the inertial rigid body kinematics is carried out in a large number of time intervals, typically in the region of 500 steps for each cam revolution. The time requirement for a solution of the EHL problem is typically in the region of 1-2 hours per time step. Therefore, a simultaneous solution is impractical. However, the effect of the lubricated contact dynamics in the multi-body analysis is essential. An appropriate approach is to obtain an extrapolated equation to describe the non-linear characteristics of the EHL conjunction under transient contact dynamic conditions. The solution of the EHL problem under combined entraining and squeeze film motion has been shown to yield such equations that can be used in the dynamic analysis of bearings [15] and gears [16]. The same approach is to be adopted here for the case of valve train multi-body dynamics.

Such an equation has been obtained by Rahnejat [9]. For the central oil film thickness:

$$h_0^* = 1.67W^{*0.059}U^{*0.541}G^{*0.421}e^{-96.775W_s^*} \qquad [13]$$

And for the side constriction oil film thickness:

$$h_{sc}^* = 1150W^{*-0.037}U^{*0.857}G^{*0.406}e^{-138.059W_s^*} \qquad [14]$$

The central oil film thickness equation, conforming closer to the actual numerical results is employed in the contact dynamic analysis to obtain film thickness at all the intervals of time.

4. The Multi-body Model Formulation

There are two parts in the multi-body model, the motion of each of which can be described in terms of the generalised co-ordinates, q by the Lagrange's equation for constrained systems as:

$$\frac{d}{dt}\left(\frac{\partial K}{\partial \dot{q}}\right) - \frac{\partial K}{\partial q} - F_q + \sum_{k=1}^{n} \lambda_k \frac{\partial C_k}{\partial q} = 0 \qquad [15]$$

The generalised co-ordinates are given by: $\{q\} = \{x, y, z, \psi, \theta, \varphi\}^T$, where the rotational components are the Euler angles (i.e. body 3-1-3 successive rotations).

The reaction forces in the multi-body system are given by the summation term in equation [15] along each of the generalised co-ordinates. These are introduced by non-linear algebraic scalar functions, C_k. Therefore, the assembly of parts can be represented mathematically in a manner that conforms with the required dynamic functions of the system.

Under dynamic conditions equation [15] provides six equations of motion per part in the valve train system.

The multi-body model described in section 2, has 3 parts (including the ground) and 12 imposed constraints which includes the specified cam rotation. Using Gruebler-Kutzbach expression, a kinematic model results:

DOF= 6(n-1) - Σconstraints = 6(3-1) – 12 = 0 [16]

4.1 FORMULATION OF CONSTRAINT FUNCTIONS

Note that the motions in the kinematic mechanism are prescribed by the cam rotation and valve translation, themselves linked by the combustion process as well as by the applied assembly constraints. The coupled action of cam rotation and valve timing is determined by the cam lift profile. The simultaneous solution of constraint functions and kinematic force equations resulting from equation [15] is required.

Therefore, the constraint functions must be formulated. These include:

(i)- cylindrical joint:

$$z_i.x_j = 0, z_i.y_j = 0, A_{ij}.x_j = 0, A_{ij}y_j = 0 \quad [17]$$

(ii)- translational joint:

$$y_i.x_j = 0, y_i.z_j = 0, x_i.z_j = 0, A_{ij}.x_j = 0, A_{ij}z_j = 0 \quad [18]$$

(iii)- curve-curve:

$$A_{ij} = 0, T_i.N_j = 0 \quad [19]$$

(iv)- motion

$$\psi_i = 2\pi f_q t \quad [20]$$

where: $(x, y, z)_{i,j}$ indicate the unit vector directions of local part frames of reference of the two parts that form a constraint function and A_{ij} represents the at-point constraint at the assembly location of two neighbouring parts.

Now the Euler transformation matrix is employed in order to transform the above constraint function into algebraic equations in terms of the generalised co-ordinates: $\{x, y, z, \psi, \theta, \varphi\}^T$. The transformation matrix is given as:

$$[T] = \begin{bmatrix} C\psi C\varphi - S\psi C\theta S\varphi & -C\psi S\varphi - S\psi C\theta C\varphi & S\psi S\theta & 0 \\ S\psi C\varphi + C\psi C\theta S\varphi & -S\psi S\varphi - C\psi C\theta C\varphi & -C\psi S\theta & 0 \\ S\theta S\varphi & S\theta C\varphi & C\theta & 0 \\ 0 & 0 & 0 & 1 \end{bmatrix} \quad [21]$$

where : $C \equiv Cos, S \equiv Sin$.

Now applying for the following joints:

(i)- The cylindrical joint between the cam (part 1) and the ground (part 3):

$$C_1 = A_{13}.x_3 = x_1 - L_c(S\psi_1 C\theta_1 S\varphi_1 - C\psi_1 C\phi_1) = 0 \quad [22]$$
$$C_2 = A_{13}.y_3 = y_1 - L_c(C\psi_1 C\theta_1 C\varphi_1 + S\psi_1 S\phi_1) = 0 \quad [23]$$
$$C_3 = z_1.x_3 = S\theta_1 S\varphi_1 = 0 \quad [24]$$
$$C_4 = z_1.y_3 = S\theta_1 C\varphi_1 = 0 \quad [25]$$

(ii)- The translational joint between the valve (part 2) and the valve guide, considered as ground (part 3):

$$C_5 = A_{23}.x_3 = x_2 - L_v(S\psi_2 C\theta_2 S\varphi_2 - C\psi_2 C\phi_2) = 0 \quad [26]$$
$$C_6 = A_{23}.z_3 = z_2 + L_v C\theta_2 = 0 \quad [27]$$
$$C_7 = y_2.x_3 = S\psi_2 C\varphi_2 + C\psi_2 C\theta_2 S\varphi_2 = 0 \quad [28]$$
$$C_8 = y_2.z_3 = -C\psi_2 S\theta_2 = 0 \quad [29]$$
$$C_9 = x_2.z_3 = S\psi_2 S\theta_2 = 0 \quad [30]$$

(iii)- The curve-to-curve joint between the cam (part1) and the valve (part 2):

$$N_2 = y_2 \quad [31]$$
$$T_1 = \frac{\partial y_1}{\partial \psi_1} \quad [32]$$

$$C_{10} = T_1.N_2 = \left(\frac{\partial y_1}{\partial \psi_1}\right).y_2 = 0 \quad [33]$$

where: $y_1 = s\left[\frac{\psi_1}{\beta} - \left(\frac{1}{2\pi} Sin(\frac{2\pi\psi_1}{\beta})\right)\right]$ [34]

and:

$$C_{11} = x_1 - x_2 = y_1 - y_2 = z_1 - z_2 = 0 \quad [35]$$

(iv)- The cam rotation:

$$C_{12} = \psi_1 = 2\pi f_q t \quad [36]$$

There are 12 constraint functions ($C_1 - C_{12}$) and 12 unknowns. A position kinematic analysis can then be carried out to observe the articulation of the mechanism.

4.2 FORCE KINEMATICS

However, the main interest is in a combined force kinematic and lubricated contact dynamics of the valve train system. With the kinematic mechanism (as is the case with valve train systems, with an assumed rigid camshaft), the inertial terms in the Lagrange's equation [15] are ignored, yielding a reduced Jacobian matrix that yields equation [37]. This equation is then solved simultaneously with the 12 constraint functions for the required solution vector, comprising the state variables; q and the Lagrange multipliers; λ.

$$\left[\frac{\partial C}{\partial q}\right]\{\lambda\} = -\{F_q\} \qquad [37]$$

$$C(q) = 0 \qquad [38]$$

For each constraint function there exist $\lambda \frac{\partial C_k}{\partial q}$ terms in equation [37], which provide the joint reaction forces/torques. The particular reaction of interest is related to the curve-to-curve constraints in the y-direction in the global frame of reference. This represents the instantaneous contact force between the cam and the follower.

Using equation [37], the following force kinematic relations are obtained for the parts in the multi-body model as:

For cam:

$$\begin{bmatrix} 1 & 0 & 0 & 0 & 0 \\ 0 & \frac{s}{\beta}[1-C(\frac{2\pi\psi_1}{\beta})] & 0 & 0 & 0 \\ 0 & 0 & 1 & 0 & 0 \\ L_c S\psi_1 S\theta_1 S\varphi_1 & L_c C\psi_1 S\theta_1 C\varphi_1 & 0 & C\theta_1 S\varphi_1 & C\theta_1 C\varphi_1 \\ -L_c(S\psi_1 C\theta_1 C\varphi_1 - C\psi_1 S\varphi_1) & -L_c(S\psi_1 C\varphi_1 - C\psi_1 C\theta_1 S\varphi_1) & 0 & S\theta_1 C\varphi_1 & -S\theta_1 S\varphi_1 \end{bmatrix} \begin{bmatrix} \lambda_1 \\ \lambda_2 \\ \lambda_3 \\ \lambda_4 \\ \lambda_5 \end{bmatrix} = \begin{bmatrix} 0 \\ (k_s y_2 + f) - mg \\ 0 \\ 0 \\ 0 \end{bmatrix} \qquad [39]$$

where: $C \equiv Cos, S = Sin$.

For valve:

$$\begin{bmatrix} 1 & 0 & 0 & 0 & 0 & 0 \\ 0 & \frac{s}{\beta}[1-C(\frac{2\pi\psi_1}{\beta})] & 0 & 0 & 0 & 0 \\ 0 & 0 & 1 & 0 & 0 & 0 \\ -L_v(C\psi_2 C\theta_2 S\varphi_2 + S\psi_2 C\varphi_2) & 1 & 0 & C\psi_2 C\varphi_2 - S\psi_2 C\theta_2 S\varphi_2 & S\psi_2 S\theta_2 & C\psi_2 S\theta_2 \\ L_v(S\psi_2 S\theta_2 S\varphi_2) & 0 & -L_v S\theta_2 & -C\psi_2 S\theta_2 S\varphi_2 & -C\psi_2 C\theta_2 & S\psi_2 C\theta_2 \\ -L_v(S\psi_2 C\theta_2 C\varphi_2 + C\psi_2 S\varphi_2) & 0 & 0 & C\psi_2 C\theta_2 C\varphi_2 - S\psi_2 S\varphi_2 & 0 & 0 \end{bmatrix} \begin{bmatrix} \lambda_6 \\ \lambda_7 \\ \lambda_8 \\ \lambda_9 \\ \lambda_{10} \\ \lambda_{11} \end{bmatrix} = \begin{bmatrix} 0 \\ (k_s y_2 + f) - mg \\ 0 \\ 0 \\ 0 \\ 0 \end{bmatrix} \qquad [40]$$

C553/041 © IMechE 1998

where: $\lambda_2 = \lambda_7$ and both relate to the curve-to-curve vertical contact reaction force.

For a force kinematic analysis there are 12 unknowns in $[q_j]=[x_j, y_j, z_j, \psi_j, \theta_j, \varphi_j]^T$ for $j = 1$(cam) and $j = 2$(valve) and 11 unknowns in $[\lambda_i], i = 1,11$. There are 23 equations; 12 constraint functions $C_1 - C_{12}$ and 11 relations in equations [39] and [40]. A simultaneous solution is obtained for this equation set in small steps of time.

4.3 LUBRICATED CONTACT DYNAMICS

The cam-follower contact is subjected to elastohydrodynamic lubrication. The contact load, being the instantaneous integrated lubricant pressure distribution, given by:

$$W = \frac{s}{\beta}[1 - C(\frac{2\pi\psi_1}{\beta})]\lambda_2 = (k_s y_2 + f) - mg \qquad [41]$$

where: W is a function of the oil film thickness, given by either of the expressions [13] or [14] (in dimensionless form, see notation).

Therefore, the lubricant film thickness: $h = h^* R$ can be obtained from expression [13] by replacing for the dimensionless groups (with the load obtained from equation [41]).

When the cam lift occurs, the cam-follower contact is subjected to a combined entraining and squeeze film motion. The lubricant film thickness and the corresponding squeeze film velocity are obtained in each step of time as follows:

$$\frac{\partial h_n}{\partial t} = \frac{h_{n-1} - h_n}{\Delta t} \qquad [42]$$

$$h_n = R h^*_{sc} \qquad [43]$$

The initial conditions for each time step, Δt, are:

$$h_{n-1} = h_r = R h^*_{sc} \mid \frac{\partial h_{n-1}}{\partial t} = 0 \text{ and } h_n = 1.05 h_r \qquad [44]$$

5. RESULTS OF MULTI-BODY DYNAMIC MODEL

The main task in the design of cam/follower friction pair is to ensure coherent lubrication by providing a sufficient oil film thickness guarding against wear, whilst ensuring contact loads that do not exceed the fatigue strength limits.

The problem is that conditions that yield low loads and thus reduced Hertzian pressures do not necessarily render optimal lubrication conditions. It has been shown that a cam with a higher radius of curvature, enjoying lower maximum Hertzian stresses can yield lower lubricant film thicknesses. This finding is now generally accepted and has been corroborated by both experimental and theoretical investigations (5,7,17). Quasi-static EHL solutions have indicated higher lubricant film thicknesses at high Hertzian stresses than those generated by other prevalent regimes of lubrication at low contact loads (7,8,17). The rather complex relationship between dynamic contact loads and lubrication can only be studied in an integrated solution of valve train kinematics and lubricated contact dynamics of cam/follower.

The solution provided in this paper, therefore, integrates EHL of finite line contact of cam and follower in the kinematics of the valve train system.

A cycloidal cam profile, typical of a mid-height camshaft and a flat follower arrangement has been employed. The cycloidal cam in such an arrangement can have a rather broadened flat nose in order to follow the required constant lift profile as shown in figure 2. The cycloidal curve is used often as a basis of designing cams, especially for high speed applications. It yields low noise, vibration and wear characteristics. Furthermore, there are no abrupt changes in the associated valve train accelerations and for a given rise time the acceleration is somewhat higher than that of many other types of cam. The trend in the development of modern valve train systems is towards faster acting systems with lower component masses. Therefore, modified cycloidal curves can be employed to produce the desired valve accelerations and owing to high accelerations the instances of zero entraining velocities in the contacting region are short lived, resulting in rapid lubricant replenishment. However, high values of acceleration and deceleration with cycloidal cams can result in high inertial forces which must be kept under control with correspondingly stiff valve springs. Therefore, valve train design problems are somewhat accentuated when cycloidal cams are employed. Precisely because of this reason, such a cam profile is employed in this study to illustrate the cam/follower design problems. Also that detailed kinematic, lubrication and structural integrity study of cycloidal cams have not been reported in literature to the same extent as the more conventional higher polynomial automotive cams. In addition cycloidal cams have been used extensively in other high speed machinery applications.

The kinematic relations for a flat follower is relatively simple. This allows for the differentiation of the valve lift over the cam angle which yields the velocity and the acceleration profiles (also shown in figure 2). Note should be taken that for an arrangement of

a cycloidal cam and a flat follower the follower acceleration has a sharp spike at the contact with the cam nose. This is because with the cycloidal curve, the follower acceleration is symmetrical about the cam half cycle. The advantage of this is that the acceleration variation is continuous (18). The disadvantage of this is seen in lubrication conditions within the vicinity and prior to the cam nose contact as will be discussed later. This disadvantage is alleviated due to fast contact transit time at high speeds (as previously discussed). Therefore, at low speeds of revolution the lubrication performance of this type of cam can be poor. The follower velocity profile exhibits a point of inflexion relating to the acceleration reversal at the cam nose contact. These characteristics of the arrangement employed here have been observed experimentally by Fessler and Ham (17).

Empirical formulae for lubricant film thickness for EHD line contact conditions have been obtained by various researchers (9,10,19). These formulae corrolate with the experimentally measured film reported by others (5,17), except for the regions where the speed of entraining motion is zero or subject to fast changes, indicating inlet boundary reversals. In these regions

C553/041 © IMechE 1998

the squeeze film effect is found to be a significant contributory factor in lubrication film retention (5,7,8,17). In the regions of zero entrainment velocity, the squeeze effect supports the loads and prevents surface to surface contact. It should also be noted that the time window in these regions is very short indeed and the entraining flow is rapidly replenished [8].

In the current study, finite line EHD contact under combined entraining and squeeze film motion is considered. A plot of lubricant film thickness against the cam angle is shown in figure 3 under steady state cyclic conditions for one complete cam cycle. The cam nose location is designated to be at 180°, with the flank and nose actions occurring from 90° to 270°. It should be noted that the entraining velocity crosses the zero value at four locations (see figure 4), corresponding to the instances of minimum film thickness shown in figure 3. In this cycloidal cam, zero entraining velocities occur at ±12° as well as ±36° to the cam nose location. The minimum lubricant film thickness is in the region of 0.1 μm. This is a very thin film thickness that is rapidly replenished in a short interval of time. The transit time between the occurrence of the first and the last (i.e. the fourth) minima at the engine idling speed of 1000 rpm is in fact 1 milli-second. Nevertheless, the major contribution to the fluid film formation between the cam angles +36° to -36° is by squeeze film lubrication. The contribution due to the entraining motion is minimal (in fact non-existence at the 4 minima locations).

The variation of squeeze velocity with cam angle is illustrated in figure 5, showing the large squeeze velocity variations that result in the vicinity of the lubricant film minima. Referring to figures 3 and 5, it can be observed that the film thickness remains almost constant in the region of cam base radius to flat follower contact (regions A and G in figure 3). The is no corresponding squeeze velocity in figure 5. The small fluctuations in both figures are due to a combination of numerical error and inaccuracies in the cam profile. As the contact moves on to the flank region (regions B and F) both the entraining velocity and the contact load increase (see figure 6) and a larger elastic deformation occurs. A separation effect is noted by the positively increasing squeeze velocity. When the contact moves towards the nose region the radii of curvature progressively decrease and along with it there is a corresponding reduction in the entraining velocity, the lubricated pressure distribution and the elastic deflection in the EHL conjunction. The squeeze velocity, having reached its maximum separation (i.e. positive) value is progressively reduced. It goes through a zero value and to a local maximum approach (i.e. negative) value to carry almost the entire contact load as the entraining velocity reverses direction (corresponding to an inlet reversal at 36° to the cam nose, see region C in figure 3). The negative values of squeeze velocity correspond to normal approach of contacting members, having their maxima where inlet reversals occur, followed by momentary separation of bodies which are represented by intances of positive squeeze velocity. As the inlet boundary reverses the speed of entraining motion is momentarily reduced to zero. This can be observed by rapid changes in the squeeze velocity in figure 5, corresponding to regions C and E in figure 3. As inlet reversals occur 4 times in a cam cycle, this phenomena is observed 4 times in figure 5 (for positions ±36° and ±12°). In between each set of inlet reversals pure entraining motion with an almost constant negative squeeze effect resumes. In the immediate vicinity of the cam nose (i.e. region D) the lubricant film thickness is first increased, reaching a maximum value and then immediately reduced as the contact load diminishes.

The is no squeeze effect during the entraining motion of the base circle against the flat follower, when a constant film thickness of 4.6 μm is predicted (see figure 3). The lubricant film at the nose is larger as the speed of entraining motion there is in excess of 4m/sec, giving a film thickness of 6.25μm. The lubricant film thickness is even larger along the cam flanks, but never in excess of 13μm for this analysis.

6. CONCLUSION

The results indicate that valve train kinematics, incorporating tribo-contact dynamics of cam and flat follower is a complex problem that profoundly affects the performance of IC engines. Traditional methods of dealing with lubrication, valve train kinematics, and contacting bodies structural integrity in isolation are not suitable for design analysis of modern valve train systems. The approach highlighted in this paper is seen to contribute to modern engine valve train development, where lighter components are increasingly subjected to higher duty cycles.

7. NOTATION

b	: Contact half-width
C	: Constraint function
D	: Instantaneous cam diameter
E	: Modulus of Elasticity
f	: Spring preload
F_q	: Vector of generalised forces
g	: Gravitational acceleration
G^*	: $E\alpha$
h	: Lubricant film thickness
$\dot{h}$	: Squeeze film velocity
h^*	: h / R
k_s	: Valve spring rate
K	: Kinetic energy
L	: Contact length
L_c	: Distance of cam local part reference frame from the global frame of reference
L_v	: Distance of valve local part reference frame from the global frame of reference
m	: mass
n	: Number of constraints
p	: Hertzian or EHL pressure
p^*	: p / E
p_r	: Cam contact profile
$\{q\}$	: Generalised co-ordinates
q^*	: Reduced hydrodynamic pressure
R	: Instantaneous reduced radius of the counterformal contact
s	: Follower lift
t	: Time

u : Entraining velocity
u^* : $u\eta / ER$
W : Contact load
W^* : W/ERL
w_s^* : Squeeze-roll speed ratio
x,y,z : Local Cartesian frame of reference
α : Lubricant piezo-viscosity index
λ : Lagrange multiplier
δ : Contact deflection
ρ : Lubricant density
η : Lubricant dynamic viscosity
ν : Poisson's ratio
ψ, θ, φ : Euler angles

8. REFERENCES

(1)- Taylor C.M., "Fluid film lubrication in automobile valve trains", Proc. Instn. Mech. Engrs., Part J: J. Engng. trib.,Vol.208, pp.221-234, 1994

(2)- Ciulli E., "A review of internal combustion losses, Part1: specific studies on the motions of pistons, valves and bearings", Proc. Instn. Mech. Engrs., Part D: J. Auto.Engng.,Vol.206, pp.223-226, 1992

(3)- Eichhorn U. and Schonfeld H., "The valve train of internal combustion engines as a source of vibrations-experimental results and methods of calculations", Proc. XXIII FISITA Congress, Torino (Italy), pp.389-395, May 1990

(4) Rahnejat H., Multi-body Dynamics: Vehicles, machines and mechanisms, PEP Press, June 1998.

(5)- Hamilton M., "The Hydrodynamics of a cam follower", Trib. Int., pp.113-119, June 1980

(6)- Dyson A., "Kinematics and wear patterns of cam and finger follower automotive valve gear", Trib. Int. pp.121-132, June 1980

(7)- Dowson D., Taylor C.M. and Zhu G., "A transient elastohydrodynamic lubrication analysis of a cam and follower", J. Phys. D: Appl. Phys., Vol.25,.pp.313-320, 1992

(8)- Xiaolan Ai., Haiqing Yu., "A numerical analysis for the transient EHL process of a cam-tappet pair in I.C. Engine", Trans. ASME, J. Trib., Vol.111, pp.413-417, July 1989

(9)- Rahnejat H., Influence of vibration on the oil film in concentrated contacts, PhD. Thesis, Imperial college of Science and Technology, 1984

(10)-Mostofi A. and Gohar R., "Oil film thickness and pressure distribution in elastohydrodynamic point contacts", J. Mech. Eng. Sci., Vol.24(4), pp.173-182, 1982

(11)-Johns-Rahnejat P.M. and Gohar R., "Point contact elasto-hydrodynamic pressure distribution and sub-surface stress field", in Rahnejat H. and Walley R. (Eds.) Multi-body dynamics: Monitoring and simulation techniques, MEP Press, pp.161-178,1997

(12)-Houpert L., Ioannides E., Kuypers J.C. and Tripp J., "The effect of the EHD pressure spike on rolling bearing fatigue", Trans. ASME., J. Trib.,Vol.108, 1986

(13)-Johns P.M. and Gohar R., "Roller bearings under radial and eccentric loads", Trib. Int., pp.131-136, June 1981

(14)-Jalili-Vahid D., Rahnejat H. and Jin Z.M., "Elasto-hydrodynamic lubrication of point contact under combined rolling and squeeze film motion", in Rahnejat H. and Walley R.

(Eds.) Multi-body dynamics: Monitoring and simulation techniques, MEP Press, pp.179-186,1997

(15)-Rahnejat H. and Gohar R., " The vibrations of radial ball bearings", Proc. Instn. Mech. Engrs., Vol.199(C3), pp.181-193, 1985

(16)-Mehdigoli H., Rahnejat H. and Gohar R., "Vibration response of wavy surfaced disc in elasto-hydrodynamic rolling contact", Wear, Vol.139, pp.1-15, 1990

(17)-Fessler H. and Ham R., "Lubrication and stress analysis as a basis for camshaft optimisation", Proc. XXIII FISITA Congress, Torino (Italy), pp.565-579, May 1990

(18)-Reeve J., Cams for Industry, MEP Press, 1997

(19)-Chittenden R.J., Dowson D., Dunn J.F. and Taylor C.M., "A theoretical analysis of the isothermal elasto-hydrodynamic lubrication of concentrated contacts-part II: general case, with lubrication entrainment along either principal axis of the Hertzian contact ellipse or at some intermediate angle", Proc. Roy. Soc. Lond., A, Vol.397, 1985

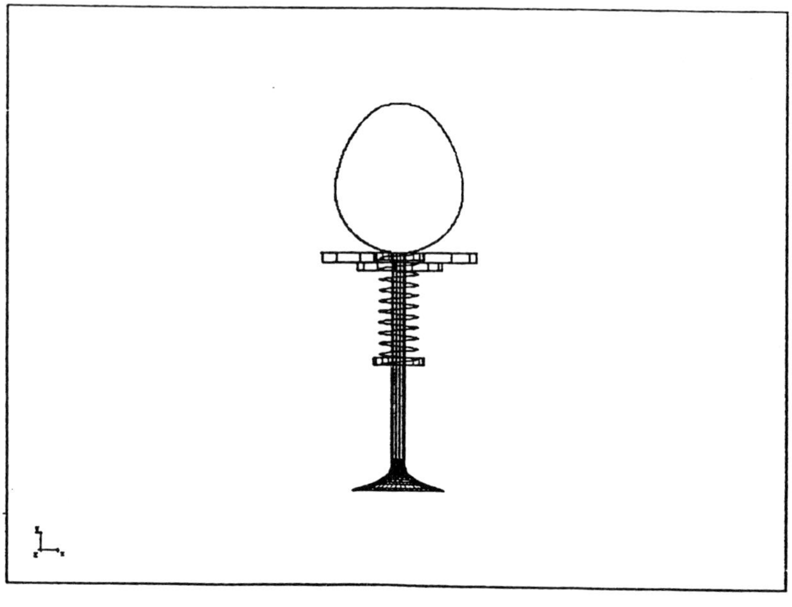

Figure 1: Cam/Tappet multi-body model

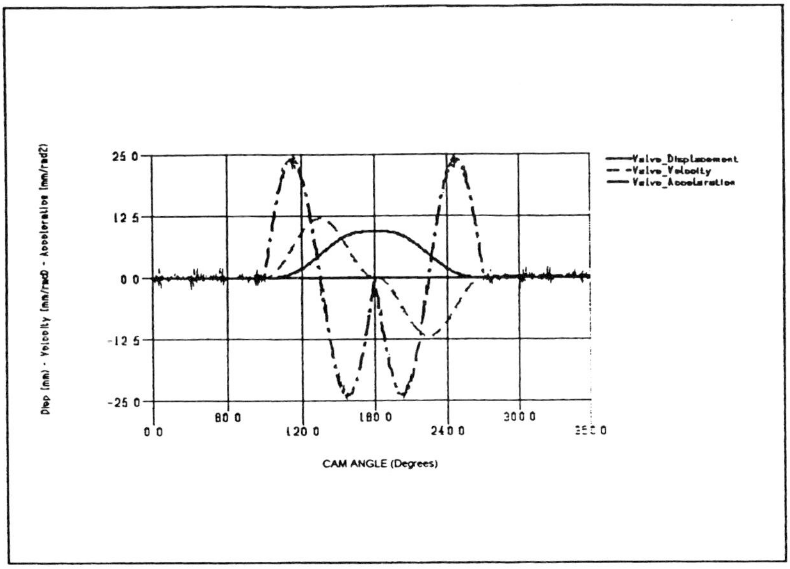

Figure 2: Lift, Velocity and Acceleration characteristics of the cycloidal cam

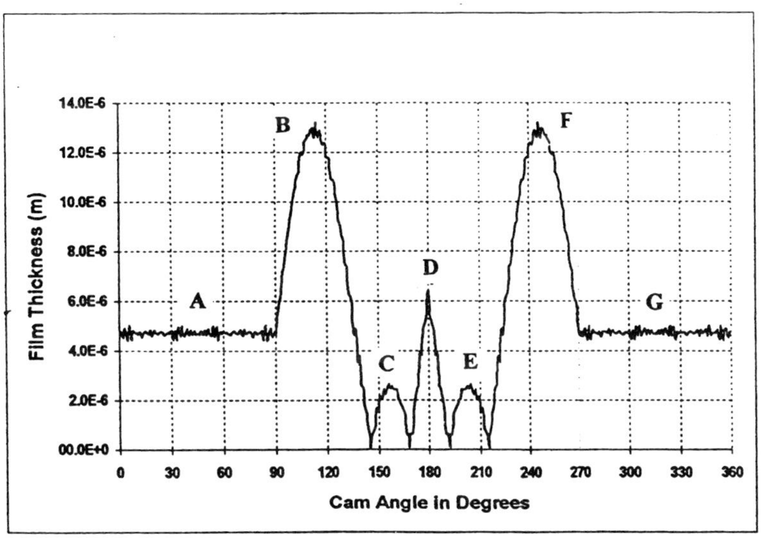

Figure 3: Contact oil film thickness

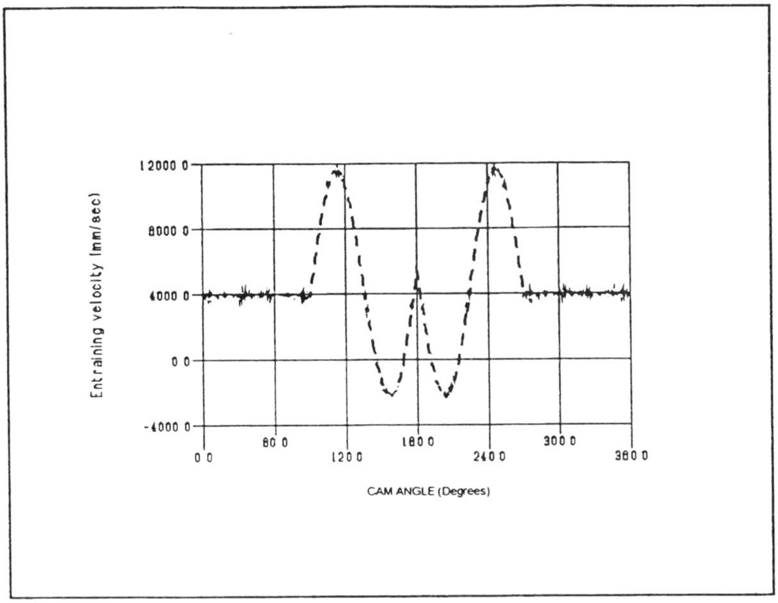

Figure 4: Entraining velocity at point of contact of cam and tappet

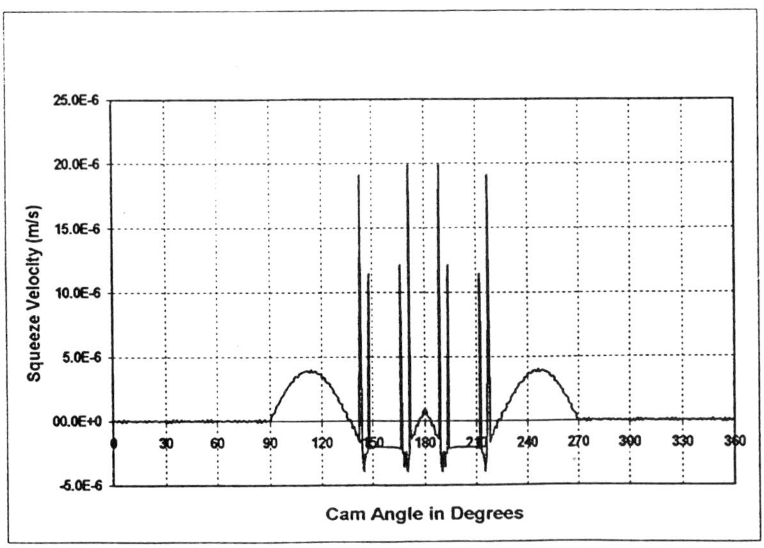

Figure 5: Squeeze velocity

C553/041 © IMechE 1998

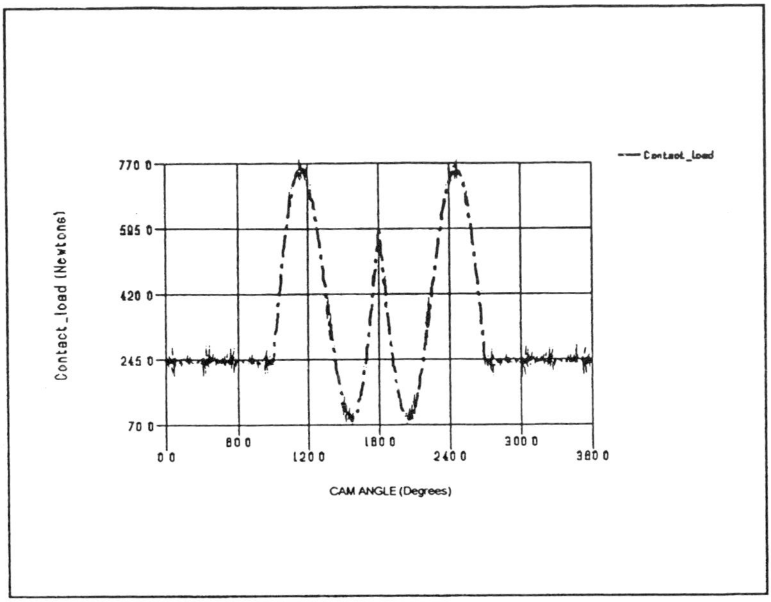

Figure 6: Contact load between cam and tappet

C553/028/98

On the modelling of system flexibilities in milling operation

F ABRARI, M A ELBESTAWI, and **A D SPENCE**
Mechanical Engineering Department, McMaster University, Hamilton, Canada

ABSTRACT

In some industrial applications, the need to use long and slender end mills in machining relatively thin and flexible components is unavoidable. For these cases, dynamic interaction between the tool and workpiece results in dimensional errors and surface irregularities. In this paper, a methodology capable of prediction of workpiece dynamic response in multi-axis side milling of flexible parts is developed. To account for the effect of metal removal on the dynamic behavior of the workpiece, a solid model based automatic mesh generation algorithm is developed which is capable of generating valid finite element models for the in-process workpiece geometries. Assuming an isotropic workpiece material properties, a special purpose 3-D finite element package is developed to model the vibrations of the workpiece while machining simulation. For each tool rotational position, predicted cutting forces are distributed (as concentrated forces) on the proper nodes of the tool/workpiece contact face. The dynamic response of the workpiece is then computed for the time period of two successive tool rotational positions. Dynamics of the tool is modeled as a system with two mutually perpendicular degrees of freedom aligned with x and y axes of the machine tool. The necessary modal parameters (stiffness, mass and damping) are experimentally determined at the tip of the tool. Effective deflection of the tool/workpiece system, at the Surface Generation Point (SGP) is used to capture the irregularities left on the machined surface. Comparison of the profile/errors of the simulated with the machined surface for the case of a flexible four-axis workpiece showed a good agreement between them.

1 INTRODUCTION

Side milling is a process that enables machining time and cost to be reduced. This type of machining is particularly suitable for ruled surfaces and all surfaces that have one principal curvature very small compared to the tool radius and that changes little over the entire surface. For some more complicated situations, for example, side milling of turbo-compressor blades, multi-axis side milling is the only possible process. This is because the tunnel between two adjacent blades is too small to accommodate the tool overhang length, so the tool-axis direction must be parallel to the generatrix of the

© IMechE 1998 C553/028

blade's ruled surface. Ruled surfaces (also called lofted surfaces) are generated by passing straight lines (generatrices) between two space curves $C_1(u)$ and $C_2(u)$ (rails) [1]. The main characteristic of a ruled surface is that there is at least one straight line passing through the surface point $P(u,v)$ and lying entirely on the surface [2]. Accordingly, the ruled surfaces can only allow curvature along one space parameter u (see Figure 1), while the curvature along the other parameter v is zero. This feature of ruled surfaces, makes them ideal candidate for peripheral milling. In fact only ruled surfaces can be *accurately* machined using peripheral milling. In this operation the axis of the tool always remains in parallel with the local generatrix of the surface at an offset distance equal to the radius of the tool minus radial depth of cut. However, in case of non-developable ruled surfaces, i.e. surfaces that can not be deformed to a planar shape without changing length measurements in them, tool orientation (setting) should be calculated for a minimum interference.

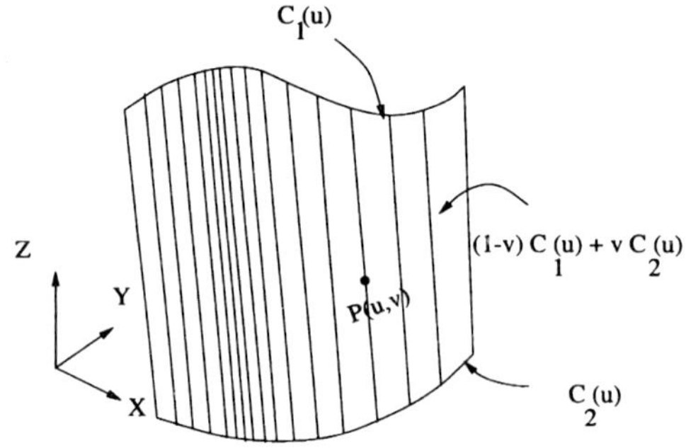

Figure 1: Ruled surface generated using two curves.

2 LITERATURE REVIEW

The static deflections of the tool and the workpiece (a thin rectangular web) was studied in [3] and [4] where a mechanistic force model of the end mill was used to calculate the cutting forces. In [4], the chip load is calculated such that it balances the cutting forces for the resulting system deflections. The tool is modeled by a cantilever beam rigidly supported by the tool holder. The workpiece is modeled by the rectangular plate elements clamped on the three edges with the fourth edge free to deflect. The developed model, provides accurate surface errors predictions as long as the system deflections are essentially static. In [5] it is suggested that if the tooth frequency is 10 times less than the lowest natural frequency of the part, then the cutting force may be assumed static. For a thin walled flexible structure, this condition can easily be violated. In the last 15 years, there have been few workers who studied the subject of flexible part machining. In [3], milling of a CCCF plate with a flexible end mill is studied. Although the finite element method is used to model the part deflections, the interaction between milling forces and the structural deflections is neglected. Accordingly, the model can still be considered as a rigid model. In [6] the model developed in [3] is improved to include the dynamic milling forces and the regeneration mechanism. The mechanism of regeneration

 C553/028 © IMechE 1998

in chip thickness calculation makes this model the most comprehensive and accurate one. Accordingly, the cutting force is modeled as:

$$F = K_T a_a (f \sin(\psi) + \delta(t) - \delta(t-T)) \quad (1)$$

where the quantity $f \sin(\psi(t))$ is the instantaneous chip thickness in the absence of any deflection in the system, $\delta(t)$ is the instantaneous dynamic deflection in the system (primary feedback) and the quantity $\delta(t-T)$ accounts for the undulations left from the cutting action of the previous teeth (secondary feedback). In [6] a $3-D$ finite element model with the brick elements is used to model the dynamics of a CFFF plate. The peripheral milling of flexible plates is also studied in [7] and [8], where the tool is assumed to be rigid and the plate is modeled by finite element analysis. Finally, milling of thin ribs is studied in [9] where an imprint of the high frequency structural vibrations is predicted by the model. In all above studies, the workpiece geometry is exclusively restricted to a thin plate and the tool path to a straight line. This clearly prevents the applicability of the developed methods and the conclusions obtained to the general cases where the workpiece has a complicated shape and the tool path is not a straight line. In this paper, however, a more general approach is adopted in which the multi-axis peripheral milling of flexible parts can be simulated.

3 SOLID BASED AUTOMATIC MESH GENERATION

Meshing may be approached from two viewpoints, either from that of a mathematician, who perceives it as a partitioning of a domain into pieces of predefined topology, or that of a practitioner who wants to use a meshing algorithm to apply Finite Element software for an engineering problem. The solid based automatic mesh generation algorithm introduced in this paper is closer to the latter case and is specifically developed for the simulation of peripheral milling operation.

3.1 CAD Versus CAM Applications

In recent years some automatic finite element mesh generation algorithms have been developed for CAD systems where geometry of the model is constructed and maintained using mainly solid models. However, very little effort has been devoted to development of an automatic mesh generation algorithm suitable for the CAM environment. In CAD applications, although the geometry of the part is not known for the meshing algorithm in advance, it is not subjected to variation either. This situation is similar to a case in which a mechanical part is meshed for a static or dynamic analysis. When the mesh is generated it is not subjected to a change later on. In CAM applications, on the other hand, not only is the part geometry not known a priori, but, because of the machining or forming process it is subjected to variation in time as well (see Figure 2). Moreover, while in CAD applications, a general mesh density control mechanism is sufficient for acceptable mesh quality, in a CAM environment a strict mesh quality control mechanism is required to assure generation of the desired number and order of elements at the contact zone of the tool/workpiece system. All of these issues complicate the task of automatic mesh generation for CAM applications.

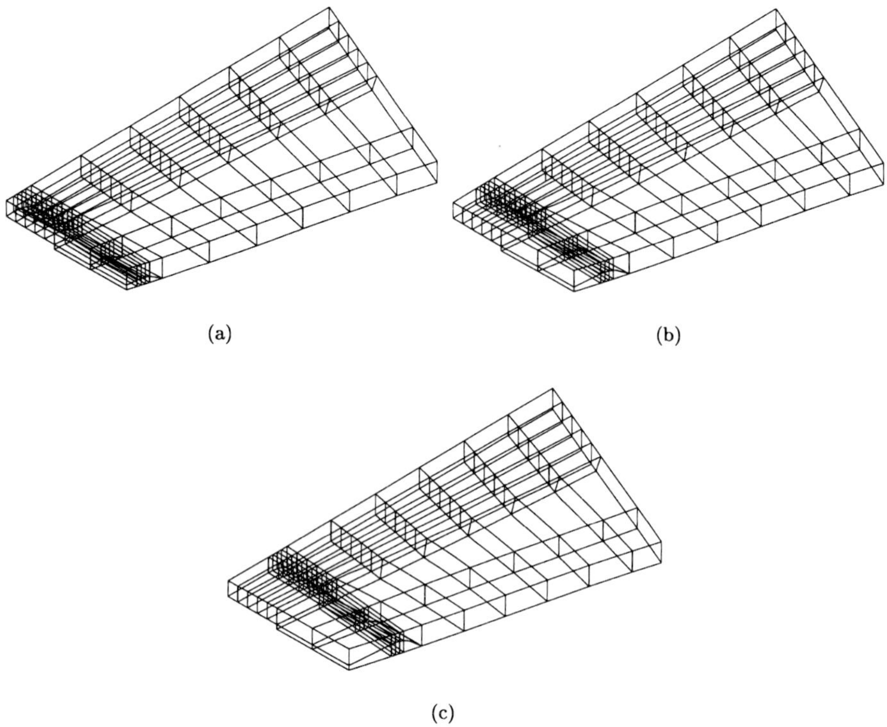

Figure 2: Automatic mesh generation in 4-axis milling simulation of a thin blade

3.2 Cellular Topology

A commercially available geometric solid modeler is used in this paper which provides an auxiliary data structure through its cellular topology husk [10]. By a proper interrogation of the data structure of a cellularized object, it is possible to access to the details of its cells geometry, including the cell vertices. These cell vertices, which later will serve as finite element nodes, are used to generate associativity table for the generated cells, (which later will server as finite elements.) Inputs to the algorithm are as follows:

- A solid model of the updated workpiece. Each updated workpiece (object) corresponds to the geometry of the workpiece at some point along the tool path. Accordingly, a sequence of these objects contains all the information required for the geometrical simulation of the metal removal process in the milling operation.

- The type of cutter used in machining, i.e. flat, tapered or ball end mill and their associated geometrical data, such as radius, helical angle and so on. These information are required for the grouping of the nodes which are located on the contact face of the tool/workpiece system.

C553/028 © IMechE 1998

- Cutter location data file. This file contains all information regarding position/orientation of the tool throughout the tool path. Information for the generation of one of the crucial partitioning layers is extracted from this file.
- Meshing quality control parameters:
  1. Number of elements along the tool axis on the contact face.
  2. Number of elements along the arc of contact on the contact face.
  3. Number of elements along the tool path (include item 2).
  4. Number of elements along the thickness of the workpiece.
  5. Number of elements between fixture level and tip of the tool.

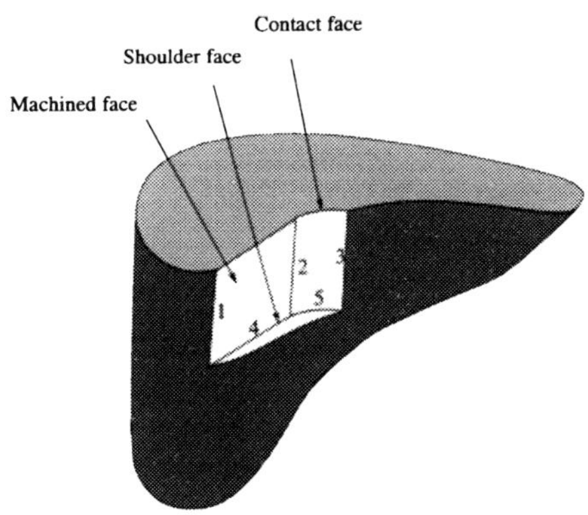

Figure 3: Geometry of the updated workpiece

3.3 Volume Partitioning

Figure 3 represents the typical geometry of a workpiece at sometime after the start of machining. Geometrical simulation of the milling operation always adds three faces to the data structure of the original workpiece. Namely, the added faces are the *machined*, *contact* and *shoulder* faces. Intersection of these faces with the boundary of the workpiece and each other creates five boundary curves labeled 1 to 5 in Figure 3. Accordingly, there always will be five partitioning sheets (planes) that are passing through these five curves. The process of cellularization of the workpiece always starts with the partitioning of the workpiece using these critically positioned planes. The orientation of these planes are also dependent on the tool path which is one of the inputs to the meshing algorithm. The critical planes 1 to 3 are generated such that not only they pass through the critical curves 1 to 3 but also such that they are normal to the local tool path. To partition the workpiece along the tool axis and along the radial depth of cut, the critical planes 4 and 5 are generated such that both of them are passing through the curves 4 and 5. One of them, however, is *normal* to the local direction of the tool axis while the other is *parallel* to it. It can be seen that the union of the critical planes followed by a Boolean subtraction from the workpiece geometry partitions the workpiece volume into 8 to 10 sub-volumes. At this stage, the critical partitioning is done and cellularization of the sub-volumes can proceed as follows. Depending on the mesh quality control parameters specified by the user, a series of copying/translation/rotation operations are performed on the critical planes to generate yet another cellularization structure for the sub-volumes. A great care should be taken to ensure that the partitioning planes of each family do not intersect each other within the workpiece volume. Figure 4 shows union of all partitioning sheets.

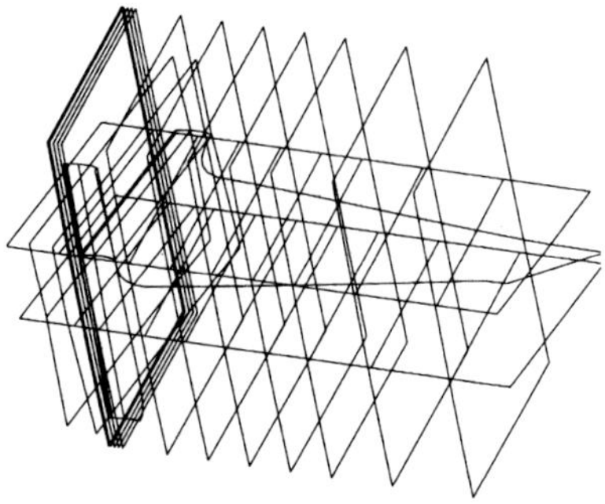

Figure 4: Partitioning sheets.

3.4 Finite Element Input Data

Although solid modeling is becoming the basis of the majority of CAD software seats, it is still fraught with different kinds of defects. These lack of integrities fall roughly into two categories; geometry and topological. While geometriç defects has to do with the issue of positions in the model space, eg. face/edge sloppiness, topological one has to do with connections or relationships between entities, eg. 'neighboring' faces not connected to the same edge (i.e. not having reversed coedges in common). To generate the required finite element input data from a solid based automatic mesh generation it is required to overcome these numerical problems. The rest of this section deals with this issue.

3.4.1 Bandwidth Minimization

In a cellularized object, the order of positioning of the 3D cells in the object space may be different from that of the *list* of 3D cells. This implies that for a large finite element job the bandwidth of the stiffness matrix generated by the FE code could be quite large. To overcome this problem, the generated mesh data should be processed for bandwidth minimization which is basically a sorting routine. This algorithm uses the location of center of gravity of an element to find its whereabouts relative to the other elements and the workpiece boundary. At the beginning, all the brick elements are sorted based upon their position along the 'height' of the workpiece. In the second sorting operation, elements of each layer, one at a time, are sorted based upon their 'width' and 'thickness' positions from the boundary sides of the workpiece. After the spatial order of all elements is corrected, the local FE node numbers of each element is set (see Figure 5). Here, nodes

C553/028 © IMechE 1998

of each element are divided into two groups. In group one elements are below and in group two they are above the center of gravity of the element. Nodes of each group are then ordered counter clockwise relative to the projection of the center of gravity on the plane containing the group.

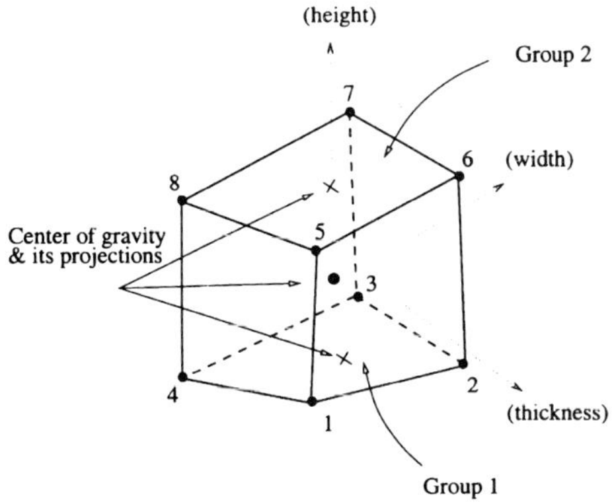

Figure 5: Local numbering of nodes of a brick element

3.4.2 Extra Edge and Zero Area Face Filters

Extraction of the required cell vertices starts with creating an empty vertex list for registering valid cell vertices and navigating through all the faces of the cell. In a healthy topological situation, a valid cell has to have 5 or 6 faces. However, in some circumstances there can be some extra faces usually with zero area. The first filtering net prevents those faces of entering into the vertex extraction procedure. Moreover, in a healthy geometrical situation, each face should have 3 or 4 edges. However, in some circumstances a valid face can have some extra edges, which will lead to extra invalid vertices for that face. The second filtering net, prevents registering a vertex not located at the corner of a face as a valid one. The algorithm for this filtering procedure is shown in Figure 6. First, all vertices of the troubled face are extracted and their center of gravity (cg) is found. Next a group of radial vectors, i.e. RV, from cg to vertices are generated. One of the RV's is picked as the reference and the others are ordered based on the ascending angle to the reference RV. Then, the end positions of three successive RV's are checked for alignment. If they are aligned then the *middle* vertex is *invalid.* If they are not aligned then the middle vertex is *valid.* In the case of detection of a valid vertex, it will be registered in the list of valid vertices only if it is *not* already present in the list due to the processing of the other faces of the cell. The outcome of this routine is the detection of all corner vertices of a brick cell.

3.4.3 Merging Coincident Nodes

Finally, it is needed to delete the coincident nodes of the adjacent finite elements. Starting from element 1 all the global node numbers are set equal to the local ones and then stored in the entity list of vertices. For the rest of the FE brick elements the FE nodes

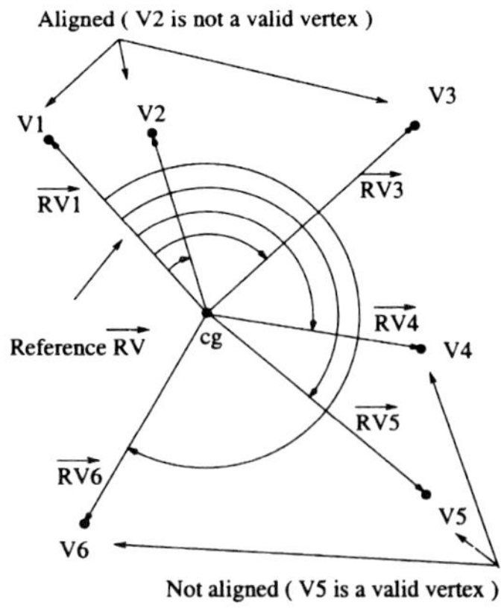

Figure 6: Detection of invalid vertices in a group

are checked, one element at a time, versus *all* the vertices in the vertex entity list. Since the consequent elements in the list are also consequent in the workpiece, the vertex equality check should always start from the *tail* of the list. If an equality happens, that FE node is regarded as a coincident node and next node of the element is examined, otherwise the vertex is *added* to the vertex entity list and its global number is set to be equal to its index in the entity list. By the time all the elements in the cellularized object are exhausted, global nodes of all the FE nodes are set properly and coincident nodes are deleted from the system. Also the entity list of vertices eventually contains all the distinct (not coincident) nodes generated by the algorithm. At this stage all the information required to generate the finite element *input file* is extracted from the list of processed brick elements. In addition, depending on the given tool type, an array of *contact nodes* is generated.

4 FINITE ELEMENT ANALYSIS OF PART DYNAMICS

In order to compute the dynamic response of the workpiece to the cutting forces, a special purpose dynamic finite element package is developed. The type of element used in the package is the eight-noded brick element with three translational degrees of freedom per node. Since the method of Modal Superposition is employed in the computation of workpiece dynamic response, a pre-specified number of mode shapes and frequencies are calculated first. To solve for the response of the workpiece to the cutting forces, during the time period of two successive cutter rotational positions, a force history file is then generated. The inputs to the nonlinear force distribution routine are the cutting edge geometry/position/orientation and the list of nodes at the vicinity of the cutting edge (see Figure 7). Using the computed mode shapes/frequencies and the generated force history file, response of the workpiece is computed and stored as the initial condition for the next time step. Accordingly, dynamic continuity between the successive dynamic analyses of workpiece is preserved, and a continuous and uninterrupted response of the workpiece is computed as the machining simulation continues. For the time period between the disengagement of one tooth and engagement of the approaching one, a free vibration

C553/028 © IMechE 1998

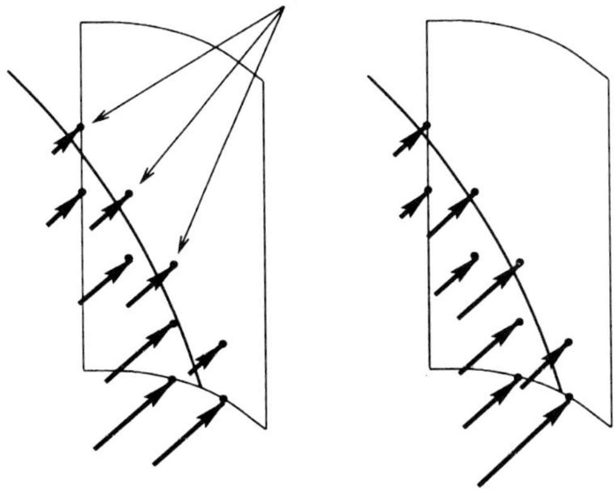

Figure 7: Nonlinear distribution of cutting forces.

analysis is performed on the workpiece structure.

4.1 Surface profile prediction

In a milling simulation that the dynamics of both tool and the workpiece are considered, the Effective System Deflection ($\overrightarrow{ESD}$) is defined as the algebraic summation of tool and the workpiece deflections. Machined surface profile, on the other hand, can be predicted by defining the Surface Generation Point (SGP) as the point in which an engaged tooth leaves a mark on the machined surface [3]. This point is always located on the *entrance* curve of the contact face. For the case that the machining system is perfectly rigid, $\mathrm{SGP_r}$ is simply the intersection of the cutting edge with the entrance curve of the contact face. However, for a flexible system the actual position of $\mathrm{SGP_f}$ is determined by translating the position of $\mathrm{SGP_r}$ by its $\overrightarrow{ESD}$ vector

$$\mathrm{SGP_f} = \mathrm{SGP_r} \cdot \overrightarrow{ESD} \tag{2}$$

The length of $\overrightarrow{ESD}$ vector at any location along the tool path and along the tool axis, is an estimation of the machining error at that location, i.e. distance between the local design surface and machined one.

5 EXPERIMENTAL VERIFICATION

To verify the developed model a thin workpiece with ruled surfaces at its front and back faces is machined using a five axis CNC machine. The generatrix curves for the workpiece are two straight lines which are angled 12.5° from each other along the rule (see Figure 8.) The structural twist and the thinness of the workpiece represent two basic characteristics of many flexible components in the aircraft and power generation industries. The initial geometry of the workpiece was machined out of a block of steel

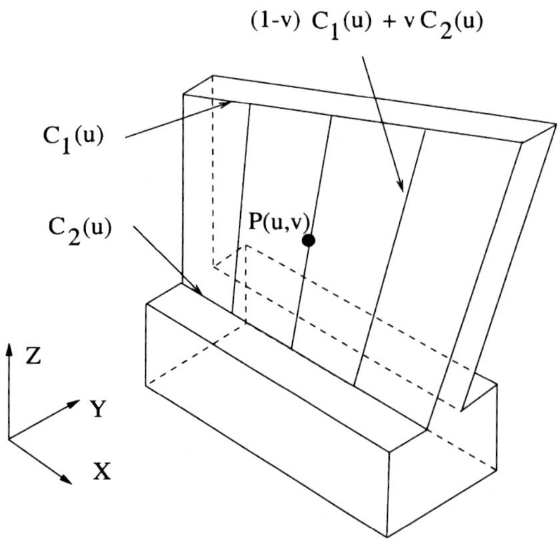

Figure 8: Ruled surface between two straight rail curves.

1045 using a four axis tool path. Single point offset (SPO) method is used in this paper for the CL data generation [11]. A five axis post-processor is then used to generate the required NC code for machining [12]. Figure 9 is a comparison between the predicted profile and the actual surface close to the end of the tool path. Also, Figure 10(a) shows an area of 1 cm^2 of the simulated surface profile. The corresponding area on the workpiece is shown in Figure 10(b). A good agreement is seen between the simulated profiles and the actual workpiece surface. The geometric inaccuracy seen on the work surface is the direct result of the workpiece flexibility. As the approaching tooth engages with the bottom of the contact face, due to the vibration of the workpiece, it encounters a deflected workpiece. As cutting proceeds, the Surface Generation Point moves towards the free edge of the workpiece where the vibration amplitude is highest. Accordingly, the surface error increases towards the free edge of the flexible workpiece (see Figure 11(a).) Also, by the time that the SGP reaches to the top of the contact face, the workpiece completes one or more of vibrational cycles; and as a result, a wavy surface is left behind. Also, as it is shown in Figure 11, comparison between the machining error across the simulated surface profile with that of the actual surface (measured using TalySurf) reveals a close agreement between them. Due to the high flexibility of the workpiece (thickness of only 2.0 mm), the geometric errors are both positive and negative relative to the design surface, i.e. zero error. However, negative errors are higher in magnitude compared to the positive errors. This is known as the undercut effect of the up-milling operation, i.e. workpiece is under-size. Moreover, closer to the workpiece free edge the higher the errors, i.e. the workpiece vibrations. As suggested in [13], for the case of long end mills this phenomenon can be avoided by using a relieved shank tool, which limits the tool workpiece contact to the intended cutting zone (see Figure 12.)

 C553/028 © IMechE 1998

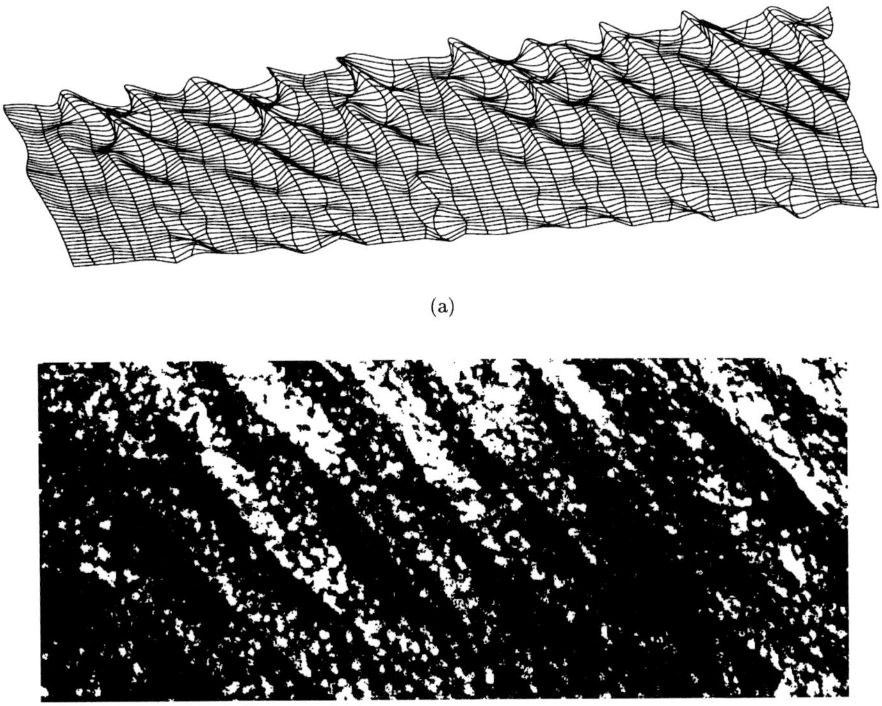

(a) (b)

Figure 9: Comparison between the actual and simulated surfaces.

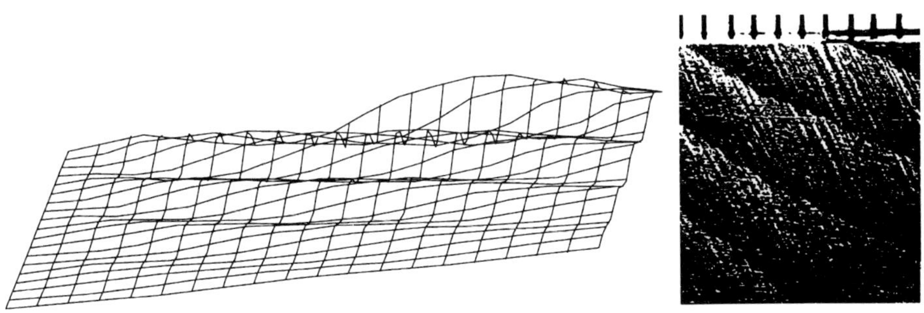

(a) (b)

Figure 10: Simulation and actual surface profile in flexible part machining.

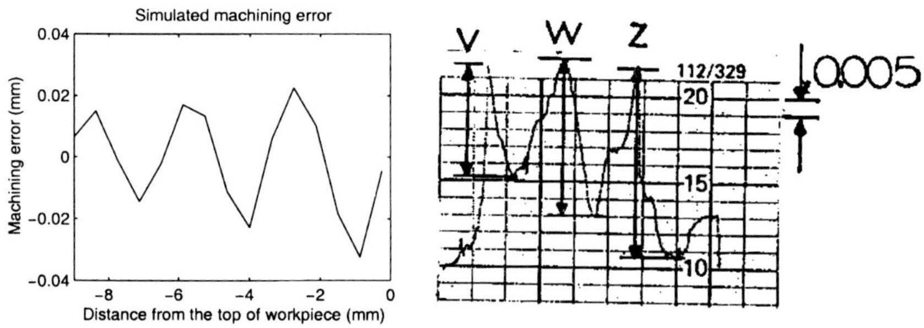

(a) (b)

Figure 11: Comparison between the actual and simulation error distribution.

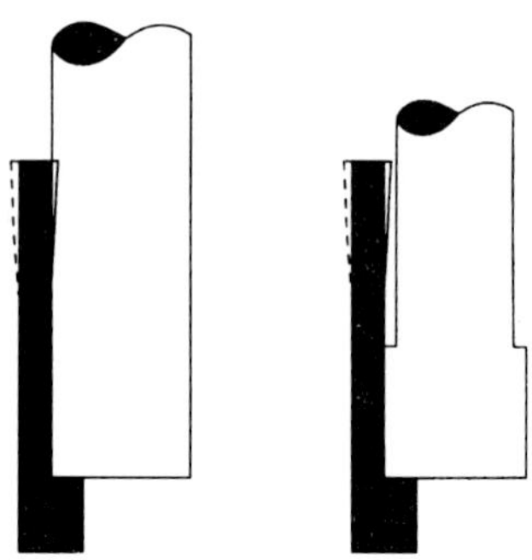

Figure 12: Relieved shank for thin workpiece milling.

6 CONCLUSION

In a flexible milling system, dynamic interaction between the tool and workpiece results in dimensional errors and surface irregularities. A methodology capable of prediction of tool/workpiece dynamic response in side milling is developed. A solid model based automatic mesh generation algorithm is developed which is capable of generating valid finite element models for the instantaneous workpiece geometries. Assuming an isotropic workpiece material properties, a special purpose 3-D finite element package is developed to model the vibrations of the workpiece while machining simulation. For each tool rotational position, predicted cutting forces are distributed on the proper nodes of the tool/workpiece contact face. The dynamic response of the workpiece is then computed for the time period of two successive tool rotational positions. The necessary modal parameters (stiffness, mass and damping) at the tip of the tool was determined experimentally. Effective deflection of the tool/workpiece system, at the point of contact between the tool and surface (normal to the direction of the feed) is used to capture the irregularities left on the machined surface. A good agreement between the simulated and machined surface profile/errors is seen.

C553/028 © IMechE 1998

REFERENCES

[1] G. Farin. *An introduction to solid modeling.* Wiley, New York, 1990.

[2] I. Zeid. *CAD/CAM theory and practice.* McGraw Hill, Inc., 1991.

[3] W.A. Kline, R.E. DeVor, and I.A. Shareef. The prediction of surface accuracy in milling. *Journal of Engineering for industry*, 104:272, 1982.

[4] J.W. Sutherland and R.E. DeVor. An improved method for cutting force and surface error prediction in flexible end milling system. *Journal of Engineering for industry*, 108(Nov.):269, 1986.

[5] J. Faupel. *Engineering design.* Wiley, New York, 1964.

[6] R. Sagherian and M.A. Elbestawi. A simulation system for improving machining accuracy in milling. *Computers in industry*, 14(1):293–305, 1990.

[7] D. Montgomery. *Milling of flexible structures.* M.Eng thesis, University of British Columbia, 1990.

[8] Y. Altintas, D. Montgomery, and E. Budak. Dynamic peripheral milling of flexible structures. *Trans. ASME Journal of Eng. for Industry*, 114(2), 1992.

[9] M. Anjanappa, D.K. Anand, and J.A. Kirk. Identification and optimal control of thin rib machining. *ASME Winter Annual Meeting*, 6:272–278, 1987.

[10] Spatial Technology Inc. *ACIS Geometric Modeler - Application Guide.* 1.7. Spatial Technology Inc., 1995.

[11] X. Liu. Five-axis NC cylindrical milling of sculptured surfaces. *Computer Aided Design*, 27(12):887–894, 1995.

[12] R.V. Fleisig and A.D. Spence. Integrated digitizing, path planning, and five-axis machining for the refurbishment of compressor and turbine blades. *Manufacturing Science and Engineering*, pages 31–38, 1997.

[13] J. Tlusty, S. Smith, and W.R. Winfough. Techniques for the use of long slender end mills in high-speed milling. *Annals of the CIRP*, 1996.

C553/004/98

The identification of shaft bow and unbalance from bearing vibration

S EDWARDS BEng, MSEM, **A LEES** BSc, PhD, MIMechE, MInstP, MINucE, and **M FRISWELL** MA, MSc, PhD, MASME, MInstP, FIMA
Department of Mechanical Engineering, University of Wales, Swansea, UK

SYNOPSIS

Bends in rotating shafts cause synchronous vibration similar to that produced by mass unbalance. It is important to be able to distinguish between these two faults in order to apply the appropriate correction procedure. In this paper, a method is presented to determine the modal components of a bend, and hence the bend geometry, from measured vibration signals during a typical run-up or run-down of the machine. The only requirement is a good numerical model of the rotor. The calculation is presented in two directions orthogonal to the rotor, and it is shown how the method can determine the extent and location of a bend in the presence of noisy data and unbalance.

NOMENCLATURE

x	axial distance along rotor
y	displacement orthogonal to rotor
z	second orthogonal direction
L	rotor length
G	rotor Green's function
F	force on rotor
β	bend modal coefficients
c	number of bend modes
e	unbalance
d	number of unbalance planes
n	number of bearings
m	bend mode number
k	bearing stiffness
$[a]$	force sensitivity matrix
G_{ij}	discrete representation of G
EI	rotor bending stiffness
$[K_s]$	rotor stiffness matrix
$[K]$	foundation stiffness matrix
$[M]$	foundation mass matrix
$[w]$	frequency dependent displacement matrix
$\{v\}$	unknown component vector
$[W]$	concatenation of $[w]$

© IMechE 1998 C553/004

$[q]$	bend influence matrix		ρ	rotor mass per unit length
$[Q]$	concatenation of $[q]$		ψ, Ψ	rotor free-free modes
$[u]$	unbalance influence matrix			
$[U]$	concatenation of $[u]$		Subscripts	
$[p]$	force influence matrix		i,j	counters
$\{P\}$	generalised force vector		bi	bearing i
δ	Kronecker delta		m	bend mode
ω	shaft rotational speed		r	rotor
ω_m	natural frequency		p	pedestal

1. INTRODUCTION

Rotors which are subject to the synchronous forcing caused by a bend, produce unwanted vibration. Bends are typically caused by creep, thermal distortion or a previous large unbalance force. Industry constantly requires machinery to run quicker and to closer tolerances, meaning that the understanding and control of vibration is ever more important. During the previous two decades, an increasing amount of research has been undertaken into the investigation of shaft bow (bend), albeit much less than the better-understood mass unbalance. A bent shaft produces a somewhat different, although not dissimilar, forcing to that caused by unbalance; the main differences being in amplitude and phase angle responses at high and low speeds. The forcing due to unbalance is proportional to the square of the shaft running speed, whilst this forcing remains constant with shaft speed in the case of a bend. There have been many cases in industry where a bend has been confused with an unbalance and balancing has been performed, resulting in an even worse problem than before balancing took place. To be able to distinguish between a bend and an unbalance in a system is therefore of great importance to engineers involved in the control of rotating machinery.

One of the first extensive investigations into shaft bow was made by Nicholas et al. (1,2), in a series of two papers. Part I (Unbalance Response) (1) discusses the unbalance response of flexible rotors due to shaft bow. Part II (Balancing) (2) proposes a balancing theory and gives experimental results for the balancing of a flexible rotor with shaft bow. One important conclusion drawn from these tests is that the change in phase angle from rest, up to the speed where maximum rotor amplitude occurs, is not usually 90°, as it is with conventional unbalance. Parkinson et al. (3) described the differences in whirl experienced by a rotating shaft subject to shaft bow and mass unbalance. By balancing the net whirl (total whirl minus shaft bow), which contains certain components due to mass unbalance and others due to shaft bow, the residual total whirl at running speed is approximately equal to the shaft bow. For practical applications this is more desirable than simply balancing the total whirl, which leaves a residual bow at non-critical speeds and is only balanced at resonance. The reader is also referred to the investigations of shaft bow undertaken by Meacham et al. (4), Flack et al. (5) and Salamone and Gunter (6,7), and to the review given by Edwards et al. (8).

Lees and Friswell (9) showed how it is possible, given an accurate rotor model, to identify the unbalance forcing acting on a system by vibration measurements taken at the bearings. This paper extends that method to include the identification of a rotor bend. The analysis is also developed to include the identification of a rotor-bearing system subject to forcing from both a bend and an unbalance.

C553/004 © IMechE 1998

2. SYSTEM FORCES

In the following analysis, the shaft bend geometry shall be expressed in terms of the free-free modal properties of the rotor, which can be readily determined, given a model of the shaft. Since the dominant damping effects are usually due to the bearings, no rotor or support damping is included. The bend profile may be described as

$$y_{bend}(x)=\sum_{m=1}^{\infty}\beta_m\psi_m(x) \qquad [1]$$

where ψ_m are the free-free mass normalised rotor modes and β_m are the corresponding modal coefficients. The resulting force per unit length due to the bend is then

$$F_{bend}(x)=\frac{\partial^2}{\partial x^2}\left(EI(x)\frac{\partial^2 y_{bend}(x)}{\partial x^2}\right) \qquad [2]$$

At an arbitrary position on the shaft, x, the displacement is given by

$$y_r(x)=\int_0^L G(\omega,x,x')F(x')dx' \qquad [3]$$

where $F(x')$ is the force per unit length acting on the rotor and $G(\omega,x,x')$ is the rotor Green's function, giving the response at position x due to a unit force applied at position x' (9).

The rotor experiences two kinds of forcing: the unknown bearing reactions $F_{b1},\dots,F_{bn}$ at n bearing stations $x_{b1},\dots,x_{bn}$ and the force due to the bent shaft profile described above. The total forcing acting on the rotor is then

$$F(x,\omega)=\sum_{m=1}^{\infty}\beta_m\frac{d^2}{dx^2}\left(EI(x)\frac{d^2\psi_m(x)}{dx^2}\right)+\sum_{i=1}^{n}F_{bi}\,\delta(x-x_{bi}) \qquad [4]$$

where δ is the Dirac delta function. Combining equations [3] and [4] then yields an equation for the displacement of the shaft at any position x along its length. Writing this displacement equation using the notation

$[G_{bend}]_{mm}=\dfrac{1}{\omega^2-\omega_m^{\,2}}$ and $\hat{k}_m=\omega_m^{\,2}$, where $\omega_m^{\,2}=\int_0^L\psi_m(x)\dfrac{d^2}{dx^2}\left(EI(x)\dfrac{d^2\psi_m(x)}{dx^2}\right)dx$ gives

$$y_r(x)=\sum_{m=1}^{\infty}[G_{bend}]_{mm}\hat{k}_m\beta_m\psi_m(x)+\sum_{i=1}^{n}G(\omega,x,x_{bi})F_{bi} \qquad [5]$$

For shaft displacement x at the bearing locations j,

$$x=x_j \qquad [6]$$

then

$$y_{rj} = \sum_{m=1}^{\infty} [G_{bend}]_{mm} \hat{k}_m \beta_m \psi_m (x_{bj}) + \sum_{i=1}^{n} G(\omega, x_{bj}, x_{bi}) F_{bi} \qquad [7]$$

It can be seen from equation [7] that the bearing forces are independent of foundation stiffness, and although the effects of foundation stiffness will obviously have a large effect on the measured response values, this is dealt with by the fact that the response is indeed measured. This method and the conditioning of the matrix $[a]$ are discussed in detail by Lees and Friswell (10) and Friswell et al. (11).

3. IDENTIFICATION METHOD

In order to successfully identify the bend geometry it is necessary to be able to identify the modal coefficients β in equation [1], the free-free modes being known. This unknown vector may be written $\{\beta\} = \{\beta_1, \ldots, \beta_c\}^T$, where c is the maximum number of modes being considered. Adopting this notation allows equation [7] to be written in vector form as

$$\{F\} = [a]^{-1}\{y_r\} - [a]^{-1}[G_{bend}][K_S][\Psi]\{\beta\} \qquad [8]$$

where $[a(\omega)] = \begin{bmatrix} G(\omega, x_{b1}, x_{b1}) & . & . & G(\omega, x_{b1}, x_{bn}) \\ . & . & . & . \\ . & . & . & . \\ G(\omega, x_{bn}, x_{b1}) & . & . & G(\omega, x_{bn}, x_{bn}) \end{bmatrix}$, $\{y_r\} = \{y_{r1}, y_{r2}, \ldots y_{rn}\}^T$, $[K_S] = diag[\hat{k}_m]$,

$[G_{bend}] = diag\left[\dfrac{1}{\omega^2 - \omega_m^2}\right]$, and $[\Psi]$ contains the shaft free-free modes.

Substituting $[p] = [a]^{-1}$ and $[q] = [a]^{-1}[G_{bend}][K_S][\Psi]$ gives

$$\{F\} = [p]\{y_r\} - [q]\{\beta\} \qquad [9]$$

$\{F\}$ is the reaction of the bearings on the rotor, its dimension being the number of bearings in the system (x2 for both perpendicular directions). The force due to the bearings acting on the supports is $-\{F\}$.

The unknown foundation forces may be expressed in terms of the foundation mass and stiffness matrices $[M]$ and $[K]$, which are assumed to be symmetric. No damping from the foundations is included since it is found in practice to be negligible in comparison to the damping produced by the bearings. The force acting on the bearings due to the foundations at each measured frequency is

$$-\{F\} = [K]\{y_r\} - \omega^2 [M]\{y_r\} \qquad [10]$$

so that
$$[p]\{y_r\} - [q]\{\beta\} = -[K]\{y_r\} + \omega^2 [M]\{y_r\} \qquad [11]$$

$[K]$, $[M]$ and $[\beta]$ may be chosen so as to minimise the residual term of equation [11], resulting in the solution of a linear problem.

For a solution over the whole frequency range, equation [11] can be re-written as

$$[W]\{v\} + [Q]\{\beta\} = \{P\} \qquad [12]$$

where $[W]$, which contains the mass and stiffness terms of equation [10], $[Q]$ and $\{P\}$ are simply the concatenation of each sub-matrix over the considered range. Equation [12] is then solved for the unknown vectors $\{v\}$ and $\{\beta\}$ using an appropriate method of solution.

Complications may arise in practice such as a lack of information available to fully determine all system parameters. It is often preferable to solve equation [12] using the Singular Value Decomposition (SVD) method (12), although this was not necessary for the example considered in this paper. It is also found that these matrices are often complex, in which case they must be divided into their real and complex parts. These matters, the conditioning of the regression matrix and the method employed in the case where some extra knowledge of the system may be available, either from further modelling or testing, are discussed in detail in (10). If damping is included in the model of the bearings then $[W]$, $[Q]$ and $\{P\}$ will become complex, although the system parameters it is intended to identify remain real.

3.1 Inclusion of Unbalance

The development of the identification method to include both a bend and unbalance is simply an extension of the above theory. The unbalance definition previously developed (9), where the unbalance forcing acts at a pre-determined set of d unbalance planes, at the locations $x_{e1},\ldots,x_{ed}$ is used. For a finite frequency range, the unbalance may be assumed to be a summation over the balance planes. Equation [12] can then be expanded, including unbalance, to

$$[W]\{v\} + [Q]\{\beta\} + [U]\{e\} = \{P\} \qquad [13]$$

where $[U(\omega)] = \omega^2 [a]^{-1} [G_{be}]$, $\{e\} = \{e_1, \ldots, e_d\}^T$, the unknown unbalance vector, and $[G_{be}]$ is the relationship between the bearings and the unbalance planes. Equation [13] now contains three forcing terms, in contrast to the two terms of equation [12], and is simply solved accordingly. For a complete treatment of the unbalance identification theory, the reader is referred to (9).

4. METHOD VERIFICATION

In order to verify the proposed method, both numerical simulations and experimental testing of a small experimental rig, in the Mechanical Engineering Department at the University of Wales Swansea, have been performed. A simplified diagram of the apparatus is shown in Figure 1. The steel shaft is of length 750 mm and diameter 10 mm, and supports three discs of diameter 40, 75 and 40 mm and width 35, 15 and 35 mm respectively, which can be positioned at various points along the length of the shaft. The shaft is supported on two spherical bearings and driven, via a direct flexible coupling, by an electric motor. Modal testing was carried out to determine both the free-free and in-situ natural frequencies of the rotor. The first two bending natural frequencies of the system with discs at 130, 320 and 520 mm are:

	Experimental (Hz)	Theoretical (Hz)
Rotor (free-free)	56.25, 161.25	57.32, 160.52
Rotor/Bearings/Foundations	21.76, 91.04	22.07, 92.68

4.1 Numerical Simulations

A series of tests using simulated data were carried out to examine the behaviour of the method under different conditions, namely frequency range, step size, noise and modelling error, as shown in Table 1. The model of the rotor system utilised four, two-noded, beam elements, each node having two translational and two rotational degrees of freedom (DoF). Noise was introduced as a factor of 2%, 5% and 10% of the high frequency response. In the cases using noisy data, each test was performed twenty times: the mean value and corresponding standard deviation are shown for each case. Tests are considered for a bend and an unbalance, individually and combined. The results for the cases described in Table 1 are shown in Tables 2, 3 and 4. For a bend with a half-sine wave profile having maximum displacement of 5 mm at the mid-point along the shaft, the first two modal coefficients are $\beta_1 = 0.04$ and $\beta_2 = 0.00$. An unbalance of e = 1.0e-6 kg.m, was defined at the second disc. To introduce an error into the numerical model of the rotor, the shaft density for the model was increased by 10% of its true value, thus increasing the mass of model. The foundation stiffnesses and masses are 100 kN/m and 1 kg respectively, at each end of the rotor. The absolute rotor displacements at the bearing positions are shown in Figure 2 (combined bend/unbalance, case 1) and Figure 3 (individual unbalance, case 3).

4.2 Numerical Results

In case 1, using the reference frequency range of 100 Hz, with no noise or error present, both foundation and excitation (bend, unbalance) are identified with either negligible or zero error

C553/004 © IMechE 1998

for all three of the models considered. The largest error is 3% in the bend estimation for the combined model.

Cases 2 and 3, with noise factors of 2% and 5% respectively, show that these levels of noise have negligible influence on the identified parameters in the cases of the individual bend and unbalance, whilst in the combined case, a more marked effect on the bend parameters is seen. It should be noted that the mean values shown are the averages of twenty individual runs, and that the standard deviations do increase with added noise. It could therefore be expected that during a single test, as would most probably occur in practice, these parameters would not remain as accurate.

Case 4, with the number of frequency steps reduced by half, and no added noise, again shows excellent predictions of all parameters for all three models. The maximum error is 4% in the bend estimation for the combined model, this error being considered insignificant. In case 5, a noise factor of 10% is added to the data used in case 4. The bend model shows little variation in the identified parameters, whilst the unbalance model becomes slightly less accurate, although the maximum error in this case remains less than 3%. Similar behaviour is exhibited by the combined model, the maximum error once again being in the estimation of the bend, now 20%. The standard deviations for the unbalance and combined models are comparably high in contrast to those for the individual bend model.

The results for a reduced frequency range of 50 Hz, with a step size of 1 Hz, are given in case 6. For the bend, the foundation parameter identification suffers from this reduction by around a 20% error in the prediction of both masses and stiffnesses. The bend parameters, however, remain virtually exact. The combined and the individual unbalance models are barely affected by this change. Case 7 is as case 6, only with 2% added noise. All three models are very obviously affected by this relatively small amount of error in the simulation data. The foundations, whilst still being of the correct orders, are incorrect by around 50% for the three different scenarios. The excitation parameters meanwhile, remain very close for the individual models, the maximum error being 6% in the bend estimation. For the combined bend and unbalance model, the unbalance parameter is correct, whilst the bend parameters can no longer be believed, since although β_1 remains close to its true value, β_2 is of approximately equal magnitude: not zero as it should be. This is because the identification of the bend relies on exciting higher frequencies than are needed for the identification of the unbalance, since the bend forcing remains constant with shaft speed, whilst the unbalance forcing is proportional to the square of the speed, thus exciting higher frequencies.

In case 8, the influence of introducing a modelling error, in the form of an incorrect density value, can be seen. Here the foundation parameters change by a maximum of around 5% in each of the models. This systematic error also affects the excitation parameters, with the unbalance slightly low in the individual case and slightly high for the combined case. In the case of the bend, β_1 increases by around 15% of its true value in the individual case, and, in contrast to the unbalance, reduces in the case of the combined model.

5. DISCUSSION

It may be drawn from the above results that, so long as certain conditions are met, then the method presented in this paper can identify, with what is considered to be a high degree of accuracy, the existence of a bend and/or unbalance in a rotor-bearing system. Cases 6 and 7 highlight the importance of selecting a suitable frequency range over which the method should be applied. If the range does not fully cover both of the foundation natural frequencies, then the identified foundation parameters become very inaccurate with added noise. The excitation parameters, however, do remain accurate, and the prediction of these is indeed the aim of the method. The introduction of a modelling error shows that as much care as possible should be taken in establishing any *a priori* knowledge of the numerical model, as this could have a potentially large effect on the error of the results. In the case where some bearing information is available, either from modelling or testing, then vibration measurements at the bearing pedestals can be taken, and this information then incorporated into the identification method.

No damping has been included in the above investigations. This has been done for the sake of clarity, and although no difficulty is envisaged in its introduction, it was omitted in this case simply in order to highlight the most important features of the method itself.

6. SUMMARY AND CONCLUSIONS

There have been many cases in industry where a bend in a rotor has been confused with an imbalanced rotor, leaving large vibration responses after balancing has been carried out. In this paper, a method has been presented which, under certain conditions, can accurately estimate the bend geometry and foundation mass and stiffness properties of a rotor-bearing system. The identification of a bend and unbalance acting concurrently has also been successfully included in this method, which has shown itself to be suitably robust to variations in solution conditions.

ACKNOWLEGEMENTS

The authors are pleased to acknowledge the support of BNFL Ltd. (Magnox Generation). Dr. Friswell gratefully acknowledges the support of the EPSRC through the award of an Advanced Fellowship.

REFERENCES

(1) Nicholas, J.C., Gunter, E.J., and Allaire, P.E., Effect of Residual Shaft Bow on Unbalance Response and Balancing of a Single Mass Flexible Rotor Part 1 - Unbalance Response, Journal of Engineering for Power, Vol. 98, pp. 171-181, 1976.

(2) Nicholas, J.C., Gunter, E.J., and Allaire, P.E., Effect of Residual Shaft Bow on Unbalance Response and Balancing of a Single Mass Flexible Rotor Part 2 - Balancing, Journal of Engineering for Power, Vol. 98, pp. 182-189, 1976.

(3) Parkinson, A.G., Darlow, M.S., and Smalley, A.J., Balancing Flexible Rotating Shafts with an Initial Bend, AIAA Journal, Vol. 22, No.5, pp. 683-689, 1984.

(4) Meacham, W.L., Talbert, P.B., Nelson, H.D., and Cooperrider, N.K., Complex Modal Balancing of Flexible Rotors Including Residual Bow, Journal of Propulsion and Power, Vol.4, No.3, pp. 245-251, 1988.

(5) Flack, R.D., Rooke, J.H., Bielk, J.R., and Gunter, E.J., Comparison of the Unbalance Responses of Jeffcott Rotors with Shaft Bow and Shaft Runout, Transactions of the ASME, Vol. 104, pp. 318-328, 1982.

(6) Salamone, D., and Gunter, E.J., Synchronous Unbalance of a Multimass Flexible Rotor Considering Shaft Warp and Disk Skew, M.Sc. Thesis No. UVA/464761/ME77/142, University of Virginia, Charlottesville, Virginia, 1977.

(7) Salamone, D.J., and Gunter, E.J., Synchronous Unbalance Response of an Overhung Rotor with Disk Skew, Journal of Engineering for Power, Vol. 102, pp. 749-755, 1980.

(8) Edwards, S., Lees, A.W., and Friswell, M.I., Fault Diagnosis of Rotating Machinery, The Shock and Vibration Digest, Vol. 30, No.1, 1988.

(9) Lees, A.W., and Friswell, M.I., The Evaluation of Rotor Imbalance in Flexibly Mounted Machines, Journal of Sound and Vibration, Vol. 208, No.5, pp.671-683, 1997.

(10) Lees, A.W., and Friswell, M.I., Estimation of Forces Exerted on Machine Foundations, International Conference on Identification in Engineering Systems, Swansea, UK, pp. 793-803, 1996.

(11) Friswell, M.I., Lees, A.W., and Smart, M.G., Model Updating Techniques Applied to Turbo-Generators Mounted on Flexible Foundations, NAFEMS Conference on Structural Dynamics Modelling: Test Analysis and Correlation, Cumbria, UK, 1996, pp. 461-472, 1996.

(12) Golub, G., and Van Loan, C., Matrix Computations, John Hopkins University Press, London, 1989.

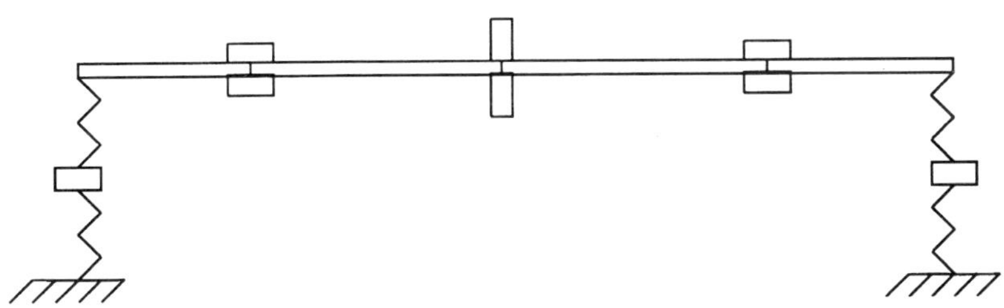

Figure 1: Schematic diagram of rotor rig

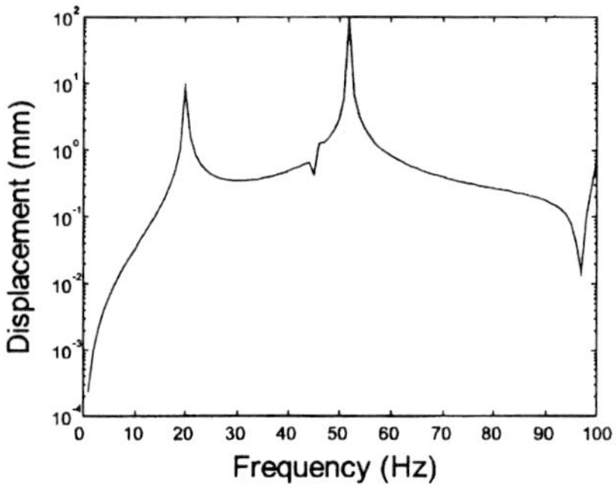

Figure 2: Absolute rotor displacement at bearing 1, case 1, combined model

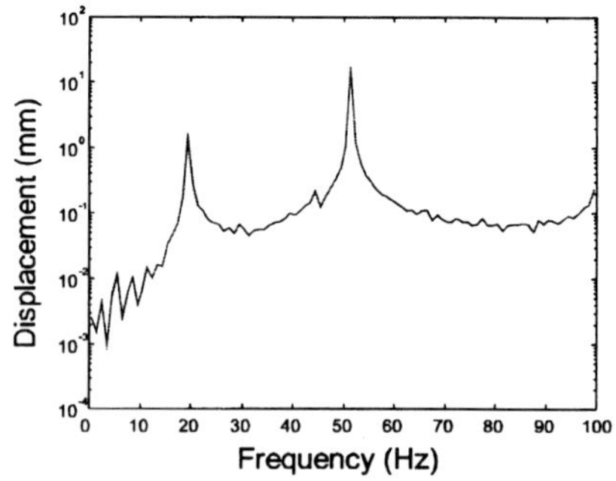

Figure 3: Absolute rotor displacement at bearing 2, case 3, unbalance model

Table 1: Test cases considered

Case	Frequency Range (Hz)	Number of steps	Noise (%)
1	100	100	0
2	100	100	2
3	100	100	5
4	100	50	0
5	100	50	10
6	50	50	0
7	50	50	2
8	100	100	modelling error

Table 2: Results for bend model, standard deviations for noisy cases in brackets

Case	K_{11} (kN/m)	K_{22} (kN/m)	K_{12} (kN/m)	M_{11} (kg)	M_{22} (kg)	M_{12} (kg)	β_1 *e-02	β_2 *e-02
1	100.22	100.24	-0.29	1.00	1.00	0.00	4.00	0.00
2	100.04	100.01	-0.08	1.00	1.00	0.00	4.00	0.00
	(0.56)	(0.66)	(0.61)	(0)	(0)	(0)	(0)	(0)
3	100.78	100.85	-0.90	1.00	1.01	-0.01	4.00	0.00
	(1.66)	(1.97)	(1.79)	(0.01)	(0.01)	(0.01)	(0)	(0)
4	100.15	100.16	-0.22	1.00	1.00	0.00	4.00	0.00
5	100.09	100.08	-0.15	1.00	1.00	0.00	4.00	0.00
	(0.64)	(0.72)	(0.68)	(0.01)	(0.01)	(0.01)	(0)	(0)
6	83.80	83.00	16.51	0.80	0.79	-0.20	4.00	0.02
7	50.88	48.42	50.19	0.39	0.37	0.62	3.99	0.06
	(0.20)	(0.19)	(0.29)	(0.01)	(0.01)	(0.01)	(0.10)	(0.04)
8	105.82	105.25	-0.23	1.03	1.03	0.14	4.62	0.01

Table 3: Results for unbalance model, standard deviations for noisy cases in brackets

Case	K_{11} (kN/m)	K_{22} (kN/m)	K_{12} (kN/m)	M_{11} (kg)	M_{22} (kg)	M_{12} (kg)	e*e-06 (kgm)
1	100.15	100.22	-0.23	1.00	1.00	0.00	1.00
2	100.49	100.46	-0.65	1.00	1.00	0.00	1.00
	(1.03)	(1.27)	(0.96)	(0)	(0.01)	(0.01)	(0.04)
3	100.60	100.67	-0.91	1.00	1.00	0.00	1.00
	(3.38)	(2.25)	(3.85)	(0.01)	(0.03)	(0.02)	(0.10)
4	100.11	100.21	-0.09	1.00	1.00	0.00	1.00
5	100.54	98.76	-0.59	1.00	0.99	0.00	0.98
	(6.40)	(4.08)	(6.01)	(0.06)	(0.04)	(0.07)	(0.11)
6	99.93	99.93	0.00	0.99	0.99	0.00	1.00
7	51.83	48.99	49.45	0.41	0.36	0.61	1.02
	(0.47)	(0.5)	(0.44)	(0.01)	(0.01)	(0.01)	(0.04)
8	98.32	94.61	15.90	1.00	0.97	0.12	0.95

Table 4: Results for combined model, standard deviations for noisy cases in brackets

Case	K_{11} (kN/m)	K_{22} (kN/m)	K_{12} (kN/m)	M_{11} (kg)	M_{22} (kg)	M_{12} (kg)	β_1 *e-02	β_2 *e-02	e*e-06 (kgm)
1	100.08	100.10	-0.17	1.00	1.00	0.00	4.13	-0.02	1.00
2	100.15	100.20	-0.27	1.00	1.00	0.00	4.41	-0.12	0.99
	(1.66)	(1.29)	(1.46)	(0.01)	(0.01)	(0.01)	(0.71)	(0.38)	(0.01)
3	99.81	100.87	-0.55	0.99	1.00	0.00	3.78	-0.24	1.00
	(3.62)	(2.92)	(3.03)	(0.02)	(0.03)	(0.02)	(1.70)	(0.75)	(0.03)
4	100.05	100.08	-0.17	1.00	1.00	0.00	4.14	-0.03	0.99
5	98.96	97.86	0.26	0.99	0.98	0.00	5.02	-0.29	0.97
	(7.66)	(2.36)	(4.85)	(0.07)	(0.03)	(0.04)	(2.18)	(0.86)	(0.06)
6	99.96	99.96	-0.04	0.99	0.99	0.00	4.05	-0.01	0.99
7	50.86	48.25	50.21	0.39	0.37	0.62	3.46	3.65	0.99
	(0.21)	(0.19)	(0.16)	(0)	(0.01)	(0)	(4.88)	(0.45)	(0.11)
8	106.05	105.13	12.24	1.03	1.02	0.13	3.25	0.18	1.13

C553/011/98

Design optimization of a contouring feed drive system

L JOHNSON BSc, MSc and **E EBRAHIMI** BSc, MEng, PhD, CEng, MInstMC
Department of Mechanical and Medical Engineering, University of Bradford, UK
R AINI
Rye Machinery Limited, High Wycombe, UK

Synopsis
A contouring model of an X and Y drive axis from a CNC Milling machine is synthesised for analysis and optimisation. A physical modelling approach using block diagram technique is employed to derive the relation for the X and Y-axes. Numerical parameters for the model are found by experimentation. The parameters are varied on two levels such that Taguchi Design of Experiments can be employed. The Design of Experiments will determine which factors have the greatest effect upon positional accuracy.

1. Introduction

A simple model of contouring movement of table and saddle (X, Y axes) is presented here, however the model is generic and modular so further refinement and enhancement are easily attainable. Within this model there are a number of parameters. Each of these parameters represents a physical element within the machine, which can effect the accuracy. Therefore the axes drive system can be looked upon as a system with many different variables, that could feasibly be at any number of different settings.

Broadly sources of error in machining can be categorised as mechanical error, cutting effects and controller error. Koren (**1**) lists some possible sources of error. Mechanical error can come from backlash in drive mechanism arrangements, non-straightness of table and/or guides, lack of rigidity in drive system and poor operating procedures regarding warm-up periods and maintenance. Error due to cutting can come from tool deflection, tool wear, poor pre-setting regarding trueness and offset values. Controller and drive dynamics errors are most frequently overlooked and can be reduced by improving the servo-control algorithms. These sources of error, although most frequently overlooked by machine tool builders may be the dominant sources, particularly with high-speed machines.

Much has been written regarding control strategies for CNC controlled machine axes, both for milling and lathe machines. The issues arising when choosing the strategies to use are cost knowledge of machine dynamics, computer/processing capabilities and machine speed (**2**). Koren (**1**) identifies a number of controller algorithms as Feedback, Feed-forward, Cross-coupled, optimal control, predictive control, adaptive control and repetitive control. The most common type of control is feedback control, which exists mainly as a PID (Proportional, Integral and Derivative) controller. Feedforward controllers are employed to reduce tracking error (**3**). The two basic types are Zero Phase Error Tracking Controller (ZPETC) as first proposed by Tomizuka (**4**) and the second type Inverse Compensation Filter (IKF) proposed by Weck (**5**). The ZPETC system was adapted to form the IKF, which was found to be better at tracking sharp corners.

Excellent tracking characteristics have been observed by the cross-coupled biaxial controller first proposed by Koren (**6**) and developed further by Koren and Lo (**7**). The cross-coupled controller consists of two major parts, a contour error model and a control law. The control law attempts to place the cutter on the nearest point of the reference contour, thus coupling the axes. Simple Proportional controllers on the other hand just use individual axes errors to correct the position and for some contours, like circles, it is not the most accurate. Steethaler (**2**) describes a method of regulating position error using phase lag of an axis behind the reference input to slow down the entire system. Its strength is thought to be the systems ability to cope, in real time, with system constraints and non-linearities. Bickel and Tomizuka (**3**) examine a repetitive controller. An example of errors due to repetitive action is given by Tung et. al. (**8**). Many error in manufacturing are of a periodic nature, such as sticktion (low velocity friction) and can be observed at 90° intervals around a circular profile. The repetitive controller can take advantage of this characteristic after a few cycles to eliminate the error.

Concentrating on the component parts, in particular on the machining operations, it is clear a need exists to develop a model of the machining process. In metal cutting there is a trade off between speed and accuracy; and it is desirable to predict optimum machining criteria in the early stages of design. The goal is to produce a model where the input is a part program and the output is the finished product. A CNC machining centre can be modelled by integration of Computer Aided Control System Design (CACSD), Computer Aided Multi-body Dynamics (CAMD) and FE modelling tools for Machine Tool / Process Modelling (MPM). The basis of the model can either be Knowledge Based or a mathematical model based upon the physics of the system (**9**).

Once such models are produced, Design of Experiment (DoE) can be utilised to indicate the best option to determine optimal settings for a system, or which factors have the greatest effect upon system performance. Here a multi-body, non-linear model of an X and Y drive axes of a CNC Milling machine is developed for simulation and optimisation purposes. The model is capable of generating the contour profile in X, Y plane. Electro-mechanical characteristics of the servo systems and actuation mechanism of the drive systems are included in system realisation. By varying the model parameters the contour accuracy can be studied. The parameters within the model are varied on two levels such that Taguchi Design of Experiments can be employed. The Design of Experiments (DOE) will determine which factors have the greatest effect upon positional accuracy. Applying DOE allows a pre-determined arrangements of settings to be tested. Conclusions are drawn from the analysis about the parameters involved in the drive systems model.

2. Modelling and Simulation

Figure 1 is a schematic representation of a typical drive system. The part program is the input to the CNC controller which generates the reference command signal to X and Y position loop controllers. The position loop controller after comparing the reference command signal with actual position of axis generates a velocity command, which is the input to the drive actuation mechanism amplifier.

Koren (7) modelled the X-axis drive system as a first order lag, accounting also for the discrete and continuous parts of the system. Backlash of leadscrew and nut arrangements is of great interest to CNC drive control. Backlash can be eliminated by pre-loading re-circulating ball nuts. Kao et. al. (**10**) have introduced a backlash study on CNC lathes and found that not only does static backlash occur during direction reversal, but also dynamic backlash caused by the imperfect transient response of the driving mechanism. Burdekin (**11**) describes the standard CNC test of two axes, which is the generation of a circle. Essentially this means the two axes follow a sine reference and one axis is 90^0 out of phase.

The contouring model of the X-Y axes was generated by connecting the X and Y drives axis models through the cutting process. Figure 2 shows the block diagram representation of the system. The input to the model is the position command signal, which is function of the axis feed rate. A detail mathematical analysis of the model is given in reference (**10**). The derived block diagram model of the system was programmed in Simulink™ for simulation and analysis. The numerical values of the parameters within the model were found by experimentation. Some of the parameters such as the length and mass which has a physical value can be calculated from the technical specification of the system, other parameters such as friction and backlash were identified by measurement (current measurement for friction and position measurement for backlash).

The chosen modelling environment represents a graphical modelling structure, which facilitates model parameter manipulations with minimum effort. The model was simulated after the individual elements within the block diagram model of the axes are validated against the experimental results. Figure 3 shows the simulated results and the parameter variation (stiffness, friction and mass) in the model.

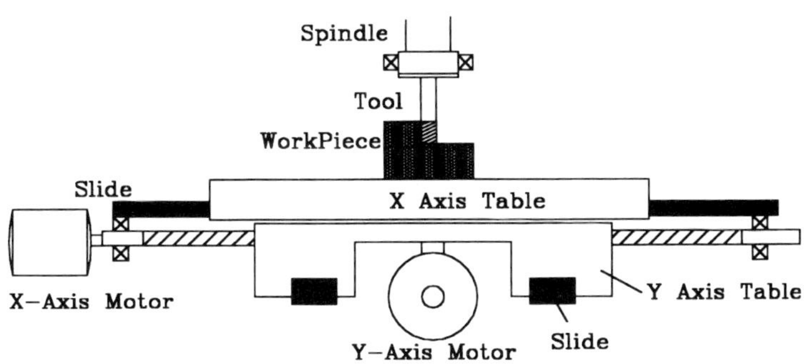

Fig 1 Schematic Representation of a Typical Axes Drive System

3. Design of Experiments

Once the simulation and modelling environment is set up the experimentation can commence. Fowlkes (**12**) gives a broad categorisation to the present types of experimental design currently used by engineers. The orthogonal array is a method of setting up experiments that only requires a fraction of the full factorial combinations. The orthogonal array imposes an order on the way the experiment is carried out. Orthogonal refers to the balance of various combinations of factors so that no one factor is given more or less weight than the others.

Orthogonal also refers to the factors effects being determined mathematically independently of the other factors (independence and balance makes an array orthogonal). Using Analysis Of Variance (ANOVA), the relative contribution each control factor makes to the overall measured response is obtained and expressed as a percentage. A mathematical technique, the "*sum of squares*", is used to quantitatively examine the deviation of the control factor effect response averages from the overall experimental mean response. The technique allows the engineer to gain insight into which factors are most significant, and which has little effect at all. For example, control factor X may be found to be contributing only 7% to the response, and similarly the noise factors are contributing 5%. Thus the engineer can conclude that the control factor X is having no more effect to the response than the noise factors.

Care needs to be taken in choosing the control factors since DoE will only analyse what is in place. Application of engineering knowledge to determine the control factors is a very important early step in the process.

4. Results and Analysis

The DoE can be applied to generated model to analyse the effect of parameter variation. First the control factors and noise factors are determined, followed by a suitable quality characteristic. Then the experiment is designed, conducted and the results analysed.

During model testing it was observed that a reference input to the drive mechanism shows an error between the actual position of the table and its desired position. The error between the desired position and actual position of the table for the contour tracking at any time t is given by the vector e (error) as shown in figure 4. This vector has components dx and dy, which can be determined:

$$e = \sqrt{dx^2 + dy^2} \quad (1)$$

$$dx = (x_r - x)^2 \text{ and } dy = (y_r - y)^2 \quad (2)$$

where x and y are the actual position of the table and x_r and y_r are its reference (desired) position at time (t). The reference position is attained from the circle generator output in the simulation model.

4.1. Determining the control and noise factors.

From the simulated results shown in figure 3, it is evident that certain parameter changes within the model have more significant effects on the overall accuracy of the drive system

C553/011 © IMechE 1998

than others. On this basis three factors were initially chosen to be used in the experiment. These factors were the drive system stiffness, pre-load (coulomb friction) and the backlash in each axis. The full list of factors taken into account during the experiment is listed in Table 1. To implement the DoE, the orthogonal array was chosen and the response data from the simulation entered and analysed. In order to handle the chosen list of parameters an orthogonal array that could accommodate eleven control factors was used. Table 2 shows the orthogonal array used to test all the control factors, which is the L12 array. An L12 array doesn't have any interaction columns, which means results can be attributed more to factors and not interactions between factors.

A summary of the responses from the experimental runs is shown in Table 3. The response plots of the mean and standard deviation of the positional error are shown in Figure 5, and the response table giving the exact values of the plot points is shown in Table 4. The ANOVA procedure is followed and the results displayed in Table 5, which shows the percentage contribution of each control factor.

Results from the model have been structured into two parallel frameworks. Firstly, an L4 orthogonal array was used and secondly, an L12 orthogonal array was used. This approach allows some comparison of results when different numbers of control factors are used, but are subjected to exactly the same analysis. The object is to confirm that the analysis agrees for the common control factors. Considering figure 5, the plot shows that the greatest effect on response is given by changing the stiffness in the drive systems, also it shows that the biggest effect on mean error comes from the rotary to linear conversion change of both axes. The percentage effects of rotary conversion x-axis (Rx) and the rotary conversion y-axis (Ry) are 52% and 40% respectively. However, the spread of response indicates the stiffness components have the greatest effect with a 32% and 25% contribution.

One would expect that the effect contributions of the linear to rotary conversion should be identical for each axis. Similarly, so should the table mass (M_x & M_y) and other matching effects. The explanation for the discrepancy probably arises because the last half of table position data from the simulation was used to calculate error. The first half of the data points were discarded to eliminate the transient response of the drive system to a step input going from radius zero to radius 100mm instantaneously.

Therefore, the response plot of figure 5, warrants as being best to minimise error, for the algebraically larger S/N values. Comparing the mean error plot to the SD error plot. The settings agree except for the rotary to linear conversion of the X and Y drive axes (R_x and Ry). The mean plot indicates the best settings are 1 and 1. The SD plot shows the best settings are 2 and 2 respectively. In this case the SD recommendation would be taken since the variation in position response is more difficult to control electronically than a steady state error. Comparing the rest of the settings shows a general agreement by both the mean measure and SD measure. The stiffness factors are generally recognised as being correct. They are shown to be main effects and should both be set to the highest stiffness value, setting 1. It is difficult to conclude which setting the table slide viscous friction, F_{gtx} and F_{gty}, should be set at. According to both sets of results F_{gtx} and F_{gty} should be set at level 1 and level 2 respectively. Furthermore, using the whole circle revolution for data may influence the way a change in an X drive parameter could appear to have a different effect from identically changing the corresponding Y drive parameter. The same can be stated about the table mass parameters (M_x & M_y), which are also contradictory.

5. Discussion and Conclusions

A contouring model of feed drive axes was created for simulating the response of the drive system under different parameter conditions. The changing of parameter conditions has been structured and analysed using Design of Experiments (DoE). It has been shown how engineering knowledge and model testing have been combined to determine which parameters are worthy of investigation. Model testing also provides a method of validating the model to ensure its response is realistic and represents a true enough account of the physical world.

Since the procedure has been set out, its application to the rest of parameters within the model can be implied, thereby a model reduction can be achieved. Furthermore, by reducing the number of control factors from eleven to around six, the opportunity for the smaller L8 array to be used is envisaged. To perform this, control factors for each axis can be considered together as in the L4 array. Finally, the L8 array does have provisions for interactions to be studied, enabling the backlash / nut friction combination to be examined.

References:

1. Koren Y, Lo C, Advanced controllers for feed drive. Annals of the CIRP. Vol. 41, No.2, 1992, pp. 689-698.
2. Seethaler R J, Yellowley I, The regulation of position error in contouring systems. International journal of Machine tool manufacturers. Vol. 36, No. 6, 1996, pp. 713-728.
3. Bickel R J, Tomizuka M, Synchronization of multiple-axis systems subject to periodic disturbances. Japan/USA symposium on flexible automation ASME. Vol. 2, 1992, pp. 821-827.
4. Tomizuka M, Zero Phase Tracking Algorithm for Digital Control. ASME transactions : Journal of dynamic systems measurement and control, Vol. 109, March, 1987, pp. 66-68.
5. Weck M, Ye G, Sharp corner tracking using the IKF control strategy. Annals of the CIRP. Vol. 39, No. 1, 1990, pp. 437-441.
6. Koren Y, Cross-coupled bi-axial computer control for manufacturing systems. ASME transactions : Journal of dynamic systems measurement and control. Vol. 102, No. 4, December, pp.265-272.
7. Koren Y, Lo C, Variable gain cross-coupling controller for contouring. Annals of the CIRP. Vol. 104, August, pp. 371-374.
8. Tung E, Anwar G, Tomizuka M. Low velocity friction compensation and feed forward solution based on repetitive control. Proceedings of the 1991 American Control Conference, Vol. 3, Boston MA 1991, pp. 2615 - 2620.
9. Martin K F, Ebrahimi K, Modelling the X-Axis drive and control of a milling machine. Modelling and simulation. Proceedings of the IASTED international conference. Columbo Sri-Lanka, July 26-28, 1995, pp. 30-33.
10. Kao J Y, Yeh Z M, Tarng Y S, Lin Y S, A study of backlash on the motion accuracy of CNC lathes, International journal of Machine tool manufacturers, Vol. 36, No. 5, 1996, pp. 539-550.
11. Burdekin M, Park J, Contisure - A computer aided system for assessing the contouring accuracy of NC machine tools, Proceedings 27th International MATADOR Conference, 1988, pp. 197-203.
12. Fowlkes W, Creveling C M, Engineering methods for robust product design. Using Taguchi Methods in Technology and Product Development. Addison-Wesley, 1995.

C553/011 © IMechE 1998

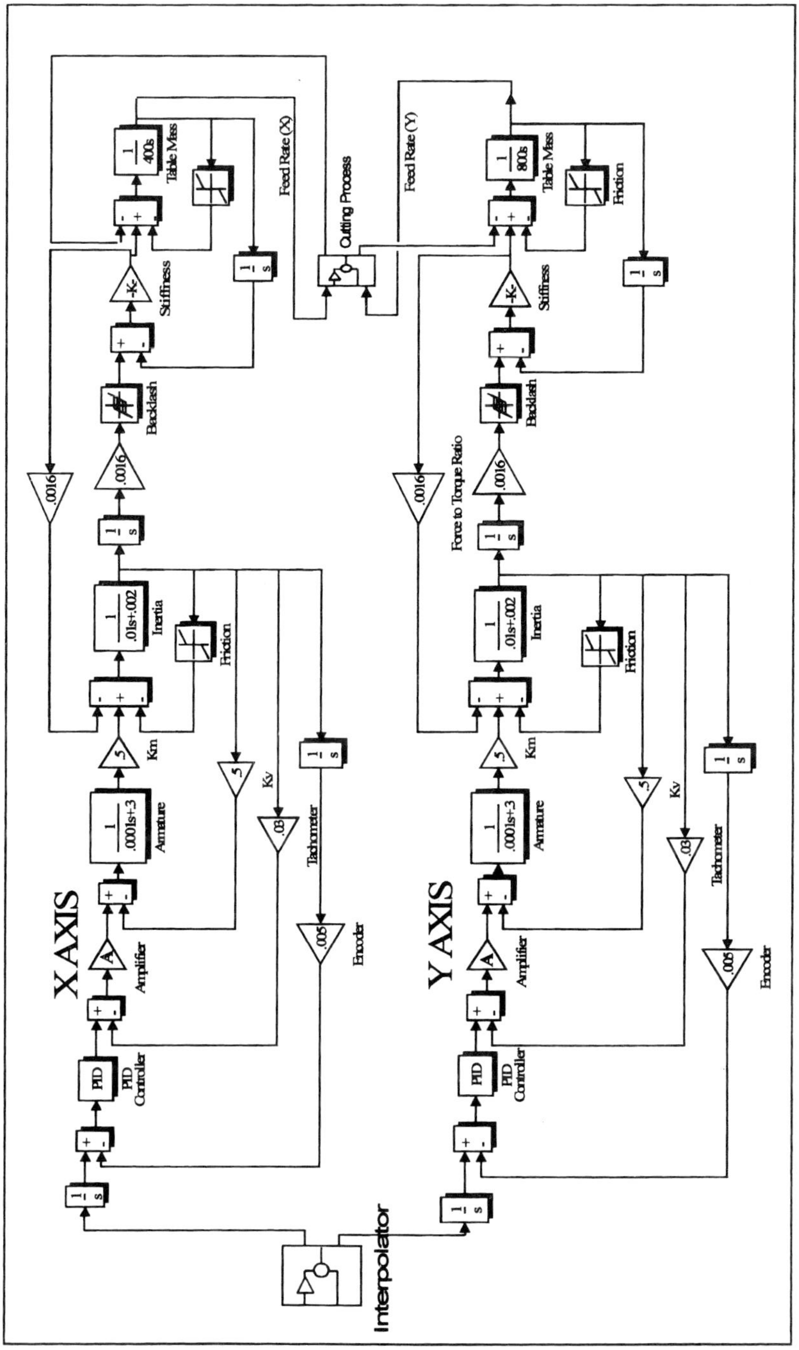

Fig. 2 Block Diagram Model of the Axes Drive system

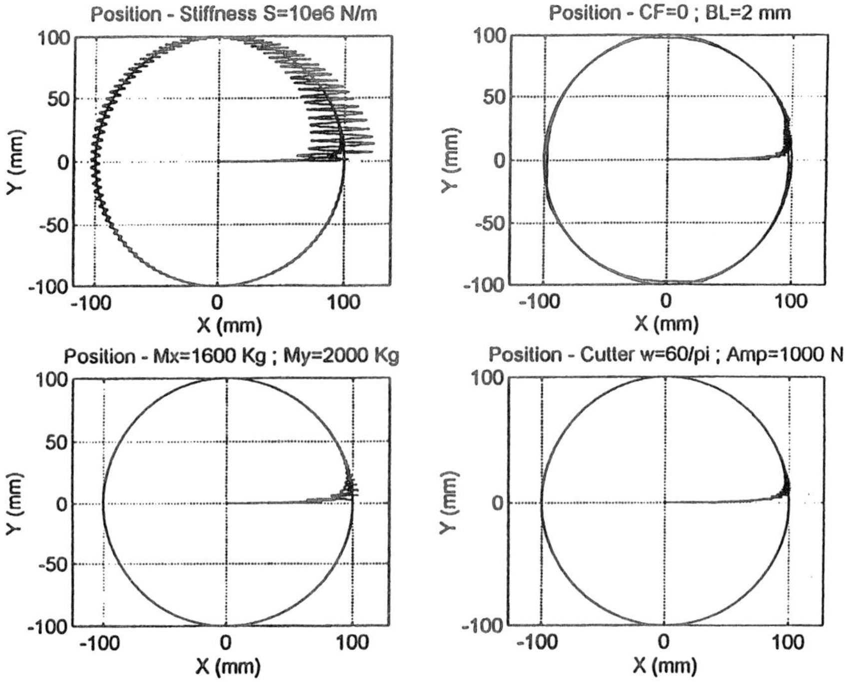

Fig. 3 Simulated Response of the Axes

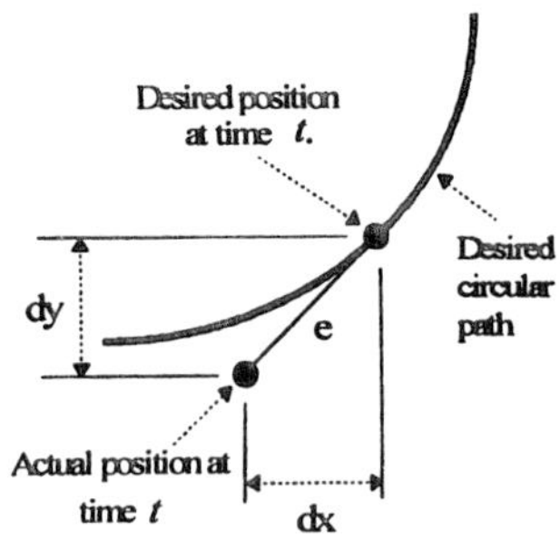

Fig. 4 Positional error around circle

No	Factor	Symbol	Level 1	Level 2
1	Backlash	Db	0	12.5×10^{-5}
2	Stiffness x-axis	Sx	10^7	10^6
3	Stiffness y-axis	Sy	10^7	10^6
4	Coulomb friction x-axis	F_{dsx}	0.5	0.0
5	Coulomb friction y-axis	F_{dsy}	0.5	0.0
6	Viscous friction x-axis	F_{gtx}	0.001	0.1
7	Viscous friction y-axis	F_{gty}	0.001	0.1
8	Rotary to linear converter x-axis	R_x	0.0016	0.0020
9	Rotary to linear converter y-axis	R_y	0.0016	0.0020
10	Table mass x-axis	M_x	400	800
11	Table mass y-axis	M_y	800	1200

Table 1 Control factors

C553/011 © IMechE 1998

Run	1 Db	2 Sx	3 Sy	4 Fdax	5 Fday	6 Fgtx	7 Fgty	8 Rx	9 Ry	10 Mx	11 My
1	1	1	1	1	1	1	1	1	1	1	1
2	1	1	1	1	1	2	2	2	2	2	2
3	1	1	2	2	2	1	1	1	2	2	2
4	1	2	1	2	2	1	2	2	1	1	2
5	1	2	2	1	2	2	1	2	1	2	1
6	1	2	2	2	1	2	2	1	2	1	1
7	2	1	2	2	1	1	2	2	1	2	1
8	2	1	2	1	2	2	2	1	1	1	2
9	2	1	1	2	2	2	1	2	2	1	1
10	2	2	2	1	1	1	1	2	2	1	2
11	2	2	1	2	1	2	1	1	1	2	2
12	2	2	1	1	2	1	2	1	2	2	1

Table 2 The L12 array.

Run	Feed 1 mean (m e-3)	Feed 2 mean (m e-3)	S/N	Feed 1 SD (m e-3)	Feed 2 SD (m e-3)	S/N
1	3.50	7.00	-14.9	8.47	8.47	-18.6
2	25.20	25.90	-28.1	8.86	8.73	-18.9
3	15.00	16.30	-23.9	840.00	830.00	-58.4
4	17.20	18.10	-24.9	800.00	820.00	-58.2
5	17.20	18.20	-25.0	810.00	830.00	-58.3
6	15.30	16.50	-24.0	810.00	820.00	-58.2
7	17.20	18.10	-24.9	780.00	800.00	-58.0
8	3.70	7.10	-15.1	95.90	97.70	-39.7
9	25.90	25.90	-28.3	9.42	9.33	-19.4
10	25.30	26.10	-28.2	160.00	150.00	-43.8
11	4.40	7.50	-15.8	190.00	190.00	-45.6
12	15.60	16.70	-24.2	780.00	800.00	-58.0

Table 3 L12 Response summary

Factor	Level	Mean (s/n)	SD (s/n)
Db	1	-23.47	-45.10
	2	-22.75	-44.08
Sx	1	-22.53	-35.50
	2	-23.68	-53.68
Sy	1	-22.70	-36.45
	2	-23.52	-52.73
Fdax	1	-22.58	-39.55
	2	-23.63	-49.63
Fday	1	-22.65	-40.52
	2	-23.57	-48.67
Fgtx	1	-23.50	-49.17
	2	-22.72	-40.02
Fgty	1	-22.68	-40.68
	2	-23.53	-48.50
Rx	1	-19.65	-46.42
	2	-26.57	-42.77
Ry	1	-20.10	-46.40
	2	-26.12	-42.78
Mx	1	-22.57	-39.65
	2	-23.65	-49.53
My	1	-23.55	-45.08
	2	-22.67	-44.10

Table 4 L12 Response Table.

Mean - Positional Response				SD - Positional Response			
Factor	S/N f1	S/N f2	SS f	Factor	S/N f1	S/N f2	SS f
1	-23.47	-22.75	0.6%	1	-45.10	-44.08	0.1%
2	-22.53	-23.68	1.4%	2	-35.50	-53.68	31.9%
3	-22.70	-23.52	0.7%	3	-36.45	-52.73	25.6%
4	-22.58	-23.63	1.2%	4	-39.55	-49.63	9.8%
5	-22.65	-23.57	0.9%	5	-40.52	-48.67	6.4%
6	-23.50	-22.72	0.7%	6	-49.17	-40.02	8.1%
7	-22.68	-23.53	0.8%	7	-40.68	-48.50	5.9%
8	-19.65	-26.57	52.2%	8	-46.42	-42.77	1.3%
9	-20.10	-26.12	39.5%	9	-46.40	-42.78	1.3%
10	-22.57	-23.65	1.3%	10	-39.65	-49.53	9.4%
11	-23.55	-22.67	0.8%	11	-45.08	-44.10	0.1%
			100%				100%

Table 5 L12 ANOVA - Factor contribution

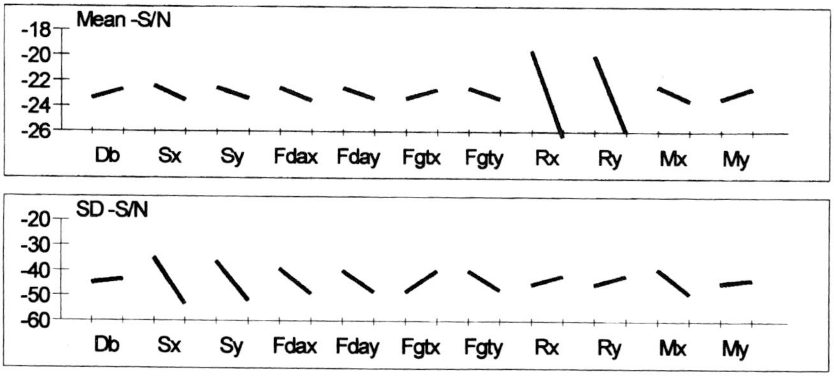

Fig 5 L12 Response plot

C553/011 © IMechE 1998

C553/036/98

Multi-body dynamics in complex objects control: an internal combustion engine dynamometer test

D E VENTZAS SMISA, MIEEE and **P TSIAKARAS**
University of Thessaly, Greece

The principles and the methods to control complex, multi-body, objects of diverse physical nature, subjected to complex forces and comprise inertial components having structural compliance under constraints are described. Non-linear control results in improved response, extended torque frequency range, system dynamics and overall speed. The paper presents the model and simulation of an internal combustion engine coupled with an eddy current dynamometer system. Steady-state and dynamic models are presented.

INTRODUCTION

The performance of an internal combustion engine is evaluated and a coupled eddy current dynamometer system is described (see fig. 1). Lumped modeling techniques are used in order to establish the inter-connections between the physical components. The coupled eddy current dynamometer absorbs kinetic power by adjusting its field excitation current.

Under steady state conditions, the engine is simulated as a voltage source with resistance, while the dynamometer is modeled as a current source with impedance. Under dynamic conditions the engine is simulated as a voltage source with resistance and capacitance, while the dynamometer is considered as a current source with impedance, capacitance and reactance. The shaft is a passive element with impedance-reactance characteristics. The shaft stiffness is measured physically, and the internal damping is derived by transient response analysis (1).

DYNAMOMETER SELECTION

Dynamometers are mechanical, hydraulic, electric, eddy current, air brake, friction and hybrid ones (9, 10). Eddy current dynamometers are capable of rapid load change, well adapted to computer control, simple and robust with low inertia, but they are vulnerable to poor cooling supply, not sustaining overload (11). Engine and dynamometer characteristics are matched (see fig. 3). An incorrect choice, installation and use of the system might lead to torsional oscillations, system vibration and shaft whirling and eventual failure. The resonant (critical frequency) of torsional oscillation is (2):

$$n_c = \frac{60}{\pi}\sqrt{\frac{C.(I_{engine} + I_{brake})}{I_{engine} \cdot I_{brake}}}$$

© IMechE 1998 C553/036

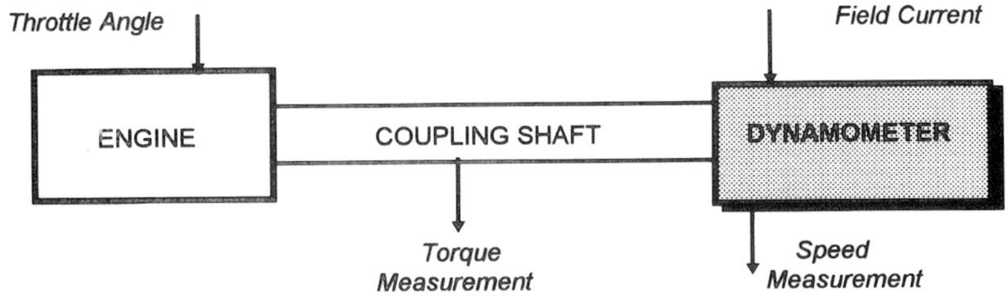

Fig. 1. Block diagram of the plant

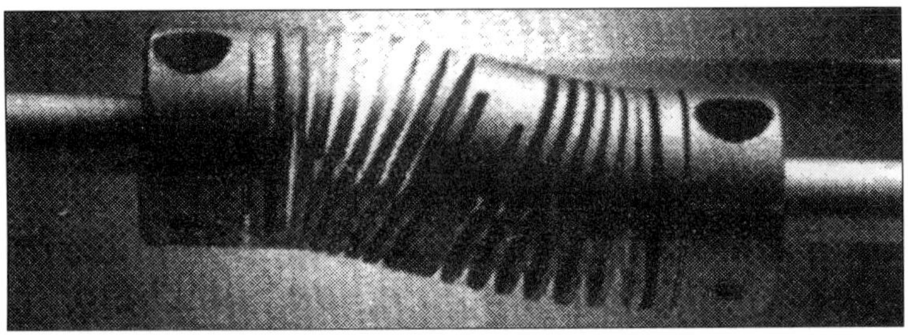

Fig. 2. Experimental flexible link

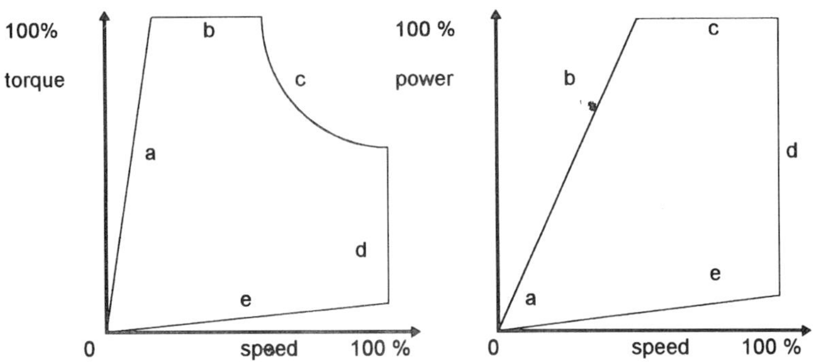

Fig. 3. Performance curves of eddy current dynamometer

For a three body model of a load-motor-tachometer system:

$$\begin{bmatrix} sL_a & sK_E & 0 & 0 \\ -k_T & s^2 J_m + sD & s^2 J_1 & s^2 J_2 \\ 0 & sD_1 + k_1 & -\left(s^2 J_1 + sD_1 + k_1\right) & 0 \\ 0 & sD_2 + k_2 & 0 & -\left(s^2 J_2 + sD_2 + k_2\right) \end{bmatrix} \cdot \begin{bmatrix} I_a(s) \\ \theta_m(s) \\ \theta_1(s) \\ \theta_2(s) \end{bmatrix} = \begin{bmatrix} V(s) \\ 0 \\ 0 \\ 0 \end{bmatrix}$$

C553/036 © IMechE 1998

where,

L_a = motor equivalent inductance
k_E = generator voltage / angular velocity proportionality constant
k_T = motor torque / current proportionality constant
k_1 = load (brake) stiffness factor
k_2 = tachometer stiffness factor
J_m = motor moment of inertia
J_1 = load (brake) moment of inertia
J_2 = tachometer moment of inertia
D = damping factor
D_1 = load (brake) damping factor
D_2 = tachometer damping factor
I_a = motor current excitation
V = motor voltage excitation
θ_m = motor angular position
θ_1 = load (brake) angular position
θ_2 = tachometer angular position

EXPERIMENTAL SET - UP	
Internal combustion engine	*1300 cc lead less oil*
Dynamometer heat absorption	*water cooled*
Internal combustion heat absorption	*heat exchanger*
Lubricating oil heat absorption	*heat exchanger*
Dynamometer casing	*swing in trunion bearings, universal joints*
Measurements	*torque, angular velocity*
Torsional strain	*resistance strain gauges*
Dynamometer speed	*ac tachometer*
Cooling water temperature	*thermistor*
Lubricant temperature	*thermistor*
Safety circuits	*shut down*
Control	*angular position (15 Hz)*
Throttle control	*step motor*
Mechanical properties	*inertia, viscous damping, compliance*

Table 1. Experimental set-up

The dynamometer field current is controlled with a rise time of 4 ms. Closed loop torque and speed control permits marginally greater bandwidth (3).

STATIC and DYNAMIC TESTS

The hardware under test and simulation is presented. The plant presented consists of the engine, the coupling shaft and the dynamometer. Parameters, such as the throttle angle, torque, field current and the speed measurement are given.

The model is developed initially to represent the system under steady state conditions, about a chosen operating point in the middle of the torque and speed ranges (5).

The model under dynamic conditions, the parameters interactions and magnitudes are evaluated by structural and dynamics considerations. Thermodynamic and mechanical losses are due to reciprocating and rotating parts (6). Excess

© IMechE 1998 C553/036

attenuation at high frequencies of the plant is due to the flux build-up (contributing first order lags) associated with the dynamometer rotor (4).

Sine wave, step, impulse and Pseudo Random Binary Signal excitation responses are studied. PRBS excites many frequencies simultaneously, its power is spread over the lower harmonics of the spectrum, but it introduces noise in the system spectrum identification results.

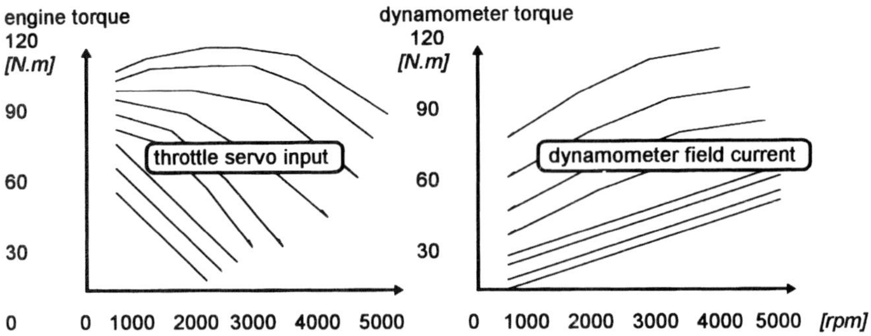

Fig. 4. Engine and dynamometer torque - speed characteristic

The throttle and dynamometer steady state transfer characteristics are presented, together with the throttle-speed frequency response of the engine (see fig. 4). The response of torque and speed to throttle and load perturbations are simulated (see fig. 8) (12).

ENGINE - DYNAMOMETER MODEL

The state variables for the electromechanical model of the engine - dynamometer system, gives a model with a valid structure throughout the torque / speed range of the engine, as:

$$\begin{bmatrix} \frac{dV_f}{dt} \\ \frac{dV_D}{dt} \\ \frac{dI_s}{dt} \end{bmatrix} = \begin{bmatrix} -\frac{1}{C_E.R_E} & 0 & -\frac{1}{C_E} \\ 0 & -\frac{1}{C_D.R_D} & \frac{1}{C_D} \\ \frac{1}{L_s}-\frac{1}{C_E.R_E.R_S} & \frac{1}{C_D.R_D.R_S}-\frac{1}{L_s} & \frac{1}{L_s}-\frac{1}{C_E.R_E.R_S} \end{bmatrix} . \begin{bmatrix} V_f \\ V_D \\ I_s \end{bmatrix} + \begin{bmatrix} \frac{1}{C_E.R_E} & 0 \\ 0 & -\frac{1}{C_D} \\ \frac{1}{C_E.R_E.R_S} & \frac{1}{C_D.R_S} \end{bmatrix} . \begin{bmatrix} V_E \\ I_D \end{bmatrix}$$

where,

V_F	=	*engine speed*	V_D	=	*dynamometer speed*
I_S	=	*shaft torque*	V_T	=	*throttle servo input voltage*
V_E	=	*engine source voltage (a function of throttle opening)*			
$-1/R_E$	=	*slope of the torque / speed curve*			
V_L	=	*load - dynamometer current amplifier input voltage*			
I_D	=	*dynamometer source current*			
I_F	=	*field current*			

C553/036 © IMechE 1998

The model (a first order system with a time delay proportional to the inverse of engine speed, i.e. the time taken to inhale and ignite a charge) decouples the engine flywheel from the shaft connecting it to the dynamometer and obtains a series of transfer functions, accepted over a range of mean speeds. The engine inertia is lumped, while the dynamometer presents a large rotor inertia (7).

The responses of shaft torque and dynamometer speed to inputs to the dynamometer field current control show poor agreement, while for throttle servo disturbance they show good agreement.

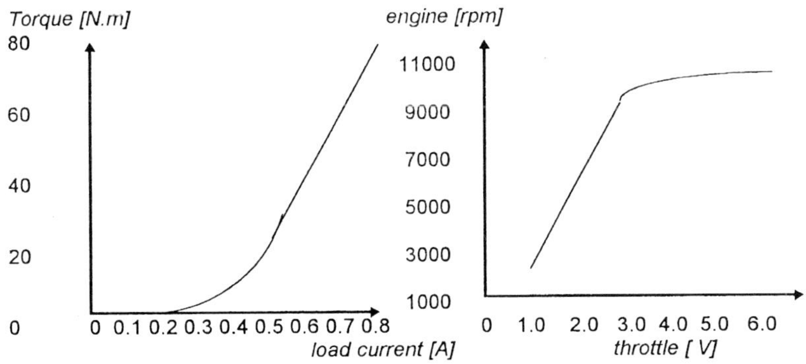

Fig. 5. Throttle and dynamometer steady state transfer characteristic

Difficulties are encountered in reproducing the curve in the high torque and low speed regime, where R_E becomes negative. At low speeds with large throttle openings the slope of the characteristic changes sign. Stability is achieved by proportional control upon the dynamometer field current. The slope of the torque / speed curves remains substantially constant for the full range of the throttle settings.

The relations between V_T and V_E are given in fig. 5, showing that for low speed high torque regime it is substantially linear.

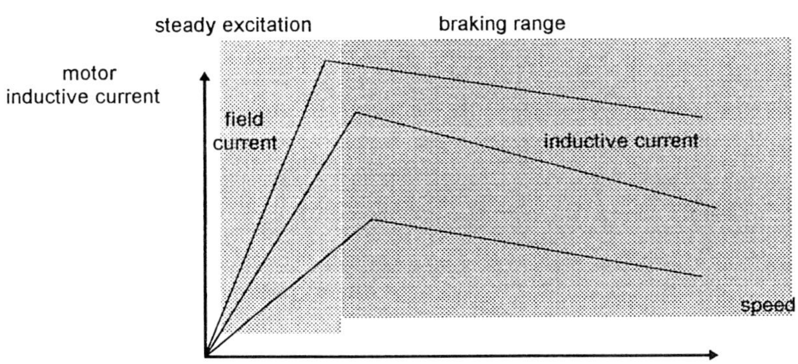

Fig. 6. Characteristic curve of electrodynamic brake

The slope of the torque speed characteristics of the engine alter significantly in certain regimes of operation (i.e. the parameters change, see fig. 6). Instabilities cause the engine to stall, while resonance imposes limits on rig performance; feedback speed control overcomes such instabilities (see fig. 2).

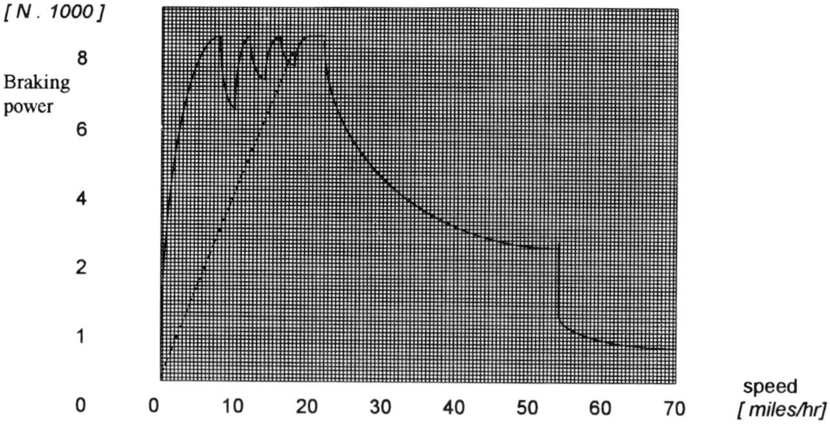

Fig. 7. Electric motor train Braking power vs velocity during maximum braking

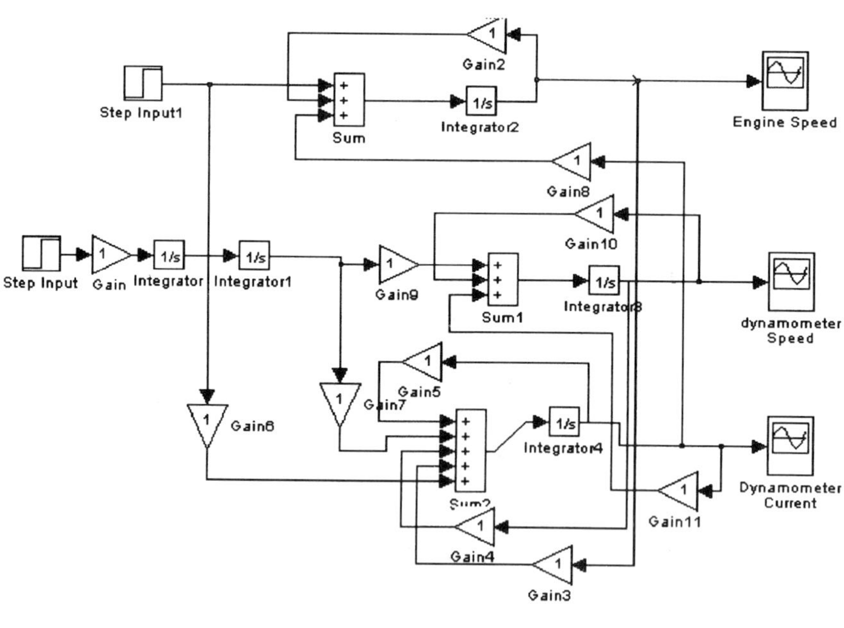

Fig. 8. Linear simulation of engine - dynamometer system

C553/036 © IMechE 1998

The throttle servo suffers from velocity limiting, as an average filter. Perturbations of the ignition advance instead of the throttle angle, results in an improved impulse response for the shaft torque (8).

CONCLUSIONS

The model of an internal combustion -dynamometer and its response is given. Simulation offers obvious advantages for testing the system without the danger in damaging it by over speeding. The system time constant is moderately short.

REFERENCES

(1) Chilikin, M, *Electric Drive*, MIR Publishers, Moscow, 1976

(2) Electro-Craft Corp, *DC Motors Speed Controls Servo Systems*, USA, 1984

(3) FROUDE Consine Ltd, *Installation, Operation & Maintenance Procedures, Instruction Manual IM 1232/2, AG Series Eddy-Current Dynamometers*, May 1990, FROUDE Consine Ltd

(4) FROUDE Consine Ltd, *Instruction Manual IM 1290/1, Issue 1, TEXCEL 100 Aero Version 3 Direc Digital Control System*, July 1995, FROUDE Consine Ltd

(5) KISTLER Instruments AG, *Quartz, 4-Component Dynamometer, Operating Instructions*, 1992

(6) MAGTROL Inc, *Digital Dynamometers, for Precision Motor Analysis*, 1992

(7) OSE, *Handbook of Analysis, Control and Regulation of Dynamic Brake of Diesel Traction Machines MLW S 450-500 and ALCO S 9100*, June 1981, Greek Railways, Study Group, Athens 1981 (in Greek)

(8) Pattas, K. N, Stamatellos, A. M, *The Effect of Exhaust Throttling on the Diesel Operation Characteristics and Thermal Loading*, SAE, 890399, February 27-March 3, 1989, Detroit, Michigan

(9) Plint, M, Martyr, A, *Engine Testing, Theory and Practice, Chapter 7: Choosing the Right Dynamometer*, p. 100, Butterworth-Heinemann Ltd, 1995

(10) Plint, M, Martyr, A, *Engine Testing, Theory and Practice, Chapter 8: Coupling the Engine to the Dynamometer*, p. 116, Butterworth - Heinemann Ltd, 1995

(11) Siakavellas, N, *Eddy Current Modelling in a Coupled Electromagnetic-Mechanical Problem*, COMPEL, The International Journal for Computation and Mathematics in Electrical and Electronic Engineering, Vol. 15, No. 2, 1996, pp. 52-72, MCB University Press

(12) Ventzas D, *Automation I-II*, Lecture Notes, TEI Athens, STEF, TIO, Athens 1990 (in Greek)

C553/031/98

Torsional vibration in planetary gears investigation of simplified models

S JOHANSEN MSc, MSEM, MSVIB, DAS, EAA
DELTA Acoustics and Vibration, Aarhus, Denmark

Torsional vibration in wind turbines with planetary gears can cause tonal noise between 200 Hz and 800 Hz. The torsional vibrations are transmitted to the wings, where noise is radiated to the surroundings.

The gearbox casing containing the ring gear is often resiliently mounted to reduce transmission of vibrations to the supporting structure. The transmission ratio is therefore not constant but varies in time due to the rotation of the gear casing.

Standard packages for torsional vibration calculations do not have proper facilities for this type of analysis. The purpose of this investigation was to find a simplified model that could be used in a standard torsional vibration calculation computer program.

Different simplified torsional vibration models of a planetary gear have been investigated and compared with a complete model derived using Lagrange's equation of motion. None of these simple models gave satisfactory results.

NOTATION

Basic terms

I	mass moment of inertia
T	kinetic energy
V	potential energy
b	transmission ratio
k_i	stiffness between coordinate i and ground
k_{ij}	stiffness between coordinate i and j
r_i	pitch circle radius, gear i
m	mass
t	time
z_i	number of teeth on gear wheel i
α	angular coordinate
ω	angular speed

Subscripts

i, j	index
G	global
L	local
S	planet carrier
1	ring wheel
2	planet gear
4	sun gear
5	input flange

1. INTRODUCTION

The sound power emitted from modern 600 kW wind turbines is approximately 98 dB(A). This is less than the sound power emitted from typical 100 kW wind turbines from 1987. This dramatic improvement is very much caused by the good results obtained by reducing aerodynamic noise. Consequently, transmission noise is again becoming an important factor in wind turbine noise reduction.

Planetary gears are used in the low speed stage in gearboxes for modern wind turbines. The torque is split into three pinions and the reaction forces cancel each other out. See figure 1. This reduces transmission of vibrations to the nacelle and tower. In order to reduce the transmission of vibrations even further the gearbox is resiliently mounted to the nacelle.

Figure 1. Planetary gear

Measurements of noise emitted from a wind turbine showed two evident harmonics of the tooth meshing frequency of the low speed stage in the gearbox. The harmonics were emitted from the wings. It was decided to make a detailed torsional vibration analysis of the wind turbine. The idea was to check if some of the tooth meshing frequency harmonics coincided with the natural frequencies of the torsional system.

Finding a commercially available computer program for torsional vibration calculation able to model resiliently mounted planetary gears was not possible. Therefore, an investigation of simple models of the planetary gear was started.

This work started in a corporation with Danish Technological Institute because of their vast experience in torsional vibration calculation and measurement. This investigation was made preliminary to a client project for BONUS Energy A/S.

2. TORSIONAL VIBRATION - COMPUTER PROGRAMS

Some commercially available computer programs for torsional vibration analysis have been examined. They all treat gearings in the same way. Mass moments of inertia and stiffness properties are multiplied by the transmission ratio squared.

$$I_G = I_L \, b^2 \tag{1}$$

$$k_G = k_L \, b^2 \tag{2}$$

The reason is probably historic. This was the classical way to deal with transmission ratios in the Holzer method (1). This excludes modelling of any type of transmission where the

C553/031 © IMechE 1998

speed of a shaft is determined by a linear combination of two other shaft speeds. Modelling branched systems but only with fixed transmission ratios is possible. An investigation was started to find a simplified model of a planetary gear. The idea was to use this model in a standard commercially available torsional vibration computer program.

3. EQUATIONS OF MOTION

Lagrange's equations of motion were used in the formulation of the problem. For simplicity only one planet is considered here.

The gear wheels and the planet carrier are shown in figure 2. Flexibilities of the gear meshes are modelled. Flexibility between the planet carrier and the planet wheel bearing is also modelled. Flexibility of the planet carrier itself has not been taken into account. This flexibility is easily added to model as an input shaft stiffness between the gear model and the input flange. The kinematics of the system in figure 1 can be found in standard textbooks on machine components (2). In the equations the number of teeth on the planet gear 2 is counted negative because of the internal gear 1.

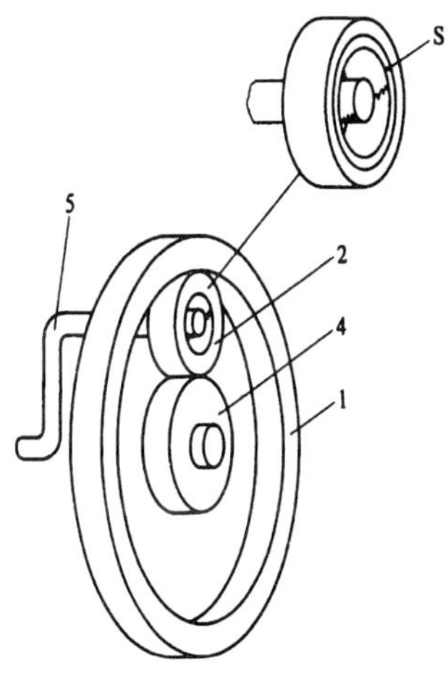

Figure 2. Planetary gear Model

$$\omega_2 = \left(1 + \frac{z_1}{z_2}\right)\omega_S - \frac{z_1}{z_2}\omega_1 \tag{3}$$

$$\omega_4 = \left(1 + \frac{z_3}{z_4}\right)\omega_S - \frac{z_3}{z_4}\omega_2 \tag{4}$$

The potential energy is:

$$V = \frac{1}{2}[k_{12}\ r_2(\alpha_2\ -\ (1\ +\ \frac{z_1}{z_2})\alpha_S\ +\ \frac{z_1}{z_2}\alpha_1)^2]$$
$$+\ \frac{1}{2}[k_{24}\ r_4(\alpha_4\ -\ (1\ +\ \frac{z_3}{z_4})\alpha_S\ +\ \frac{z_3}{z_4}\alpha_2)^2]$$
$$+\ \frac{1}{2}k_1\ r_1\ \alpha_1^2\ +\ \frac{1}{2}(k_{SS}\ r_5(\alpha_S\ -\ \alpha_5)^2 \tag{5}$$

The kinetic energy is:

$$T = \frac{1}{2}\ [I_1\ \dot{\alpha}_1^2\ +\ I_2\ \dot{\alpha}_2^2\ +\ I_4\ \dot{\alpha}_4^2$$
$$+\ I_S\ \dot{\alpha}_S^2\ +\ I_5\ \dot{\alpha}_5^2] \tag{6}$$

The free vibrations are described by insertion of Equations 5 and 6 into Lagrange's equations of motions:

$$\frac{d}{dt}\left(\frac{\partial T}{\partial \dot{\alpha}_i}\right)\ -\ \frac{\partial T}{\partial \alpha_i}\ +\ \frac{\partial V}{\partial \alpha_i}\ =\ 0 \tag{7}$$

The mass and stiffness matrices are computed from equation 7.

4. ANALYSIS OF COMPLETE GEARBOX

Analysis was done on a commercial available planetary gearbox. The total gear ratio is 54.7. The actual gearbox includes the planetary input stage, an intermediate spur gear stage and an output spur gear stage. The gear ratio for the planetary stage is 5.55. Data for the planetary gear stage are shown in table 1. The mass moment of inertia of the gear box casing supporting the ring is 493 kg m^2.

The natural frequencies of the gearbox were calculated with different values of the stiffness of the gearbox casing supporting the ring gear. The natural frequencies are shown as function of the gearbox casing support stiffness in figure 3. The system has two rigid body modes when the gear box suspension stiffness is zero. The second rigid body mode is transferred into a suspension mode when stiffness is introduced into the support of the gearbox casing. The natural frequency of this mode increases with increasing stiffness. The natural frequencies of the other modes are temporarily changed when the frequency of this rotating gearbox mode comes close. Otherwise the other natural frequencies are constant. The mode shapes do not change very much except for the gearbox casing suspension rotating mode.

It is concluded that a simple model not considering rotation of the gearbox can estimate all the modes except the gearbox casing suspension mode. The natural frequencies computed with such a model are expected to be accurate except when they are close to the natural frequency of the gearbox casing rotating mode.

A simplified model can therefore be useful for calculation of the modal properties at frequencies away from the suspension mode.

C553/031 © IMechE 1998

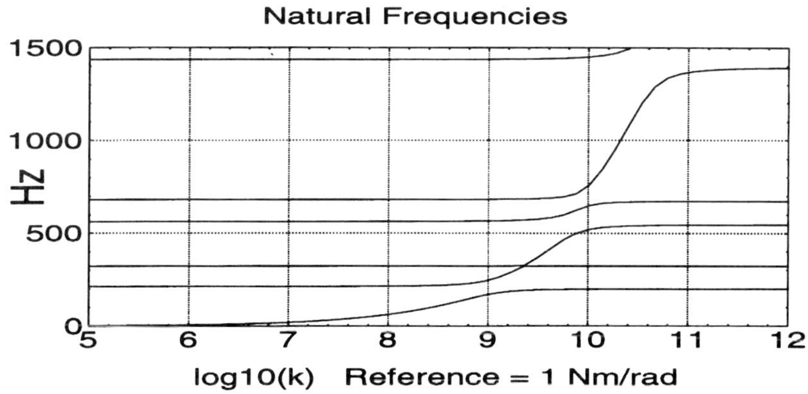

Figure 3. Natural frequencies as function of gear housing support stiffness

The actual stiffness of the gear housing support in the wind turbine is 65.2 MNm/rad. The logarithm to this stiffness is approximately 8. From figure 3 is it seen that there is good separation between the suspension natural frequency and the other natural frequencies. The reduced model could be applied here and would predict the correct natural frequencies and approximate mode shapes.

5. ANALYSIS OF PLANETARY STAGE

The reduced model that successfully modelled all the modes except the suspension mode was developed using Lagrange's equations of motions. Unfortunately it was not possible to introduce the equations of motion directly into the torsional vibration computer program.

It was therefore necessary to investigate possible simple models of the planetary gear that could be modelled in the torsional vibration computer program.

The previous calculations were made using a model of the entire gearbox. In the following only a model of the planetary stage is used. The data used are shown in table 1.

Mass Moments of Inertia kg m^2		Stiffness Nm/rad		Number of gear teeth	
I_S	20.0	k_{5S}	$1.00 \cdot 10^9$	z_1	91
I_2	0.37	k_{12}	$3.33 \cdot 10^8$	z_2	-35
I_4	5.60	k_{24}	$4.71 \cdot 10^8$	z_3	35
I_5	20.0			z_4	20

Table 1. Data for planetary stage of the analysed gearbox

The natural frequencies computed using the full model with the ring wheel grounded are shown in tables 2 and 3.

Mode number	1	2	3	4
Frequency, Hz	0	1117	1923	3782
Mode shapes				
Planet carrier	0.17	-0.28	0.28	0.01
Planet bearing	0.17	0.00	-0.53	-0.05
Planet wheel	0.27	0.44	0.67	-0.12
Sun wheel	0.93	0.85	-0.44	0.99

Table 2. Full model, natural frequencies and mode shapes

Mode number	1	2	3	4
Frequency, Hz	0	1117	1923	3782
Mode shapes				
Planet carrier	1.00	1.00	-0.52	0.03
Planet bearing	1.00	0.02	1.00	-0.30
Planet wheel	1.00	-0.98	-0.79	-0.41
Sun wheel	1.00	-0.55	0.15	1.00

Table 3. Full model, natural frequencies and mode shapes, mode shapes have been multiplied by the gear ratio

C553/031 © IMechE 1998

A simple model for the planetary gear is derived from figure 4.

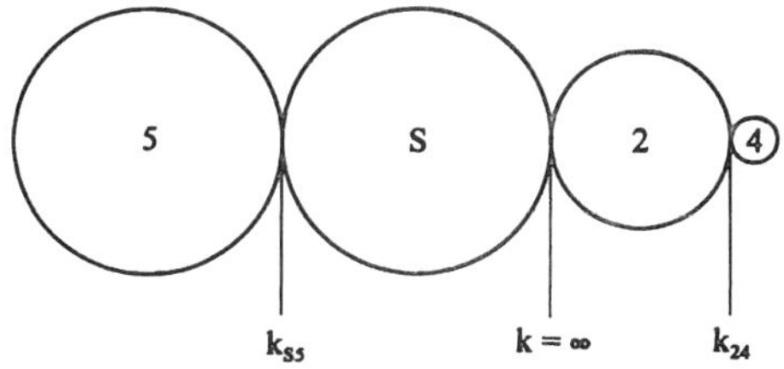

Figure 4. Simple model A. Actual gear ratios

The mass properties are the same as in the full mode, but the flexibility between the ring wheel and the planetary wheel is not included in the simple model. The number of degrees of freedom is actually only three in the simple model. The kinematic connection between the centre of the planet wheel and the sun wheel makes it possible to omit the $m \cdot r^2$ contribution of the planet wheel to the carrier mass moment of inertia and add it to the mass moment of inertia of the planet wheel instead:

$$I'_2 = I_2 + m_2 \, r_2^2 \, i_2^2 \tag{8}$$

The number of degrees of freedom is not reduced in the model used because this would make comparison to the full model more difficult.

The natural frequencies for the simple model of the planetary gear are shown in table 4.

Mode number	1	2	3	4
Frequency, Hz	0	1289	1978	$17 \cdot 10^6$
Mode shapes				
Planet carrier	1.00	1.00	0.22	0.00
Planet bearing	1.00	-0.31	-0.46	-0.72
Planet wheel	1.00	-0.31	-0.46	1.00
Sun wheel	1.00	-0.82	1.00	0.00

Table 4. Simple model A, natural frequencies and mode shapes, mode shapes have been multiplied by the gear ratio

See the modal assurance criterion for comparison with the full model in table 5.

		Full model			
	No	1	2	3	4
Simple model A	1	1.00	0.03	0.00	0.02
	2	0.03	0.73	0.14	0.14
	3	0.02	0.00	0.00	0.96
	4	0.01	0.29	0.78	0.02

Table 5. MAC, Full model and simple model A, mode shapes have been multiplied by the gear ratio

The rigid body mode is of course identical in the two models. The results for mode number two are in good agreement, but the results for mode number three are not. The natural frequency is close, but the mode shape is very bad. Mode number four in the simplified model is not expected to correspond to any mode. This is the mode that could have been removed by reduction of the equations of motion.

The simple model does not take the flexibility of the contact between the planet gear and the ring gear into account. To compensate for this the stiffness of the rest of the system was changed. It can be seen in table 6 that getting the right natural frequencies is possible.

Mode number	1	2	3	4
Frequency, Hz	0	1117	1923	$17 \cdot 10^6$
Mode shapes				
Planet carrier	1.00	1.00	0.14	0.00
Planet bearing	1.00	-0.36	-0.41	-0.72
Planet wheel	1.00	-0.36	-0.41	1.00
Sun wheel	1.00	-0.68	1.00	0.00

Table 6. Simple model A, modified stiffness, natural frequencies and mode shapes, mode shapes have been multiplied by the gear ratio

Unfortunately this does not improve the third mode shape as seen in table 7.

		Full model			
	No	1	2	3	4
Simple model A Modified stiffness	1	1.00	0.03	0.00	0.02
	2	0.02	0.76	0.15	0.07
	3	0.02	0.00	0.00	0.98
	4	0.01	0.29	0.78	0.02

Table 7. MAC, Full model and simple model A, modified stiffness, mode shapes have been multiplied by the gear ratio

An alternatively simple model, B, was investigated. The model is shown in figure 5.

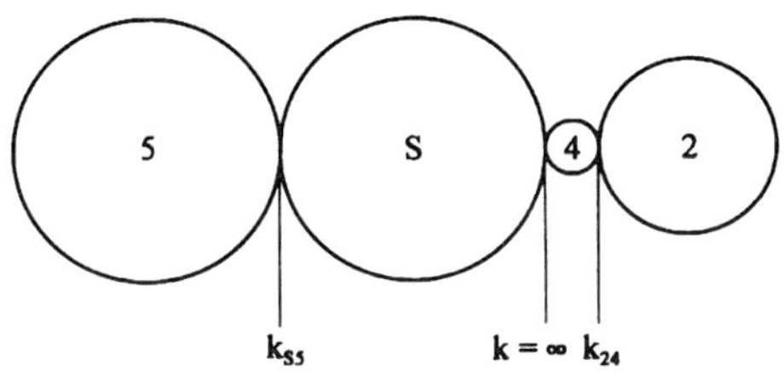

Figure 5. Simple model B. Actual gear ratios

The modal properties are shown in table 8 and compared with the full model in table 9. Model B is not better than model A.

Mode number	1	2	3	4
Frequency, Hz	0	1037	1691	$19 \cdot 10^6$
Mode shapes				
Planet carrier	1.00	1.00	-0.79	0.00
Planet bearing	1.00	0.15	1.00	-0.57
Planet wheel	1.00	-0.72	-0.45	0.00
Sun wheel	1.00	0.15	1.00	1.00

Table 8. Simple model B, natural frequencies and mode shapes, mode shapes have been multiplied by the gear ratio

	No	Full model 1	2	3	4
Simple model B	1	1.00	0.03	0.00	0.02
	2	0.05	0.75	0.02	0.09
	3	0.05	0.12	0.68	0.21
	4	0.03	0.10	0.07	0.82

Table 9. MAC, full model and simple model B, mode shapes have been multiplied by the gear ratio

The natural frequencies were overestimated by model A and are underestimated by model B. The mode shapes are not in agreement with the correct model.

6. CONCLUSION

Computations using a full model of the gearbox showed that elimination of the rotational degree of freedom of the gearbox casing preserved all modes except one. Properties for modes with natural frequencies close to the natural frequency of the gearbox casing rotation suspension mode must be discarded. A reduced model of a planetary gear would therefore be useful in some cases.

The analysed model cannot be used directly in the commercially available torsional vibration computer programs because of the restricted input format used in those programs. Two different simple reduced models satisfying the restricted input formats were examined using four degrees of freedom. The simple models were compared with the correct four degrees of freedom model. The simple models are, of course, missing a mode. The first mode found using the simple model is acceptable, but the second mode is very different from the mode found using the full model.

An acceptable simple model satisfying the restricted input format of the torsional vibration computer program was not found.

It was therefore decided not to use a standard torsional vibration program. A new program was written, where the full model of the planetary gear was implemented.

7. REFERENCES

(1) Nestorides, E. J., BICERA
A Handbook on Torsional Vibration
Cambridge University Press, 1959

(2) Köhler, G. and Rögnitz, H.
Maschinenteile 2
B. G Teubner, Stuttgart 1976

C553/021/98

Turbomachinery forced response predictions using integrated time-accurate models

M IMREGUN BSc, MSc, PhD, MIMechE, MSEE, MSEM
Department of Mechanical Engineering, Imperial College, London, UK

When the rotating blades go past flow defects created by the interaction of upstream and downstream blade rows, the ensuing unsteady aerodynamic forces can become large and may cause excessive vibration levels, hence the term forced response. Forced response can also be due to the general unsteadiness throughout the engine and the resulting vibration is usually associated with low-order nodal diameter modes. The modelling approach described in this paper is based on an integrated time-accurate analysis that includes both the stator and the rotor blades. The aerodynamic representation is viscous with a Baldwin Barth turbulence model. The flow solver is implicit and uses unstructured meshes of tetrahedra. A non-linear friction damping model, the dynamic behaviour of which is derived from measured vibration test data, is also incorporated into the numerical model. A series of computations were performed for a high-pressure turbine stage for which experimental results were available. Although good overall agreement was observed between the predicted and computed results, the calculations were found to be sensitive to a number of key parameters such as inherent structural damping and resonance shift due to friction dampers. In a second analysis, a whole-annulus model was used to predict the aerodynamic excitation arising from a variation of stator blade spacings. It was found that such variations could lead to significant aerodynamic forcing.

1. INTRODUCTION

Aeroelasticity is the study of the mutual interaction that takes place within the triangle of the inertial, elastic, and aerodynamic forces acting on structural members exposed to an airstream, and the influence of this study on design. In a turbomachinery environment, these forces act on various blade rows and can cause excessive blade vibration, a feature that continues to limit the design and performance of gas turbines. In certain high temperature environments, thermal stresses also become important, and the triangle of forces becomes a tetrahedron, giving rise to the term *aerothermoelasticity*, though it is unusual to have

© IMechE 1998 C553/021

aeroelasticity problems involving dynamic thermal stresses. Turbomachinery aeroelasticity phenomena are characterised by the interaction of two physical domains, the describing equations of which are non-linear. These are the fluid and structural domains where the dynamic equations of motion cannot be solved analytically for generalised geometries. A number of extra complications stem from the existence of large numbers of complex-shaped blades, arranged in alternate stator and rotor rows with relative motion between them. The flow therefore exhibits a number of non-linear aerodynamic effects (e.g. shock/ boundary-layer interaction, flow separation, turbulence) that are very difficult to model representatively. The structure may also bring additional non-linear features such as friction damping in blade roots, shroud interfaces and large deformations under centrifugal effects.

1.1 Modelling of structural dynamics aspects

The use of finite element (FE) techniques for the dynamic analysis of bladed disk assemblies has become standard practice. The modes of vibration are obtained via an eigensolution of constant mass and stiffness matrices. However, such an approach cannot include any non-linear behaviour. In turbomachinery applications, friction dampers are used to control the rotor blade vibration levels and their inclusion into the analysis is highly desirable. A good example of combining modal models obtained from linear FE representations with experimentally-derived non-linear elements is given in (1) by Sanliturk et al. who studied the optimum selection of friction dampers that are placed at turbine blade roots in order to prevent the vibration levels from reaching high amplitudes. These devices dissipate energy via friction and their dynamic behaviour is significantly non-linear. Turbomachinery designers must be able to predict both the natural frequency shift and the response reduction due to the non-linear contact mechanism. Using a harmonic balance approach, Sanliturk et al. were able to demonstrate good qualitative and acceptable quantitative agreement between their theoretical and experimental work **(Fig. 1)**.

1.2 CFD methods for unsteady flows

Reliable aeroelasticity predictions require an accurate modelling of the unsteady flowfield. With dramatic increases in computing power and advances in computational fluid dynamics (CFD) methods, it is now possible to obtain solutions for both inviscid (hyperbolic Euler equations) and viscous (parabolic Navier-Stokes equations) steady flows. The usual approach is to use time integration, starting from known initial conditions but unsteady flows are much more complex in nature because of their temporal variations. However, representative solutions can still be obtained if the time-dependent terms are discretised in a time-accurate fashion. Over the last two decades, time domain solutions have progressed through full-potential, linearised-Euler, Euler and, most recently, Reynolds-averaged Navier-Stokes equations. The numerical modelling of aerodynamics is reviewed in (2) while the unsteady flow modelling for turbomachinery applications is discussed in (3) and (4).

The fluid domain is usually discretized by selecting one of two distinct methods. The first one, known as finite-difference, is based on the differential form of the governing equations. However, it is not of primary interest in turbomachinery aeroelasticity, its main application area being external aerodynamics (5 & 6). The second method, known as finite-volume, uses the integral form of the governing equations of which are discretised as fluxes through control volumes of known shape but arbitrary size. No grid transformations are required and the

C553/021 © IMechE 1998

calculation takes place in the physical domain. The technique was first reported in (7 & 8) and adapted to turbomachinery applications in (9 & 10).

One of the major problems associated with 3D unsteady flow modelling is the amount of CPU time required to obtain fully-converged solutions. Although both finite-volume and finite-difference methods can be accelerated by implicit schemes and numerical techniques such as multigrid and local time-stepping, such techniques do not usually preserve time-accuracy throughout the fluid domain. So these acceleration methods are usually restricted to steady flows, a notable exception being the time-accurate multiple grid scheme reported in (11). The modelling of turbulence is probably one of the most significant challenges to the accuracy of modern CFD methods, especially in the case of turbomachinery applications. Turbulence can only be represented using simplified models which are usually based on equivalent local turbulent viscosities given by the averaged effect of the eddies at some point in the flow. There is a hierarchy of turbulence models which range from fully-empirical mixing length representations to Reynolds stress models with 9 equations in 3D.

1.3 Aeroelasticity Methods

Classical aeroelasticity methods treat the fluid and structural domains in an uncoupled fashion such that the fluid flow does not affect the structural response and vice-versa. Such methods thus split an inherently coupled non-linear phenomenon into two separate linear and uncoupled analyses. These methods have been popular for three main reasons. First, assumptions had to be made in order to solve the otherwise intractable governing differential equations with limited computing power. Second, some aeroelasticity problems, especially those involving low speed external flows, may not require coupled solutions, as the motion of structure is well defined by a particular mode of vibration that remains almost unchanged by the fluid. Third, there has been relatively little cross-fertilisation across structural dynamics and aerodynamics and the tendency has been to use existing aerodynamic or structural codes with minimum changes in order to accommodate one another. However, the methods were found to be of limited use when applied to turbomachinery cases where a number of non-linear effects and strong structure/fluid coupling may exist. Although most of the earlier uncoupled methods rely on simplified aerodynamic theories for incompressible inviscid flows, the more recent ones use linearised flow equations, a simplification that is not guaranteed to capture some important features such as shocks and flow separation. This is especially a problem in turbomachinery bladerows where an accurate representation of the shock strength and position is crucial for aeroelasticity predictions. In any case, the requirement is to use integrated methods which do not uncouple the fluid motion from that of the structure, but attempt to treat the problem of aeroelasticity in one continuous medium. The need for such an approach arises from the non-linear response of fluid flow to the motion of solid boundaries, especially at transonic flow speeds for which flutter often occurs, and the existence of non-linear structural effects such as friction damping in the blade roots and shroud interfaces. Thus the mathematical formulation needs to be devised in such a way that the fluid is allowed to modify the structural motion, and vice-versa, as the phenomenon occurs in nature. It then becomes possible to include non-linear effects for both the fluid and the structure and take into account various interactions that can take place between them. One of the most common integrated analysis techniques is the partially-integrated method which involves the solution of the fluid and structural equations separately but information is exchanged at each time step so that the solution from one domain is used as boundary condition for the other domain. In a turbomachinery context, a new blade position is calculated at each time step using the fluid forces of the previous time step and this new position is used as the new fluid/structure boundary for the next time step when the aerodynamic forces are computed first. The algorithm is shown schematically in **Fig. 2**.

2. FORCED RESPONSE MECHANISMS

When the rotating blades pass through flow defects created by the interaction of upstream and downstream bladerows, the ensuing large unsteady aerodynamic forces can cause excessive vibration levels, hence the term *forced response.* One of the first design steps for minimising the response levels is the Campbell diagram of **Fig. 3** which indicates the possibility of an assembly mode being excited at a particular rotational speed or its multiples, the so-called engine-order excitation. Resonance occurs when the engine-order and the corresponding assembly mode frequency lines cross, as indicated by circles in **Fig. 3**. In aerospace engineering, it is common practice to display engine test results in the form of a modified Campbell diagram, which then indicates both flutter and forced response behaviour. In such cases, the out-of-plane axis represents the vibration amplitude and flutter is observed at nearly constant frequency, between engine-order lines, as indicated by asterisks in **Fig. 3**. For free-standing blades, flutter is usually confined to the first family, though it is also known to occur in the second family for other designs.

What is of real interest to the designer is the prediction of the vibration amplitude under unsteady aerodynamic loading which can be due to wake passing from upstream blade-rows (wake-rotor interaction) and/or the potential field of upstream and downstream blade-rows (potential-rotor interaction). Intake distortions and fluctuating back pressures can also be significant excitation sources. Because of the numerous unknown factors such as structural damping, non-linear damping in the blade roots, and the forcing itself, forced response analyses often aim at ranking potential designs rather than predicting absolute vibration levels. Such analyses are also used to explain failure mechanisms and to evaluate the effects of possible remedial modifications. Reference (12), which presents an industrial design prediction system for forced response, also reviews much of the work in this area.

A forced response analysis system which can predict the actual blade vibration levels consists of four components: (a) unsteady flow modelling for determining the engine-order excitation, (b) structural modelling, including non-linear friction damping behaviour, (c) solution procedure to obtain the aeroelastic response of flexible blades, and (d) whole/multi stage modelling. Separate bladerow analyses are valid only for larger than practical axial gaps for which the interaction effects become minimal. The desire to reduce the engine length requires compact stages with closely-spaced stator and rotor bladerows, a feature which necessitates whole-stage, if not multi-stage, modelling. To reduce the computational effort, it may be possible to perform several "linked" single-stage calculations in order to obtain the forcing on the rotor bladerow of the last stage. In any case, the flow solver should possess the following features.

(i) The ability to capture unsteady effects such as end-wall, tip leakage, etc., as these can contribute significantly towards the three dimensionality of the flow.

(ii) The ability to deal with unsteady flows due to the interaction of pressure waves and wakes between bladerows.

(iii) The ability to model viscous effects, as they can contribute significantly towards the unsteady component of the flow.

C553/021 © IMechE 1998

(iv) The ability to allow relative motion between the rotor and the stator blades by implementing a moving grid strategy. The transfer of information between the stationary and moving grid cells must be accurate.

(v) The ability to move the fluid mesh according to the instantaneous position of the solid boundaries so that the blade vibration can be included in the calculations.

4. CASE STUDY

Recent studies of a benchmark test case report actual amplitude predictions for a HP turbine stage with 36 stator and 92 rotor blades (13 & 14). The FE model of the rotor blade is shown in **Fig. 4** while the two mode shapes of interest are shown in **Fig. 5**. The first study uses an integrated, non-linear, multi-passage, multi-row time domain analysis that also includes the flexibility of the rotor blades while the second study uses linearised aerodynamics but the same structural model. **Table 1** lists the results obtained from the integrated method, the corresponding instantaneous unsteady pressures being shown in **Fig. 6** for a typical sector. The analysis was performed for both inviscid and viscous representations of the stator domain and the resulting unsteady force is shown in **Fig. 7** for the two situations. The actual rotor blade vibration levels were also predicted and good overall agreement was reached for most of the cases studied.

One particular regime of forced response, which can occur at much lower frequencies than the blade passing frequency, is the so-called *low engine-order (LEO) forced response*. The unsteady aerodynamic forcing that causes LEO forced response is called LEO excitation, the creation mechanism of which is much more complex than that for blade passing forced response. The source of the LEO excitation is largely unconfirmed but it is believed that some loss of symmetry, such as inherent non-uniform spacing of the stator blades, is the most probable cause. However, many other factors, such as flow exit angle variations, rotor-stator axial gap differences, combustion effects such as burner blockages and non-uniform temperature distributions, combinations of blade numbers, effects of downstream and upstream stages, can also influence the LEO excitation mechanism.

The recent study was focused on the spacing variation of the stator blades and three cases of 1%, 3% and 5% variation was considered (13). The investigation was conducted using the whole-annulus two-row model of **Fig. 8**. The pressure field at the stator outflow is shown in **Fig. 9** for the 3% variation case. A Fourier transform of the pressure field indicates the harmonics that are present in the unsteady excitation. The corresponding modal displacement amplitudes, shown in **Fig. 10**, indicate that the unsteady excitation increases with increasing blade spacing.

5. CONCLUDING REMARKS

(i) Recent advances in aeroelasticity prediction tools allow the modelling of the forced response phenomenon using integrated CFD & structural dynamics methods. A main feature of such analyses is the inclusion of the blade vibration into the aerodynamic model. Similarly, the non-linear dynamic behaviour of the friction dampers is represented by using special non-linear elements together with a modal model obtained from a linear FE formulation.

(ii) Quantitative predictions of blade vibration amplitudes under forced response are beginning to look possible. However, the determination of inherent structural damping and

the tracking of the resonance shifts due to friction damper effects in turbine blades are likely to remain significant challenges for a while.
(iii) With expected increases in computing power, multi-passage, multi-stage viscous unsteady flow predictions should become increasingly common for aeroelasticity applications. The next challenge is likely to come from the area of turbulence modelling.

6. ACKNOWLEDGEMENTS

The author wishes to thanks Rolls-Royce plc for sponsoring some of the work presented in this paper. He gratefully acknowledges the work of Drs. Vahdati & Sayma at Imperial College. He also wishes to thank Drs. Hall & Marshall & Mr. Green of Rolls-Royce plc for many useful discussions.

7. REFERENCES

1) SANLITURK, K. Y., STANBRIDGE, A. B. & EWINS, D. J. 1996 Friction damper optimisation for turbine blades: experiment and theoretical predictions. IMechE paper C500/019/96
2) JAMESON, A. 1983 The evolution of computational methods in aerodynamics. *Journal of Applied Mechanics* **50**, 1052-1060
3) SHARMA, O. P., PICKETT, G. F. & NI, R. H. 1992 Assessment of unsteady flows in turbines. *Journal of Turbomachinery* **114** 79-90
4) VERDON, J. M. 1993 Review of unsteady aerodynamic methods for turbomachinery aeroelastic and aeroacoustic applications. *AIAA Journal* **31**, 235-250
5) MACCORMACK, R. W. 1976 An efficient numerical method for solving the time-dependent compressible Navier-Stokes equations at high Reynolds numbers. *NASA Technical Memorandum*, TM-X-73129
6) BEAM, R. M. & WARMING, R. F. 1978 Implicit factored schemes for the compressible Navier-Stokes Equations. *AIAA Journal* **14**, 393-402
7) JAMESON, A. 1974 Iterative solution of transonic flows over airfoils and wings including flows at Mach 1. *Comm. Pure & Applied Maths* **27**, 283-304
8) JAMESON, A., SCHMIDT, W. & TURKEL, E. 1981 Numerical solutions of the Euler equations by finite volume methods using Runge-Kutta time-stepping schemes. *AIAA paper* 81-1259
9) DENTON, J. D. 1975 A time marching method for two- and three-dimensional blade to blade flows. *A. R. C. R & M.* No. 3775
10) DENTON, J. D. 1983 An improved time-marching method for turbomachinery flow calculations. *Journal of Engineering for Power* **105**, 514-524
11) HE, L. 1993 New two-grid acceleration method for unsteady Navier-Stokes calculations. *AIAA Journal of Propulsion and Power* **9**, 272-280
12) CHIANG, H. D. & KIELB, R. E. 1992 An analysis system for blade forced response. *ASME paper* 92-GT-172
13) VAHDATI, M., GREEN, J. S., MARSHALL, J. G. & IMREGUN, M. 1998, Forced Response Predictions for a HP turbine rotor blade, PEP Symposium on Design Principles & Methods for Aircraft Gas Turbine Engines, Paper 34, Toulouse, France
14) GREEN, J. S. & MARSHALL, J. G. 1999. Forced Response Prediction within the Design Process. 3rd European Conference on Turbomachinery, IMechE, London

 C553/021 © IMechE 1998

Table 1. Measured and predicted peak-to-peak vibration amplitudes (13)

Case	Predicted amplitude (mm)	Measured amplitude (mm)
1T mode	0.28	0.28
1T mode with friction damper	0.14	0.09
2F mode	0.11	0.12
2F mode with friction damper	0.08	0.04

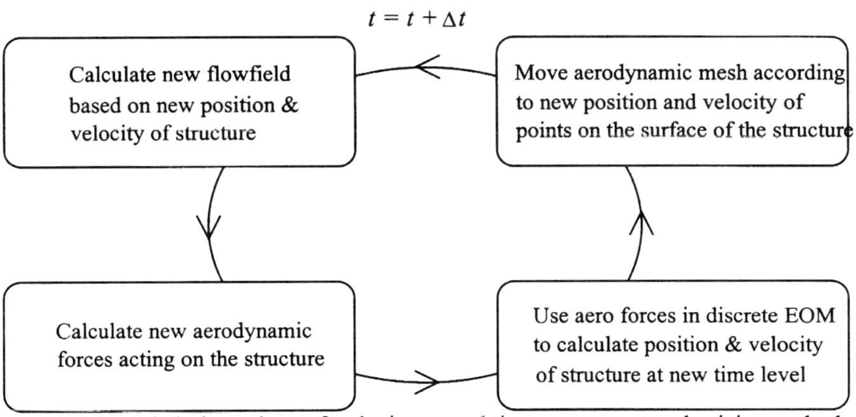

Fig. 1 Solution scheme for the integrated time-accurate aeroelasticity method

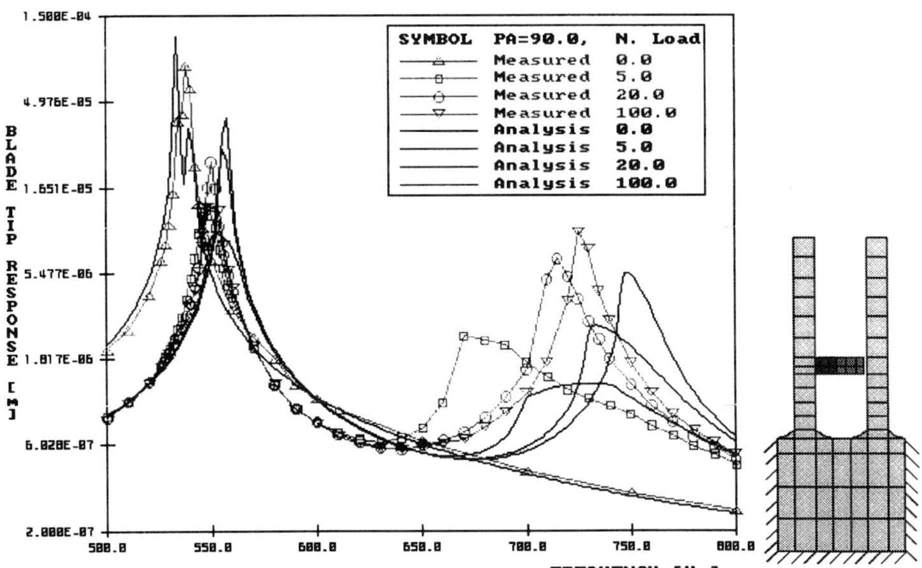

Fig. 2 Measured and predicted response levels for the idealised turbine blade & friction damper system shown on the right. The normal damper load is shown for the measured curves.

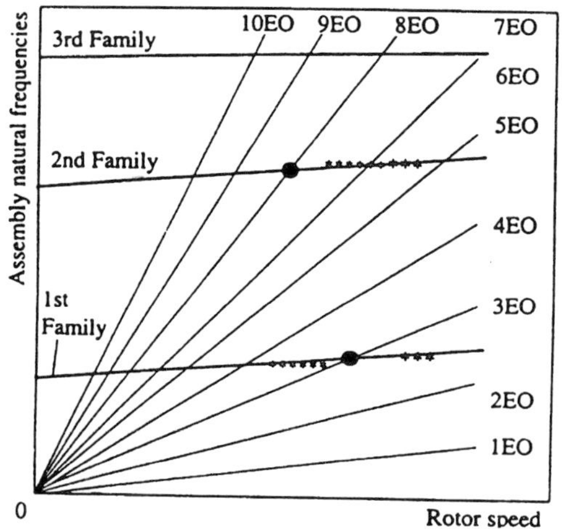

Fig. 3 Simplified Campbell diagram showing occurrence of forced response and flutter

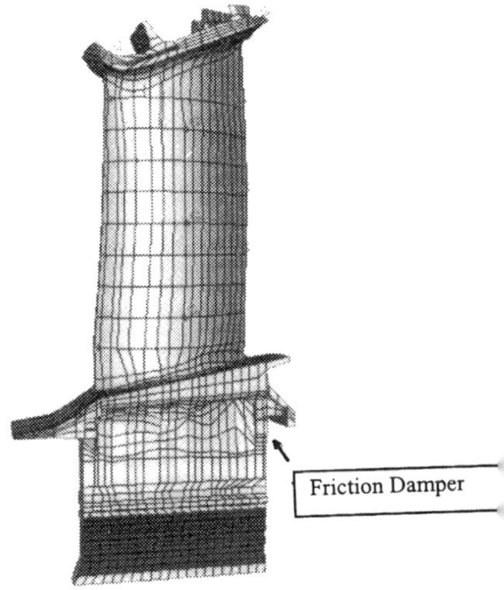

Fig. 4 FE model of the rotor blade

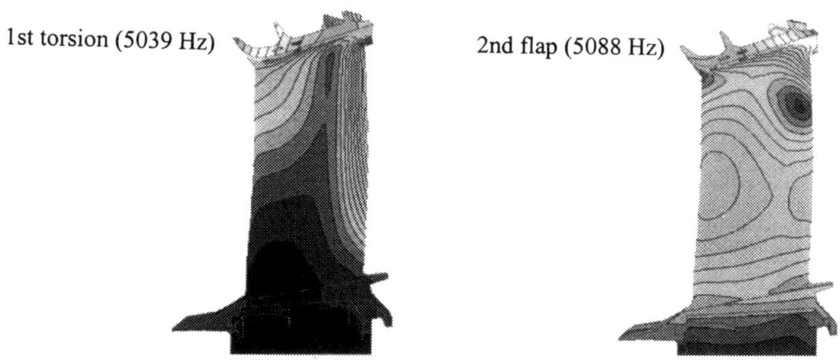

Fig. 5 Blade vibration mode shapes for 1T and 2F

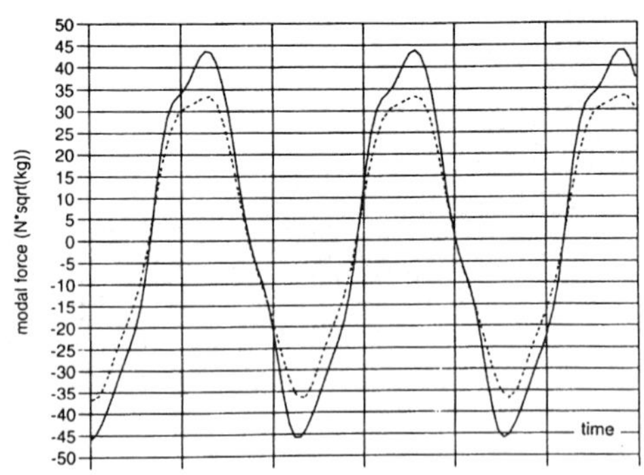

Fig. 7 Unsteady forcing due to inviscid and viscous stator domain modelling

C553/021 © IMechE 1998

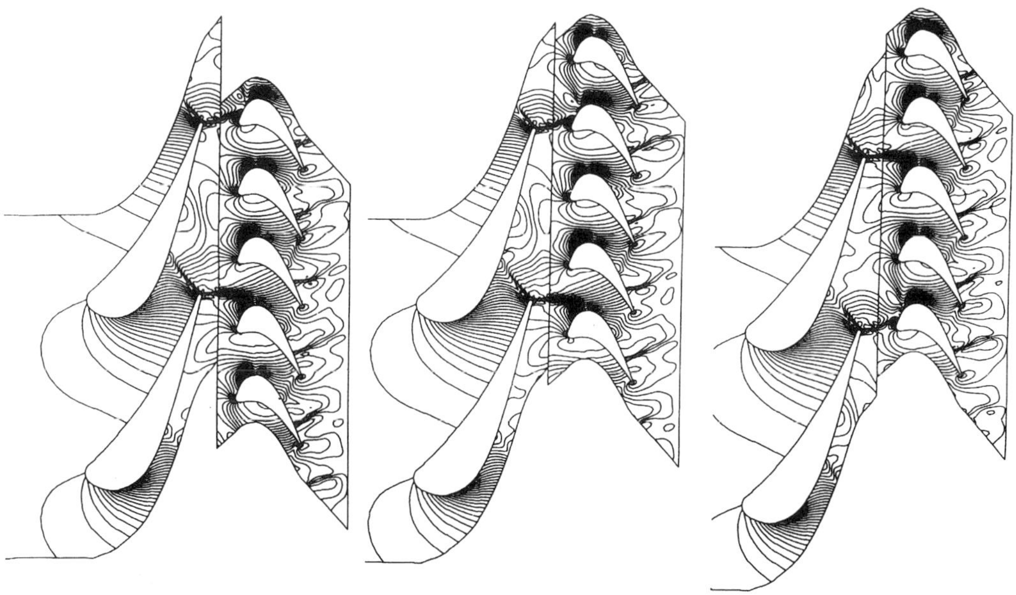

Fig. 6 Instantaneous unsteady pressures for a typical stator/rotor sector

Fig. 8 Whole-annulus stage model for low engine order calculations

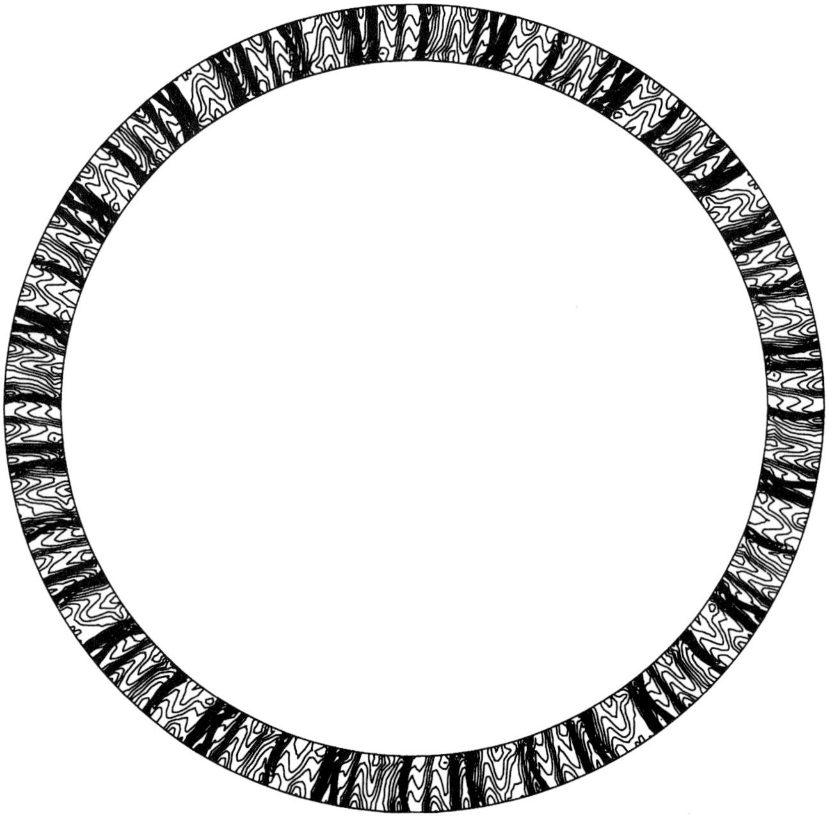

Fig. 9 Pressure at the stator outflow for the geometry shown in **Fig. 8**

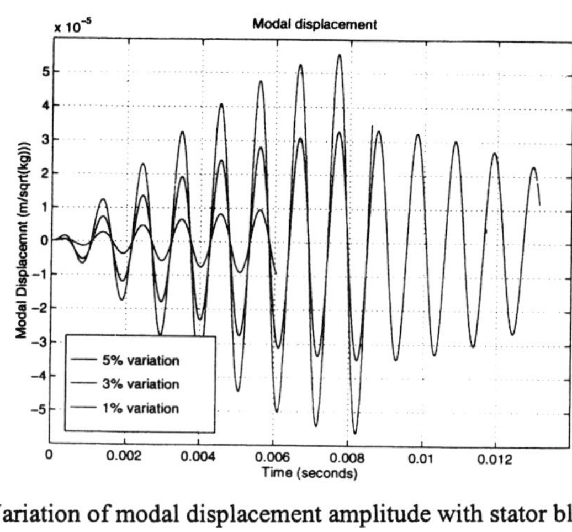

Fig. 10 Variation of modal displacement amplitude with stator blade spacing

C553/021 © IMechE 1998

C553/013/98

Multi-body dynamics investigation of clutch pedal in-cycle vibration (whoop)

P KELLY BSc, MSc, CEng, MIMechE
Powertrain Engineering, Ford Werke, Cologne, Germany
H RAHNEJAT BSc, MSc, PhD, DIC, CEng, MIMechE, FCybS
Department of Mechanical Engineering, University of Bradford, UK
J W BIERMANN
Institute for Automotive Engineering, University of Aachen, Germany

1. ABSTRACT

This paper highlights an experimental investigation and a numerical non-linear multi-body dynamic analysis of in-cycle clutch pedal vibration, referred to as "whoop". The experimental investigation has indicated that whoop is an axial vibration of the clutch mechanism, excited by "nodding" vibration of the flywheel which is caused by the elastodynamic response of the crankshaft. The torsional-deflection modes of the latter is induced by the cylinder combustion forces, particularly as the result of the firing of the 4th and 3rd cylinders at Top Dead Centres (TDC) in 4-cylinder diesel engines. The severity of whoop response is dependent upon the various sources of compliance in the clutch mechanism. Their effect upon whoop have been studied in this paper and good agreement has been obtained between the numerical predictions and the experimental findings.

2. INTRODUCTION

The term "whoop", or in-cycle vibration, is applied to a clutch pedal operating noise and vibration problem, which can be both heard as noise in the driver's footwell area and felt as vibration through the clutch pedal. This phenomenon occurs during the depression and release of the clutch pedal in a motor vehicle. The critical range for the release system of the clutch is at frequency of 250-300 Hz, when the clutch pedal is being operated. Therefore, this is a dynamic or in-cycle concern. Observations have indicated that vehicles with diesel engines and mechanically actuated clutches are more prone to this phenomenon. It is shown that whoop is excited by the elastodynamic response of the crankshaft, caused by the combustion of the engine cylinders (1,2,3). The frequency and the amplitude of the vibration depend on the design parameters of the clutch and of the clutch actuation system. Whoop has been observed experimentally to be an axial vibration problem (3,4,5).

Although whoop has been an industry-wide problem since early 1980s, it has not received the fundamental study that it clearly deserves. The traditional approach in dealing with this problem has been a palliative rather than a preventative one. Although various palliative measures have been advocated, a typical solution has been the use of a "Diehl fix". This is a mass and a rubber block damper attached to the clutch release lever, in effect shifting the natural frequency of its oscillations from the normal 250-270 Hz range to around 450-520 Hz. At these higher frequencies the amplitudes of oscillations are considerably reduced and the problem is shifted to higher engine speeds (3,6). Aside from its palliative nature there are cost and weight penalties associated with the use of the "Diehl fix". The mass of the fix is usually up to 1 Kg and its addition to the lever increases the cost of assembly typically by £3.5. With an annual production rate of several million vehicles in Europe, the cost to the industry is substantial.

Therefore, an in-depth investigation of the phenomenon of whoop is essential and a pre-requisite for preventative measures to be undertaken in this regard. This paper describes some recent experimental findings and numerical predictions.

3. DESCRIPTION OF CLUTCH SYSTEM

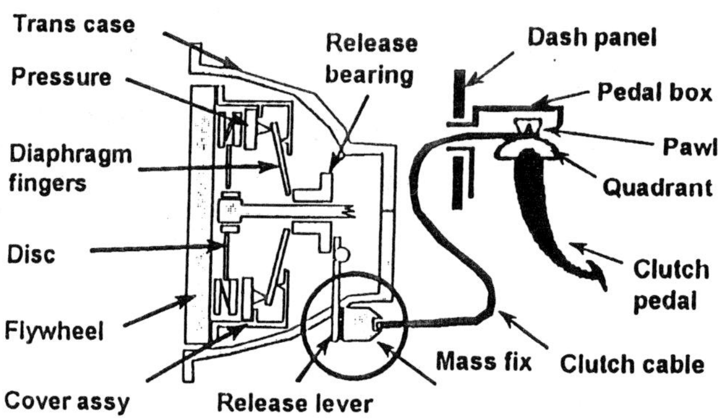

Figure 1 Complete Clutch System under investigation

Figure 1 is a schematic representation of a cable operated clutch system. It extends from the pedal, through an actuation mechanism inside the transmission bell-housing to the release bearing. During the engagement process the release bearing slides along the transmission input shaft and is pressed against the diaphragm fingers of the clutch cover. The flexion of the fingers (acting as levers) cause the pressure plate lift in the axial direction and effect the clamping of pressure plate-dry friction disc-flywheel assembly. The primary function of the clutch is to transmit and disconnect the power between the engine and the drivetrain system and to permit gear selection. Basic design issues on the engagement process have been outlined by Zimmermann et al (7).

C553/013 © IMechE 1998

4. SUMMARY OF EXPERIMENTAL FINDINGS

Figure 2 illustrates the dynamic nature of the clutch whoop. It shows the measured pedal acceleration and the monitored sound pressure in the driver's footwell area against the clutch lever motion. It can be observed that noise and vibration increase during the clutch engagment and disengagement processes. When the pedal is held stationary both noise and vibration response disappear.

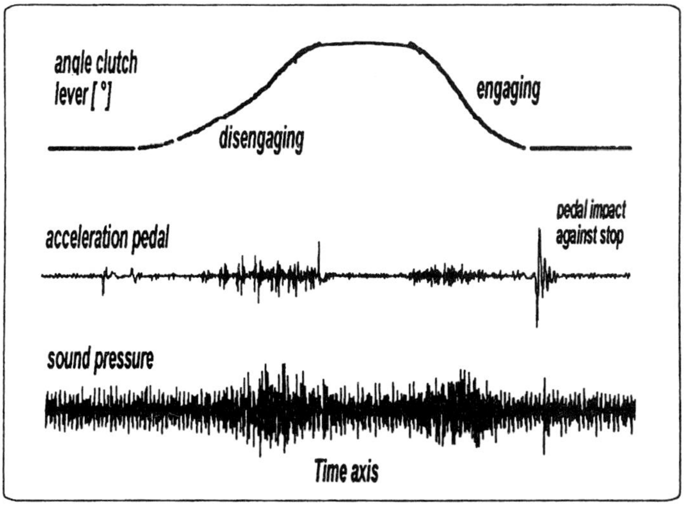

Figure 2: The dynamic nature of the whoop response

The root cause of clutch whoop excitation is found by monitoring the flywheel speed and accelerations along the release system to the pedal. Figure 3 shows the results of the flywheel speed related to the engine firing order. It is clear that acceleration levels increase along the direct path to the pedal whenever cylinders 3 and 4 fire at their TDC positions. Nakada and Tonosaki (8) have stipulated that crankshaft combined torsional-deflection mode and the proximity of the 4th cylinder to the flywheel location account for flywheel "nodding" at the combustion force frequency (i.e. half engine order for four stroke engines). This supposition has been numerically confirmed through multi-body dynamic analysis for a single cylinder engine in references (9,10).

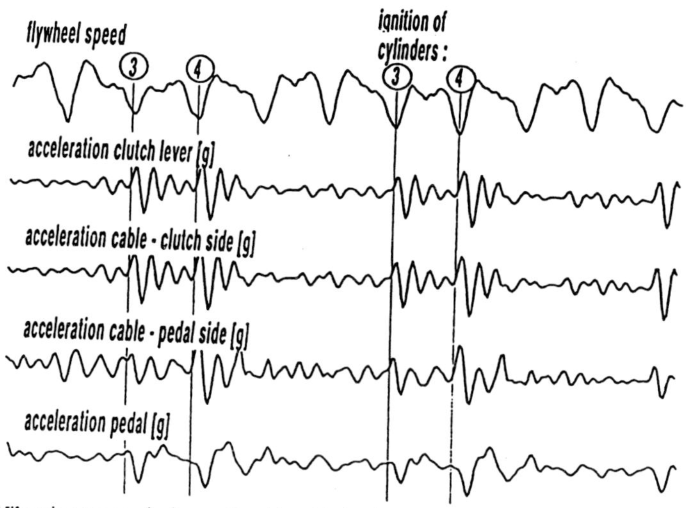

Figure 3: Vibration transmission path with cylinder ignition

To study the effect of various system parameters on the whoop response a Design of Experiments (DoE) is carried out in reference (5). The chosen measure of performance is the linear component of pedal vibration and the associated dB(A) measured noise level in the footwell area. The design of experiments in reference (5) has indicated that aside from masses of pressure plate and the flywheel, the significant system parameters affecting the whoop response are the release lever Diehl fix and the clutch cover stiffness.

During the experiment the pedal acceleration was measured for a complete pedal stroke and its maximum continuous exponential value was derived using a time constant of $\tau = 125$ ms (11):

$$\hat{a}(t) = \sqrt{\frac{1}{\tau}\int_0^t a(\xi)^2 e^{\frac{t-\xi}{\tau}} d\xi} \qquad [1]$$

The average maximum value for two pedal strokes was taken as the pedal vibration response.

These results corroborate the on-vehicle observations. Furthermore, the significance of the cover stiffness has also been reported by Hasebe et al (4) in an independent study.

To carry out further in-depth study of the whoop phenomenon in a cost effective and rapid manner necessitates the formulation of a practical multi-body dynamic model.

5. MULTI-BODY DYNAMIC MODEL

The clutch whoop model is fully described in references (1,2,3,10). Figure 4(a) is a schematic representation of the model. There are 17 rigid parts (excluding the ground frame of reference) in the multi-body model. These include: flywheel, friction disc, hub, pressure plate, clutch cover, release bearing, transmission input shaft, release shaft, release lever, Diehl fix, cable guide, cable pull-lever, pedal quadrant, cable fixture to quadrant, pedal box and the pedal. Each part accounts for 6 unconstrained degrees of freedom.

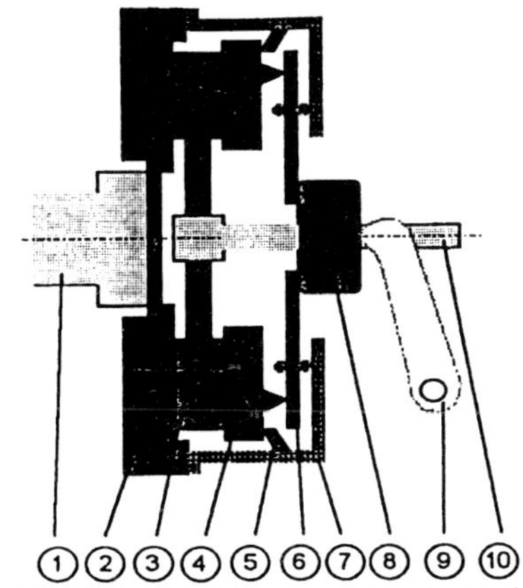

Key: 1- Crankshaft, 2- Flywheel, 3- Friction disc, 4- Pressure plate, 5- Leaf springs, 6- Diaphragm spring, 7- Clutch cover, 8- Release bearing, 9- Release lever, 10- Transmission input shaft

Figure 4(a): The main clutch components

In the system assembly the constraints between adjacent parts restrict the unwanted degrees of freedom. A list of these is provided in references (2,6,10). Each joint type introduces its own motion constraints, each of which is represented by an algebraic constraint function. A total number of 98 constraints exist in the clutch system model. Each degree of freedom is represented by a differential equation of motion (102 in this multi-body model). Therefore, there are 4 unconstrained independent degrees of freedom in the model. These are the translational motion/floating of the flywheel/pressure plate and clutch cover assembly (as a clamped assembly), the axial motion of the release bearing, the articulation of the release lever and the movement of the clutch pedal. Figure 4(b) is a simplified representation of the multi-body dynamic model, showing the 4 independent degrees of freedom described above.

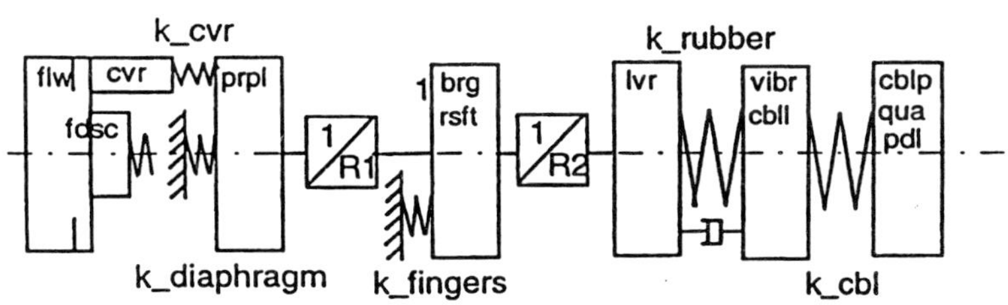

Figure 4(b): A simple mass-spring representation of the multi-body model

The sources of compliance in the model include the axial stiffness of the cover, the diaphragm spring, the "Diehl fix" rubber mass-damper, the clutch cable and the initial preload introduced by the clutch quadrant. Each of these elements is represented by a linear or non-linear algebraic function.

The input to the model is a Dirac force function of the duration equivalent to half engine order combustion process and with a nominal amplitude of 100 N. This is applied to the flywheel as an impact force in the horizontal axial direction.

5.1 MODEL FORMULATION

For each part in the multi-body clutch model parts, a set of equations of motion is established, using the Lagrange's equation for constrained systems:

$$\frac{d}{dt}\{\frac{\partial K}{\partial \dot{q}}\} - \frac{\partial K}{\partial q} - F_q + \sum_{n=1}^{m} \lambda_n \frac{\partial C_n}{\partial q} = 0 \qquad [2]$$

Where: $\{q\} = \{x,y,z,\psi,\theta,\phi\}^T$.

The constraint functions C in equation [2] depend on the types of joints that are employed to represent the physical constraints in the clutch system. For example, for the axial displacement of the pressure plate along the transmission input shaft, a translational constraint is employed. There are, in the case of a translational joint, 5 constraint functions, allowing only one linear degree of freedom along the axis of translation. These constraints are given by the scalar vector products:

$$\vec{z}_i.\vec{x}_j = 0,\ \vec{z}_i.\vec{y}_j = 0,\ \vec{x}_i.\vec{y}_j = 0,\ \vec{s}_{ij}.\vec{y}_j = 0,\ \vec{s}_{ij}.\vec{x}_j = 0 \qquad [3]$$

$\vec{s}_{ij}$ is the position vector of the point i in the frame of reference of point j:

$$\vec{s}_{ij} = (\vec{R}_i + \vec{r}_i) - (\vec{R}_j + \vec{r}_j) \qquad [4]$$

$\vec{s}_{ij} = 0$ is the at-point or point coincident constraint which introduces 3 scalar constraint functions, restricting the translational excursions of two coincident points. A spherical joint (such as cable end to release lever) is represented by this constraint. Other joint types in the model include, for example, cylindrical joint between clutch cable guide and the bell housing. The constraint functions for a cylindrical joint are given by equation [3], excluding the constraint: $\vec{x}_i.\vec{y}_j = 0$

The contact between the release lever and the release bearing is given by a curve-to-curve constraint, such that:

$$\vec{s}_{ij} = 0 \qquad [5]$$

C553/013 © IMechE 1998

$$\{\frac{\partial q_k(q_l)}{\partial q_l}\}_i \cdot \{\frac{\partial q_k(q_l)}{\partial q_l} X q_l\}_j = 0 \qquad [6]$$

where the first term gives the slope of the first curve, whilst the second cross product term provides the normal to the second curve.

Whilst, the selection of appropriate constraints to ensure the ideal functions of the clutch is of paramount importance, care must be taken that the choice of joints does not culminate in the formation of "repeat" constraints or an over-constrained mechanism. Repeat constraints occur when more than a single function is specified to restrict a given movement.

The applied forces in the model, as well as the sources of compliance are modelled as generalised forces $\vec{F}_q$ as shown in equation [2]. These include the impulse translational force used in the whoop model, representing the knocking action on the flywheel at half engine order and related to the firing frequency of the fourth cylinder.

Other applied forces include the clamp force and the restoring forces imposed by the system compliant elements such as cushion spring, cover stiffness and clutch cable stiffness and damping. Further information is provided in references (2,3,6,10).

The set of differential-algebraic equations, described above are solved simultaneously in small variable step sizes. The vector of unknowns includes the system state variables and their first derivatives, and the Lagrange multipliers, when the following substitutions are made:

$$\{\dot{q}\} = \{\zeta\} \qquad [7]$$

Thus, in the matrix form the set of equations is represented as:

$$[J]\{q,\lambda\}^T = \{F_q\} \qquad [8]$$

where $[J]$ is the Jacobian matrix, $\{q,\lambda\}^T$ is the required solution vector in all small time steps, *dt*, and $\{\vec{F}_q\}$ is the vector of applied forces. The Jacobian matrix is of the following form:

$$[J] = \begin{bmatrix} \left[\frac{s}{dt}\frac{\partial K}{\partial \dot{q}} + \frac{\partial K}{\partial q}\right] & \left[\frac{\partial C}{\partial \lambda}\right] \\ \left[\frac{\partial C}{\partial q}\right] & [0] \end{bmatrix} \qquad [9]$$

The solution procedure is fully described in reference (10).

6. SIMULATION RESULTS

The experimental and theoretical investigations point to the dynamic nature of whoop. It is clear that the changes in the effective dynamic stiffness of the system during the clutch engagement and disengagement play the most prominent role in occurrence of whoop. It is, therefore, not surprising that an increase in the mass of the lever assembly, pressure plate or flywheel will affect the system dynamics. The results of the DoE reported in reference (5) have shown marked influences from the elements with significant masses. An "optimal" solution, however, would be one that does not add to the overall mass of the system as this would affect other clutch and powertrain NVH concerns. Alternatively, the dynamic variation in the stiffness of the system during the engagement/disengagement process should be minimised, whilst guarding against any losses that may ensue in the desired clutch function. This points to further investigation of the effect of Diehl fix stiffness and the clutch cover compliance. The effect of the former has been observed experimentally in reference (3) and corroborated in this study (see figure 5).

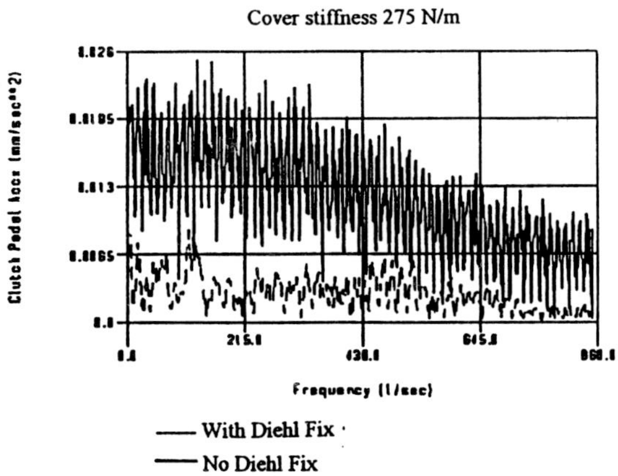

Figure 5: Clutch pedal acceleration vibration spectrum

In figure 5, clutch pedal acceleration spectra are shown for two identical clutch configurations, with and without a Diehl fix. It can be observed that the spectrum of vibration for the case without a Diehl fix exhibits contributions at the combustion frequency (at 16.7 Hz, for engine speed of 2000 rpm) and all its multiples (i.e. half engine order and all its multiples) with considerably larger amplitudes compared to the case with a Diehl fix. In fact the amplitude contributions are reduced by 75% with the inclusion of the Diehl fix. This agrees

C553/013 © IMechE 1998

well with vehicle observations and the marked reduction in the clutch pedal acceleration effects plot in reference (5).

The spectrum of vibration without the Diehl fix has its main contributions at second, fourth and eighth engine orders (i.e. $\approx$65, 130, 260 Hz). The second engine order is the main powertrain contribution in 4 cylinder in-line diesel engines, the contribution at the 4th engine order is pronounced as it coincides with the clutch cable response and the 8th engine order response is coincident with the release lever natural frequency in the range 250-270 Hz. When the Diehl fix is added to the lever its natural response shifts to the region 450-520 Hz. This can be seen as one of the three significant regions in the vibration spectrum with the Diehl fix, although with much reduced amplitudes when compared with the case without the mass-damper. The other two regions correspond to the half engine order and the 4th engine order, the former being the forcing frequency.

In all the models reported in references (2,5,6,10) the clutch cover is considered to be rigid. This makes the analysis somewhat less arduous. The clutch cover stiffness is included in the current model. For the results shown in figure 6 the clutch cover stiffness is set at 275 N/m (10 times the normal practical value), making it almost rigid and in line with the investigations in the aforementioned references. The results obtained here conform to those reported in references (2,5,6,10). A realistic cover stiffness, however, would be in the range 24-30 N/m. Figure 6 shows the pedal acceleration when a Diehl fix is included, and with a clutch cover stiffness of 27.5 N/m. Comparing the spectrum of vibration with that in figure 5 under the same conditions indicate that the vibration levels are considerably reduced, in fact by 40%-60%. This is in line with the findings of Hasebe et al (4) who have pointed out the significance of clutch cover stiffness.

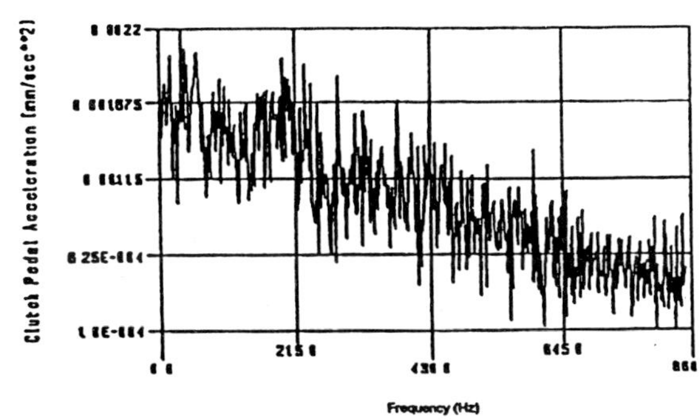

Figure 6: Vibration spectrum with a Diehl-Fix, an elastic clutch cover and cable damping

Another significant finding relates to the key role of cable damping, usually about 5 Ns/m. With effective cable damping the amplitudes of vibration around the resonant frequency peaks are reduced. These are the whoop response frequencies in the range 250-520 Hz. To confirm the experimental observations in reference (5), cable damping is removed in the multi-body model. Comparison of amplitude contributions in figures 6 and 7 shows a corresponding rise in the whoop response region 250-520 Hz. No significant change is observed in lower frequencies, including the peak around the 4th engine order as these are induced by the repetitive impacts due to the half engine order combustion process and all its higher harmonics.

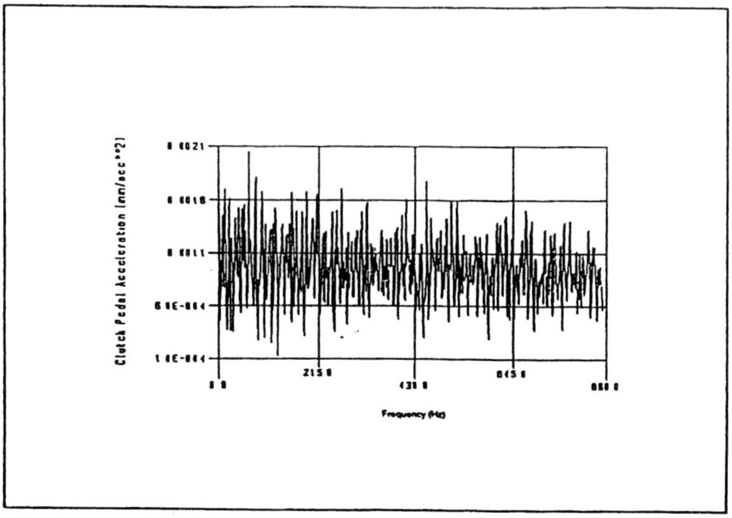

Figure 7: Vibration spectrum with a Diehl-Fix, an elastic clutch cover and no cable damping

7. CONCLUSION

Good agreement has been obtained between the experimental results and the numerical predictions in the study of clutch pedal whoop problem. It has been shown that whoop is a dynamic phenomenon, taking place during the clutch engagement and disengagement processes. The main control factors that affect clutch pedal whoop have been identified and the effect of the most appropriate of these, being Diehl fix and the clutch cover stiffness, have been studied further. It is concluded that a combination of the Diehl fix and appropriate elastic mounting of the clutch cover to the bell housing can be employed in order to substantially reduce the effect of clutch pedal whoop.

C553/013 © IMechE 1998

8. ACKNOWLEDGEMENTS

The authors would like to express their gratitude to Ford Motor Company for the sponsorship of the clutch whoop research project.

9. NOTATION

$\hat{a}$	: Measured pedal acceleration
C	: Constraint function
dt	: Integration time step size
F_q	: Generalised forces
J	: Jacobian matrix
K	: Kinetic energy
q	: Generalised co-ordinates
q_k	: The independent co-ordinate in the curve-curve constraint
q_l	: The dependent co-ordinate in the curve-curve constraint
s	: A scaling factor
t	: Time
x, y, z	: Cartesian co-ordinates
$\vec{x}, \vec{y}, \vec{z}$	: The orientation of local part frame of reference
λ	: Lagrange multiplier
ψ, θ, φ	: Euler angles
τ	: Time constant of accelerometer response

Superscript:

	: Time derivative

Subscripts:

i, j	: Constrained points
k	: Denotes a dependent co-ordinate
l	: Indicates an independent co-ordinate

10. REFERENCES

(1)- P. Kelly and H. Rahnejat, "Clutch pedal dynamic noise and vibration investigation" in H. Rahnejat and R. Whalley (Eds.) Multi-body Dynamics: Monitoring and Simulation Techniques, MEP (IMechE), March 1997.

(2)- H. Rahnejat, D. Centea and P. Kelly, "Non-linear multi-body dynamic analysis for the study of in-cycle vibrations (whoop) of cable operated clutch systems", Proc. 30th ISATA, Florence, Italy, June 1997.

(3)- P. Kelly, Clutch Pedal Noise and Vibration Analysis, MSc by Research Thesis, University of Bradford, 1997.

(4)- T. Hasebe and A. Seiki, "Experimental study of reduction methods for clutch pedal vibration", SAE Pap. No. 932007, 1993.

(5)- P. Kelly, H. Rahnejat, J.W. Biermann and B. Hagerodt, "Combining design of experiments and modelling techniques to resolve complex clutch pedal noise and vibration issues", Proc. European Vehicle Noise and Vibration Conf., IMechE, London, May 1998.

(6)- D.N. Centea, Non-linear multi-body dynamic analysis for the investigation of torsional and axial vibrations in automotive clutches, MPhil Thesis, University of Bradford, 1997.

(7)- F. Zimmermann, H. Oetting, P. Heidemeyer and R. Haack, "An automatic dry friction clutch for passenger cars and light duty trucks", SAE Pap. No. 860383, 1986.

(8)- T. Nakada and H. Tonosaki, "Excitation mechanism of half order engine vibrations", Int. Conf. on Vehicle NVH and Refinement, Paper C487/017/94, MEP(IMechE), 1994.

(9)- A. Boysal and H. Rahnejat, "ADAMS generic engine: a tribo-elastohydrodynamic analysis tool", Proc. ADAMS Int. Users' Conf., Ypsilanti, Michigan, USA, May 1996.

(10)- H. Rahnejat, Multi-body Dynamics: Vehicles, Machines and Mechanisms, PEP (IMechE), June 1998.

(11)- B. Hagerodt, "Investigation of clutch whoop", Report 6472, IKA, Aachen, July 1997.

C553/013 © IMechE 1998

C553/009/98

Performance of an aero-servo-elastic control system for varying dynamic pressure

T RAHULAN BSc, PhD
Department of Aeronautical and Mechanical Engineering, University of Salford, UK

SYNOPSIS

Development of control law design techniques utilising acceleration feedback as part of the overall non-interacting control architecture, attempted to reconcile the differences between the theoretical and the practical approaches, and addressed some of the implementational issues. This technique was successfully applied to suppress flutter by active means in a wind tunnel model at a given flight condition. This paper discusses the effect the dynamic pressure has on the stability aspects as the forward speed is varied from very low levels to values well beyond the flutter boundary for the case where the controller parameters are held constant. It is shown that this controller possesses exceptional robustness characteristics by maintaining stability at all operating points in the increased flight envelope due to active control.

NOMENCLATURE

B	principalised aerodynamic damping matrix
C	principalised aerodynamic stiffness matrix
E	principalised structural stiffness matrix
e_1	first principalised elastic displacement co-ordinate
e_2	second principalised elastic displacement co-ordinate
e_3	third principalised elastic displacement co-ordinate
F	principalised control input matrix
g	overall feedback gain parameter (scalar)
G	principalised control input rate matrix
h	vertical displacement at the centre of gravity, positive downwards (m)
J	principalised structural damping matrix
M	principalised inertia matrix

© IMechE 1998 C553/009

NZG	normal acceleration at the fuselage centre of gravity (g)
NZLM	normal acceleration at wing mid-span, leading edge (g)
NZLT	normal acceleration at wing tip, leading edge (g)
NZTM	normal acceleration at wing mid-span, trailing edge (g)
NZTT	normal acceleration at wing tip, trailing edge (g)
$q\ (= \dot{\theta})$	pitch rate (rad/sec)
q_1	principalised aircraft displacement co-ordinate vector
R	principalised disturbance input vector
s	Laplace operator
U	aircraft control input vector
U_i	symmetric inboard flaperon deflection (rad)
U_o	symmetric outboard flaperon deflection (rad)
U_t	symmetric taileron deflection (rad)
v	forward airspeed (m/s)
$w\ (= \dot{h})$	heave rate (m/s)
w_g	vertical gust velocity (m/s)
θ	pitch angle (rad)

Dot notation denotes the time derivative.

1 INTRODUCTION

Aircraft flutter is an explosive phenomenon in which the aircraft structure absorbs excess energy from the aerodynamic forces through interacting natural modes. If uncontrolled, flutter instability can rapidly destroy the aircraft structure. Passive control of flutter usually results in expensive structural modifications along with weight penalties and reduced speed capabilities. Active methods incorporating feedback principles may be used by exploiting existing control surfaces to improve the dynamic performance with negligible weight penalties. The artificial stability thus provided also has the effect of reducing structural and fatigue loads and results in greater ride comfort due to attenuation of aircraft rigid-body and flexural motions.

There has been a clear division between the theoretical and the implementational aspects of active flutter suppression. Theoretical work, although mathematically rigorous, has tended to be rather too refined for implementation in real systems. Little consideration has been given by the theoreticians towards the functional attributes. Almost invariably, the implications of the dynamic characteristics of the actuators and the sensors are ignored, and, control algorithms which require the use of non-measurable parameter identifiers (such as state observers,...) are proposed. In addition, quite often, a number of additional hydraulic and electronic components are added in the circuitry during the development process to cure 'unforeseen characteristics' problems. Inclusion of these items in the mathematical models often result in additional phase lags which significantly affect the dynamic performance of the control system as a whole.

Most of the successful wind tunnel and flight test demonstrations on the other hand have employed control system design techniques based more on intuition and experience rather than purely theoretical concepts. As a result, simpler basic control strategies have been derived which are based on a wider repertoire of reasoning and logic. In many cases, the

C553/009 © IMechE 1998

implementational process is accelerated by the use of off-the-shelf standard equipment such as dynamic compensators, phase adjusters, and notch filters. It is also quite common to feed back the rate of change of the states in practice whereas, theoretical analyses have concentrated almost exclusively on state feedback. This has arisen chiefly because rate feedback can be readily measured, and, it also automatically introduces phase lead which tends to provide compensation for the delays which are introduced by various items of signal conditioning equipment in the feedback loop. These effects are most apparent in time simulation studies where the rapid action of the controller 'around the feedback loop' can be seen to improve the transient response characteristics.

A variation in the design of control laws was presented earlier (1) in which some of the rate of state variables (the accelerations) were fed back for dynamic control purposes. This approach shows that whilst this form of control architecture possesses phase lead advantages, it also drives at least one of the poles (for any operating condition) towards the origin of the complex plane as the overall system gain is increased in order to 'tighten' performance of the system as a whole.

This paper is concerned with the study of the performance of a feedback control system (1) whose control laws are held constant whilst a major parameter inherent in the system dynamics is varied resulting in the introduction of large scale non-linearities. The control laws which were designed using the techniques developed earlier (1), were installed in the mathematical model of the Group for Aeronautical Research and Technology in EURope (GARTEUR) wind tunnel model (Figure 1) and the overall performance was optimised for the airspeed of 50 m/s (well above the flutter speed of 43 m/s). Only analogue based controllers have been considered in this paper although discretisation is relatively straightforward with this type of controller which facilitates the ready implementation of digital devices. The control architecture was then simplified using simple actuator and sensor reconfiguration (1) strategies. It is shown using root loci and time response plots that stability is maintained as the airspeed is varied from very low levels to values greater than the critical speed.

2 MATHEMATICAL MODELLING

The GARTEUR wind tunnel model (2) is described in terms of the rigid body heave and pitch degrees of freedoms (associated with h and θ respectively), and, three symmetric elastic modes (associated with e_1, e_2 and e_3). The equations of motion were provided in matrix form as

$$M.\ddot{q}_1 + (v.B + J)\,\dot{q}_1 + (v^2.C + E)\,q_1 = v.R.w_g + v.G.\dot{U} + v^2.F.U \qquad \ldots (1)$$

where

$$M = \begin{bmatrix} 6.6600e{+}01 & 0 & 0 & 0 & 0 \\ 0 & 2.3300e{+}01 & 0 & 0 & 0 \\ 0 & 0 & 2.2500e{+}00 & 0 & 0 \\ 0 & 0 & 0 & 5.7200e{+}00 & 0 \\ 0 & 0 & 0 & 0 & 4.4600e{+}00 \end{bmatrix}$$

$$M^{-1}B = \begin{bmatrix} 3.1276e-02 & 9.2106e-03 & 4.0090e-03 & 1.9670e-04 & 2.8529e-03 \\ 2.2117e-02 & 1.3933e-02 & 2.7853e-03 & -2.3145e-03 & 9.0131e-03 \\ 1.1733e-01 & 1.7390e-02 & 6.9778e-02 & 1.9556e-02 & 8.1778e-03 \\ 5.8392e-02 & 7.5053e-04 & 1.9878e-02 & 1.8304e-02 & 4.4755e-03 \\ 2.3318e-02 & 1.4576e-02 & 9.6413e-03 & 9.1031e-03 & 3.2691e-02 \end{bmatrix}$$

$$M^{-1}J = \begin{bmatrix} 4.7000e-01 & 4.2065e-02 & 0 & 0 & 0 \\ 1.2024e-01 & 8.5761e-02 & 0 & 0 & 0 \\ 0 & 0 & 9.9500e-01 & 0 & 0 \\ 0 & 0 & 0 & 1.2478e+00 & 0 \\ 0 & 0 & 0 & 0 & 1.2642e+00 \end{bmatrix}$$

$$M^{-1}C = \begin{bmatrix} 0 & 3.1276e-02 & 1.5090e-02 & -3.6231e-02 & -3.4985e-03 \\ 0 & 2.2117e-02 & 9.4913e-03 & -2.5964e-02 & -1.3942e-02 \\ 0 & 1.1733e-01 & 1.1600e-01 & -2.4222e-01 & -8.0889e-02 \\ 0 & 5.8392e-02 & 4.7378e-02 & -1.0350e-01 & -3.5140e-02 \\ 0 & 2.3363e-02 & -1.0157e-02 & -2.3991e-02 & 2.7130e-02 \end{bmatrix}$$

$$M^{-1}E = \begin{bmatrix} 2.2207e+01 & 1.9875e+00 & 0 & 0 & 0 \\ 5.6810e+00 & 4.0615e+00 & 0 & 0 & 0 \\ 0 & 0 & 8.5730e+02 & 0 & 0 \\ 0 & 0 & 0 & 1.7301e+03 & 0 \\ 0 & 0 & 0 & 0 & 3.9954e+03 \end{bmatrix}$$

$$M^{-1}R = \begin{bmatrix} 3.1276e-02 \\ 2.2117e-02 \\ 1.1733e-01 \\ 5.8392e-02 \\ 2.3318e-02 \end{bmatrix}$$

$$M^{-1}G = \begin{bmatrix} -9.1892e-04 & 1.2312e-04 & 3.8889e-04 \\ -2.2179e-03 & -4.3609e-05 & 2.8060e-04 \\ 4.8000e-03 & -2.2756e-03 & 1.2533e-03 \\ 1.5490e-03 & -3.3217e-05 & 1.3059e-03 \\ -7.2197e-03 & -6.8161e-04 & 9.0807e-04 \end{bmatrix}$$

C553/009 © IMechE 1998

$$M^{-1}F = \begin{bmatrix} -6.3363e-03 & -2.3874e-03 & -2.0420e-03 \\ -1.1230e-02 & -2.7223e-03 & 3.5742e-04 \\ 1.5644e-02 & -3.2844e-02 & -1.3156e-02 \\ 4.4056e-03 & -8.0944e-03 & -2.2203e-03 \\ -1.7489e-02 & -8.4529e-03 & 5.1570e-03 \end{bmatrix}$$

$$q_1 = \begin{bmatrix} h & \theta & e_1 & e_2 & e_3 \end{bmatrix}^T$$

$$U = \begin{bmatrix} U_t & U_o & U_i \end{bmatrix}^T$$

The wind tunnel is equipped with a gust vane located upstream from the working section. It can generate a single-pulse 1-cosine type vertical discrete gust which is modelled as

$$w_g(t) = (1/2).v.(2.\pi/180)\{(1 - \cos(2.\pi.t/0.2)\} \quad 0 <= t <= 0.2$$

$$w_g(t) = 0 \quad t >= 0.2 \qquad \ldots (2)$$

When this gust is scaled to a real aircraft, it is equivalent to a rare and severe case.

The vector describing the accelerometer outputs is defined by the equation

$$\begin{bmatrix} NZLT \\ NZTT \\ NZLM \\ NZTM \\ NZG \end{bmatrix} = \begin{bmatrix} 1 & 0.330 & 0.814 & 0.739 & 0.701 \\ 1 & 0.400 & 0.967 & 0.364 & 0.620 \\ 1 & 0.182 & 0.409 & 0.445 & 0.230 \\ 1 & 0.411 & 0.579 & 0.073 & 0.101 \\ 1 & 0.000 & -0.059 & 0.000 & -0.310 \end{bmatrix} \ddot{q}_1 \qquad \ldots (3)$$

Each of the control surface actuator hydraulics was modelled by the transfer function

$$1 / \{ 1 + 2 (0.743 s / 707) + (s / 707)^2 \}$$

The maximum flaperon deflection was specified as 10 degrees from the neutral position. Power limitations in the hydraulic system prevented any of the deflection rates to exceed 180 deg/sec. The filter dynamics associated with the accelerometers were described by the transfer function

$$\{ s / (s + 2) \} / \{ 50 / (s + 50)\}$$

The accelerometers were not included in the model as they possess a very high bandwidth. The pitch rate gyro and its second-order low pass filter were described by the transfer function

$$[1 / \{ 1 + 2 (1.2 s / 100) + (s / 100)^2 \}] . [25 / (s + 25)]$$

3 OPEN-LOOP CHARACTERISTICS

Figure 2 shows the migration of the poles of the uncontrolled aircraft as the airspeed is varied from zero to 60 m/s; each tick indicates a step increase of 5 m/s. It can be seen that the two rigid body modes labelled phugoid and SPO (Short Period Oscillation) along with the third elastic mode (labelled e_3) remain stable throughout the speed range. It is also clear that flutter in this case occurs as a result of the frequency coalescence of the first two elastic modes; whilst the first elastic mode (labelled e_1) becomes excessively damped, the second elastic mode (labelled e_2) crosses the stability boundary at the critical (flutter) speed of about 43 m/s.

The lowest frequency mode is not a proper phugoid as in a real aircraft since the dynamics is associated more with the way the model is suspended in the middle of the working section of the wind tunnel rather than that which is exhibited during free flight. However the locus trajectory behaviour is relatively unaffected when compared to real aircraft. Even though the model includes all poles associated with equipment dynamics, only the aircraft relevant ones are displayed in all root loci figures in order to enhance clarity around the stability boundary.

4 SYNTHESIS OF CONTROL LAWS

The control laws were designed using procedures previously laid down (1). These laws were compromised in order to avoid adverse build-up of structural loading of the airframe, considerations of ride comfort, and incorporation of performance limiting factors in the feedback control equipment at airspeeds upto 50 m/s. Even though the equipment dynamics did not play an active role during the synthesis of the control laws, the controller was installed in the mathematical model as in the real system. That is, only the filtered outputs were from the sensors were made accessible to the controller whereas, the accelerations displayed in the time response curves were evaluated directly from the state equation.

The controller requirements specified minimum use of actuators and sensors in the control architecture to minimise costs and maximise reliability. During the synthesis procedure, it was discovered that the inboard flaperons were surplus to flutter control requirements, and hence were eliminated from the feedback structure. Even though the design is based on a centralised controller, the method quite often yields simplified solutions; this is especially true in aircraft design where the system as a whole is generally well control configured from the outset. The initial laws (one for regulation and one for command tracking) were obtained as

$$U_t = +0.1\,q \quad \ldots (4)$$

$$U_o = +g.\ddot{e}_1 \quad \ldots (5)$$

Using the previously developed sensor reconfiguration technique (1) based on iterative procedures, one pair of accelerometers (out of a total of nine) were identified as being the most crucial for this study; hence the regulator feedback law (equation (5)) was simplified to

$$U_o = +g.NZTM \quad \ldots (6)$$

C553/009 © IMechE 1998

5 CLOSED-LOOP CHARACTERISTICS

Figure 3 shows the airspeed loci of the closed-loop poles for the case where both control laws are held constant. The initial set of studies were centred around transient responses at an airspeed of 50 m/s in order to determine a suitable value for g. It transpired that at g = 0.05 the overall performance could be optimised without too many compromises. Hence the feedback setting was frozen in this configuration for all closed-loop cases in this study.

As before, the airspeed was varied from zero to 60 m/s in steps of 5 m/s and it is clear that stability is maintained for all flow conditions - from still air to values beyond the design point (which itself was about 7 m/s beyond the flutter speed). The increase in the damping levels for the SPO and the phugoid modes is due to the action of the command tracking controller.

The behaviour of the third elastic mode has hardly changed, but the tendency of the second elastic mode to reduce damping and become unstable (at about 30 m/s) has been arrested at the expense of limiting the level of damping in the first elastic mode. In fact, it is the migration of the first elastic mode that displays the largest difference in terms of change of direction when compared to the open-loop loci.

6 TRANSIENT RESPONSE

The closed-loop transient responses when the model is subjected to the discrete vertical gust (equation (2)) at an airspeed of 60 m/s, is shown in figure 4. The first part (subscripted a) shows h, θ, NZG and q plots whilst the second part (subscripted b) shows the U_0, U_t, rate of U_0 and NZTM plots. Although the responses are limited to the initial second from the commencement of the gust, it was ensured that all responses decayed with time, and that all variables reached a maximum during this period. It can be seen that the peak values are about

h	3.5 cm	θ	0.8 deg
q	6.3 deg/sec	NZG	2.5 g
U_0	6.8 deg	$\dot{U}_0$	172 deg/sec
U_t	0.52 deg	NZTM	2.9 g

The specified limits for the maximum control surface displacement and the rate were 10 degrees (from the neutral position) and 180 degrees per second respectively. The displacement limit is necessary to ensure spare capacity to generate (pilot-demanded) manoeuvre control forces. The rate limit is imposed by the maximum performance of the servohydraulics. It can be seen that the former requirement can be met for both pairs of control surfaces, but that the demanded rate for the outer flaperons peaks to about 96% of the maximum allowed.

The altitude perturbation of the model (with a wing-span of about 2m) inside the working section of the wind tunnel is very small even though the centre of gravity experiences a level of acceleration nearly as high as that near the wing tip; the gust duration was too small to appreciably excite the SPO mode. The transient responses at the lower airspeeds (with exactly the same control laws) have not been shown as they were less critical in terms of model parameter perturbations and feedback controller energy demands.

7 CONCLUSIONS

An analogue controller was designed to suppress flutter in the GARTEUR wind tunnel model whose flutter speed is 43 m/s. The controller was simplified by eliminating one pair (out of three pairs) of control surfaces, and, seven (out of nine) sensors from the control architecture. The control laws were then 'tuned' at an airspeed of 50 m/s for optimum overall performance. The impressive robustness characteristics of the controller was demonstrated by varying a fundamental system parameter (the airspeed) by large amounts without any changes in the controller settings. There were no instances of instability at any of the airspeeds (from zero to 60 m/s), and, the demanded performance from the controller never exceeded the specified limits even when the airspeed was increased to the highest achievable value in the wind tunnel.

REFERENCES

1. Bradshaw, A. et al, "Active Flutter Suppression", Proc.15th Congress of the International Council of the Aeronautical Sciences, vol. 1, pp. 262-273, Sept. 1986

2. Kaynes, I.W., Skingle C.W. and Copley, J.C., "The Design and Wind Tunnel Demonstration of an Active Control System for Gust Load Alleviation and Flutter Suppression", 2nd International Symposium on Aeroelasticity and Structural Dynamics, Aachen, April 1985.

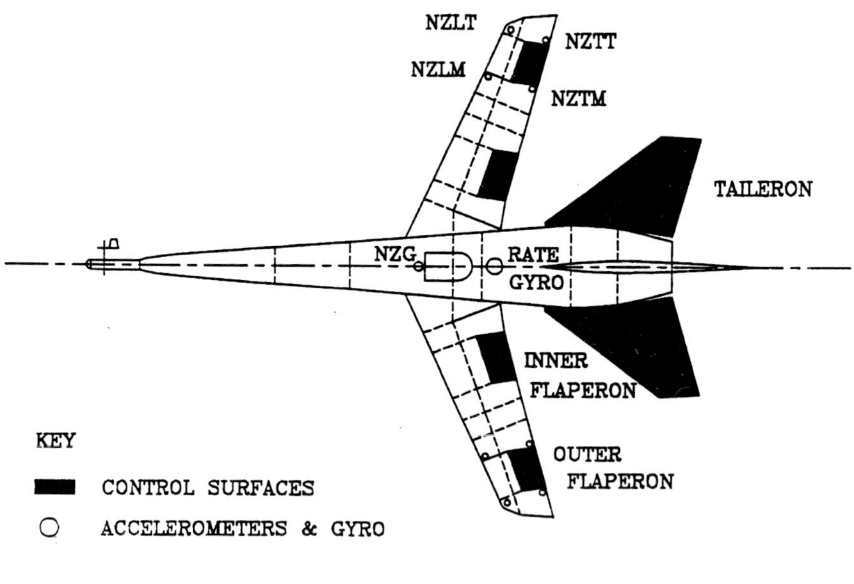

Fig. 1 The GARTEUR wind tunnel model

C553/009 © IMechE 1998

Fig. 2 Open Loop Speed Loci (0-60 m/s)

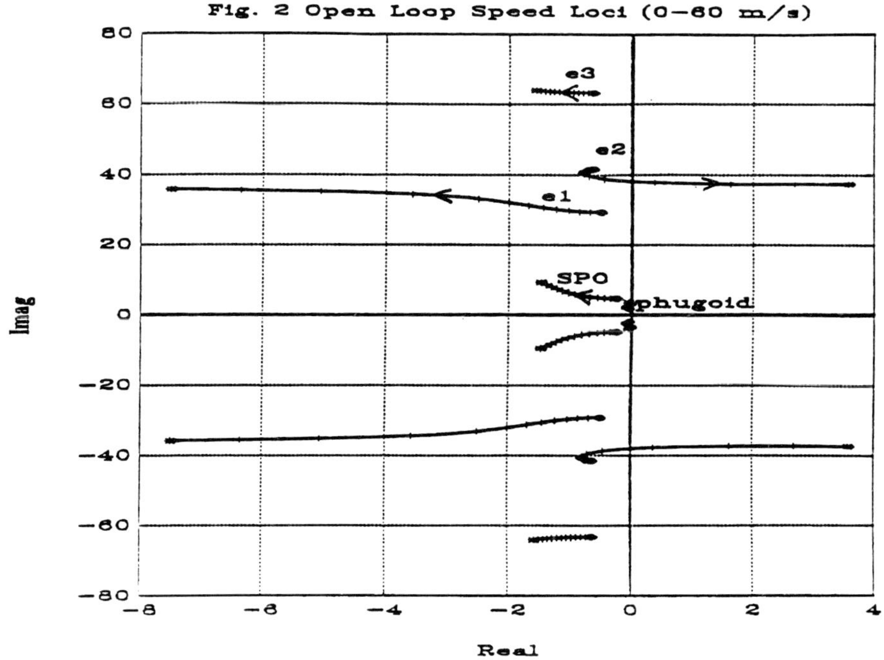

Fig. 3 Closed Loop Speed Loci

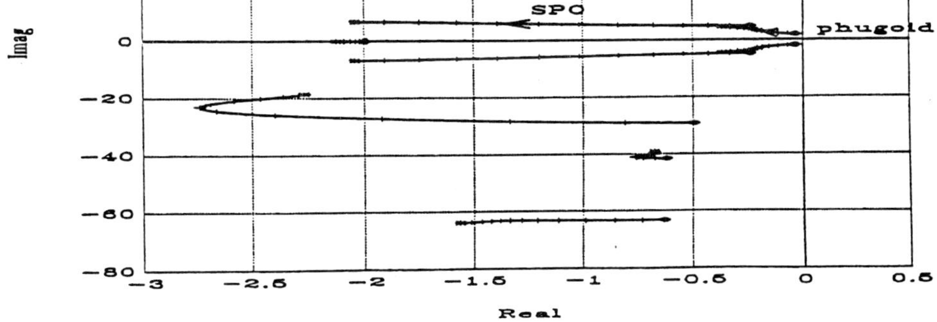

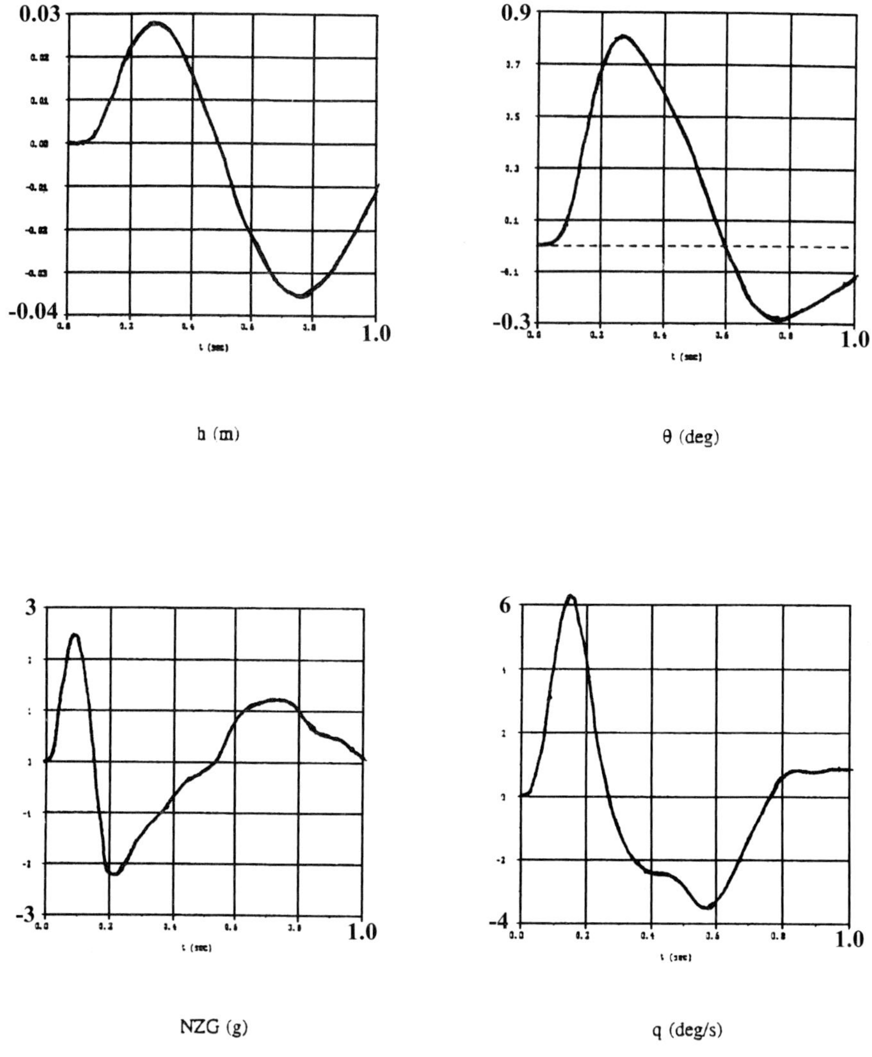

Fig. 4a Closed-loop transients at 60 /s

C553/009 © IMechE 1998

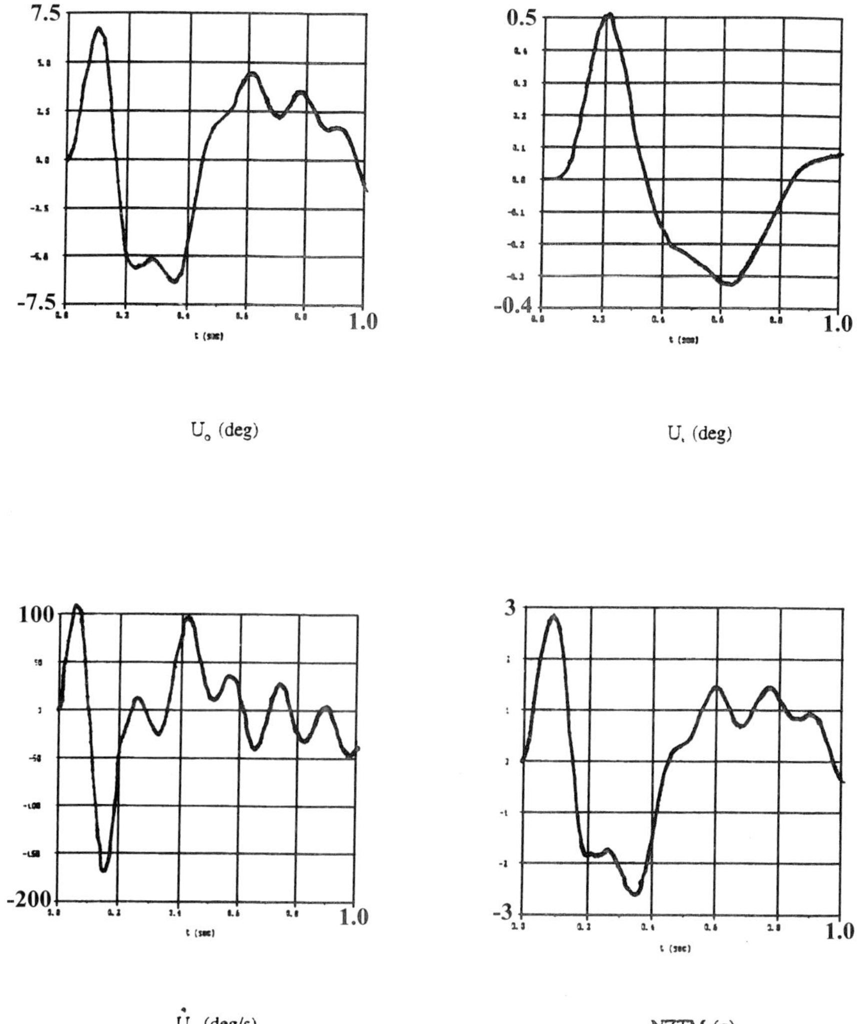

Fig. 4b Closed-loop transients at 60 /s

C553/003/98

A linear model for impact between a rigid body and an elastic ring

C H J FOX FIMechE, AMRAeSoc, **E WILLIAMS** BSc, PhD, and **Y-L CHEN**
Department of Mechanical Engineering, University of Nottingham, UK

Synopsis *Dynamic interaction between two bodies, one of which is significantly more flexible than the other, is a fairly common occurrence (for example, rotor-casing contact). This paper uses a linear model for simulating contact between rigid and flexible bodies. The motion between contacts is predicted efficiently by modal superposition, using separate models for the two components. During contact, the two models are coupled via a linear contact stiffness. The coupled equations of motion are solved by numerical integration. The results presented here relate to the short-term dynamics during contact and focus on the effects of using different numbers of degrees of freedom in the model of the flexible body.*

1. INTRODUCTION

There are many engineering applications that involve dynamic interaction between two bodies, one flexible, and the other essentially rigid by comparison. An important example is the interaction between a rotor and a flexible casing in an aero-engine. Other examples may be found in certain 'bat-and–ball' interactions. The effect of component flexibility is becoming increasingly important in the context of the drive for weight reduction. There is therefore a need to study the effect of component flexibility on contact dynamics.

This paper uses a linear model to simulate contact between a rigid body and a flexible component. The flexible body is taken to be an elastic ring and the rigid body is a spring-restrained pendulum. These constitute a simple model that can be extended to simulate rotor-casing interaction. A linear spring (contact stiffness) is used to represent local deformation of the bodies while they are in contact. The motion between contacts is predicted efficiently by modal superposition, using separate ring and pendulum models. During contact the two models are coupled by the contact stiffness and the resulting equations are solved by numerical integration. This paper deals with the short-term response of the system during

© IMechE 1998 C553/003

contact and focuses on the complexity of model that is required to give converged predictions.

2. THE SYSTEM

Figure 1 shows the system to be analysed, consisting of a two components. The rigid pendulum is restrained symmetrically by four springs at the lower bearing, is pivoted at O and can move with one degree-of-freedom, ϕ_Y, about axis OY (Figure 2). The rotor disc on the end of the pendulum is located in a clearance space within a flexible ring, symmetrically supported by eight springs. The parameter values for the system considered in this study are shown in Table 1.

Shaft radius R 5 mm	Shaft length L 1.5 m	Shaft spring position L_S 1.4 m	Shaft spring stiffness k_{SHAFT} 20 kN/m	Rotor radius R_D 100 mm	Rotor length L_D 10 mm	Rotor mass M_D 2.47 kg
Young's modulus E 206 GPa	Density ρ 7850 kg/m^3	Damping ratio ζ 0.01	Support spring stiffness k_{RING} 10 kN/m	Ring radius r 110 mm	Ring length l 140 mm	Ring thickness h 3 mm

TABLE 1 System parameters

The displacement of the ring will be described using N eigen-functions of a free circular ring as the generalised coordinates. The radial and tangential displacements, w_R and w_T respectively, of a point on the ring at angle θ from the OX direction are therefore

$$w_R(\theta) = \sum_{n=1}^{N} [q_{nc}\cos(n\theta) + q_{ns}\sin(n\theta)] \quad ; \quad w_T(\theta) = \sum_{n=1}^{N} \frac{1}{\alpha_n}[q_{nc}\sin(n\theta) - q_{ns}\cos(n\theta)]$$

q_{nc} and q_{ns} are the generalised coordinates associated with ring displacements which are, respectively, symmetric and anti-symmetric with respect to $\theta = 0$ and α_n ($\approx n$) is the ratio of the maximum radial and tangential displacements for each value of n. Note that $n = 1$ represents rigid body displacement of the ring and $n > 1$ represents elastic deformation of the ring (the symmetric deflection for $n = 2$ is illustrated in Figure 2). The question as to what value of N should be used when considering interaction between the pendulum and the ring is considered in detail in section 3.4.

The individual equations of motion for the pendulum and the ring can be derived using any suitable means. Reference (1) provides a derivation using Lagrange's equations and presents the mass and stiffness matrices in detail. These equations can be used to predict the non-contact motion of the individual components in terms of a summation of individual modal contributions. For the results presented in this paper, the pendulum is restricted to movement in the XZ-plane. Contact between the pendulum and ring therefore only occurs in the XZ-plane, i.e. at $\theta = 0°$ or $\theta = 180°$. Consequently, due to the assumed symmetry of the ring

C553/003 © IMechE 1998

support springs, only those generalised coordinates associated with symmetrical displacements of the ring (i.e. q_{nc}) will have non-zero values.

Clearly, contact between the pendulum and ring will occur when the algebraic difference between the pendulum displacement and the radial displacement of the ring at $\theta = 0$ or $\theta = 180^{\circ}$ exceeds the initial clearance. In general, the displacement of the ring will consist of a combination of rigid body displacement and elastic deformation. It will be assumed here that the elastic deformation of the ring is small relative to the initial clearance so that the point of contact will be determined with good accuracy by considering the rigid body displacements only. If however the elastic deformation of the ring is relatively large compared to the initial clearance then a different, more complex approach to finding the contact point is required. This is discussed in more detail in reference (1).

3. CONTACT BEHAVIOUR

3.1 Contact Stiffness

The force-displacement relationship between elastic bodies in contact is inherently non-linear. Nevertheless, when considering dynamic interactions, it is beneficial to linearize the relationship in the interests of computational efficiency. For the system considered in the present paper, the Hertzian force-displacement relationship for the circular contact between a sphere and a flat plate, given in reference (2), was used as the basis for calculating the stiffness of a linear spring to model the local contact deformation between the pendulum and ring. Figure 3 shows the non-linear force displacement relationship for circular contact, for a range of contact forces relevant to the present study. The linear approximation to the circular contact curve (stiffness K_C = 250 MN/m) used in the present study is also shown.

3.2 Predicting Motion During Contact

The individual mathematical models of the ring and pendulum, mentioned above, fully describe the non-contact dynamics of each component. When contact occurs, the two models are coupled via the contact stiffness and a different approach, outlined below, is used. If the contact stiffness is K_C, the contact force will be given by

$$F = K_C [v(t) - w_R(t) - c] \qquad (1)$$

where $v(t)$ and $w_R(t)$ are the radial displacement of the pendulum and the ring at the point of contact and c is the initial radial clearance. It is assumed that the contact velocities are normal to the surfaces. The equations of motion for the ring can then be written in the following form

$$[M]\{\ddot{q}\} + [C]\{\dot{q}\} + [K]\{q\} = \{Q(t)\} \qquad (2)$$

where $\{Q(t)\}$ is the vector of generalised forces associated with the contact force F and given by $\{Q(t)\} = F\{\cos\theta, \cos 2\theta, \dots \cos n\theta\}^T$. Introducing normal mode coordinates, $\xi(t)$, for the ring [reference (3)], we may write $\{q\} = [\Phi]\{\xi\}$ where $[\Phi]$ is the modal matrix. Substituting into equation (2) and pre-multiplying by $[\Phi]^T$ to transform to modal space, the rth modal space equation for the ring can be written as

$$\ddot{\xi}_r + 2\gamma_r \omega_r \dot{\xi}_r + \omega_r^2 \xi_r = \left\{ [\Phi]^{\mathrm{T}} \{Q(t)\} \right\}_r \tag{3}$$

where ω_r and γ_r are the natural frequency and damping ratio of the rth mode. Similarly, for the single degree-of-freedom rigid pendulum, which has natural frequency Ω, damping ratio δ and moment of inertia I_{Y}

$$\ddot{\varsigma} + 2\delta\Omega\dot{\varsigma} + \Omega^2\varsigma = -\frac{F L^2}{I_{\mathrm{Y}}} \tag{4}$$

where $\varsigma = L\phi_{\mathrm{Y}}$. From equation (1), the contact force can therefore be written as

$$F = K_{\mathrm{C}} \left[\varsigma - \sum_{r=1}^{N} [[\Phi]\{\xi\}]_r - c \right] \tag{5}$$

Introducing the state vector $\{u\} = \left\{ \{\dot{\xi}\}, \{\dot{\varsigma}\}, \{\xi\}, \{\varsigma\} \right\}^{\mathrm{T}}$ and eliminating the contact force, the modal equations of motion, equations (3) and (4) can be assembled in the form

$$\{\dot{u}\} = \begin{bmatrix} A & B & C & D \\ E & F & G & H \\ I & 0 & 0 & 0 \\ 0 & I & 0 & 0 \end{bmatrix} \{u\} \tag{6}$$

These equations can be integrated numerically provided that the velocities and displacements of the two structures at the start of impact are known. The contact period will end when the force reduces to zero. The modal state vector, $\{u\}$, at the end of the contact period gives the initial conditions for the subsequent period of non-contact motion.

The modal properties (natural frequencies and mode shapes) of the coupled system, which are the eigenvalues and eigenvectors of equation (6), have an important bearing on the duration of contact events, and will be discussed in section 3.4.

3.3 Numerical Considerations

It is necessary to consider the computational time-steps to be used during both the non-contact and contact phases, along with the accuracy with which the start and end of contact need to be determined. During non-contact the required time-step depends on the frequency content of the motion and a value of 100 μs was found to be appropriate in this case.

The time at which contact begins and ends is determined by the inequality relationship $v(t) - w_{\mathrm{R}}(t) > c$. The inequality was tested at each time step and an iterative interpolation scheme was employed to find the actual start of the contact more accurately. Investigations were performed to determine an appropriate interpolation tolerance. For illustration, consider tolerances of 10 μs, 1μs and 0.1 μs. Results are presented for the case where the system is initially at rest and motion of the pendulum is initiated by a radial impulse in the OX direction. This leads to three closely spaced impacts between the pendulum and the ring, all at the same point ($\theta = 0°$). The predicted force functions are shown in Figure 4. With a tolerance of 10 μs, the predicted start and end of each force pulse is significantly different from zero, in either the compressive or the tensile sense. When the tolerance was reduced to 1μs, the start and end of each force pulse are close to, but visibly different from, zero. With a

C553/003 © IMechE 1998

tolerance of 0.1μs, the start and end of each force pulse is sensibly zero. Smaller tolerances produced no discernible change and a value of 0.1μs was therefore used throughout.
The choice of time step to be used for numerical integration of the equations of motion during contact is governed by the need to have sufficient steps during one cycle of the highest mode of the coupled system. The natural frequency of this mode depends on the number of generalised coordinates used to model the ring. Suffice it to say that a time-step of 1μs was found to be satisfactory when up to 20 generalised coordinates were used for the ring.

3.4 The effect of the number of ring generalised coordinates on contact duration.

The aim of the following discussion is to explain the observed pattern of behaviour as predicted by the mathematical model and it must be acknowledged that the detailed predictions will be model dependent. Figure 5 shows the effect on the coupled natural frequencies of varying the number of generalised coordinates used to model the ring.
With N ring coordinates, there are $N + 1$ coupled modes (the extra degree of freedom being that of the pendulum). Mode 1 is characterised by coupled rigid-body motion of the ring and pendulum, the two members moving in phase with each other causing relatively little deformation of the contact spring. Since this mode depends on the rigid body behaviour of the ring, its frequency is sensibly independent of the number of flexible ring coordinates used. For modes 2 to N, the main contribution to the rth coupled mode shape is found to be from ring coordinates r and r-1. If fewer than 6 coordinates are included, the motion of the ring and pendulum at the point of contact is predicted to be in-phase with the pendulum mass in all coupled modes. When 6 or more ring coordinates are used, modes 6 and above exhibit out-of-phase movement between pendulum and ring at the point of contact. When such modes are excited during contact, there is significant compression of the contact spring, which in turn influences the contact behaviour. This is discussed later.
As an example, Figure 6 shows the mode shapes for modes 4, 11 and 12 for the model with 11 ring coordinates. Figure 7 shows the corresponding contributions of the generalised coordinates of the ring and the pendulum (marked P on the axis). For mode 4, the largest contribution is from ring coordinate 4, with a smaller, negative contribution from coordinate 3. At the point of contact, the displacements of both the ring and pendulum are positive and the mode shape of the ring exhibits the expected four lobes.
For mode 11, the largest contributions are from coordinates 11 and 10, and the mode shape has eleven lobes. At the point of contact, the displacement of the ring is negative while that for the pendulum is positive. This results in a larger deformation of the spring than for mode 5.
Mode 12 (the highest frequency mode) has significant contributions from all coordinates, with coordinate 11 being the largest. At the point of contact, the displacement of the ring is large and negative while that for the pendulum is positive. This again results in a large deformation of the spring.
Figure 8 shows the effect of the number of ring coordinates on the predicted contact duration. Increasing the number of coordinates from 1 to 12 suggests that convergence of the contact duration might be expected at a value of about 0.075 ms. In this region, the behaviour of the system during contact is dominated by the highest frequency mode. As noted earlier, this mode has significant out-of-phase movement of the ring and the pendulum at the point of contact. Excitation of this mode results in the compression and extension of the contact spring over approximately half a cycle and, once the spring returns to its unstrained length, separation occurs.

As an example, consider the case of 4 ring coordinates for which the frequency of the highest mode is 4.5 kHz (Figure 5). The duration of half a cycle is 0.111 ms and the actual contact duration is found to be 0.112 ms. In comparison, all of the other modes have much longer periods and contribute little in the way of oscillatory motion. As might be expected from this behaviour, the corresponding variation of the contact force (the 4-coordinate case in Figure 9) approximates to a half-sine wave. The corresponding displacements of the pendulum and ring at the point of contact are shown in Figure 10. (I'm not sure that I would include Fig 10)

A similar correlation between the contact duration and the half periods of the highest frequency modes exists when from 1 to 12 ring coordinates are used. These are shown as circles in Figure 8. With fewer than 6 coordinates included, the discrepancy is less than 2%. It increases if more coordinates are used and is 33% with 12 coordinates. This occurs because it is found that the highest frequency mode alone no longer dominates the contact behaviour. With 12 coordinates, for example, modes 10, 11 and 12 make significant contributions in addition to mode 13. It will be seen from the frequencies in Figure 5 that these modes have half-periods which are shorter than the contact duration. As a result, they are able to contribute additional oscillatory behaviour. This effect continues as more ring modes are included. With 15 ring coordinates for example, the contribution of these other modes prevents the contact spring from returning quickly to its unstrained length and therefore delays separation. The presence of these additional oscillatory components can be clearly seen in the force plot in Figure 9 (15 coordinates).

There are thus two distinct regions created by the model. With fewer than 12 ring coordinates, the linear spring models the contact deformation as a local effect in which the global deformation of the ring is relatively small. When more coordinates are included, the deflection of the ring becomes similar to the compression of the spring and, in this case, it is doubtful that the continued use of the linear spring to model the local deformation remains valid. Further work, including experimental evidence, is required to clarify this.

Thus, if the ring is sufficiently stiff that its radial deflection is small in comparison with the local deformation at the point of contact, the linear spring model provides a realistic approximation to the behaviour. This work forms part of a larger study of the rotor dynamics (1) and, while the detail of the individual impacts is important, the behaviour over many revolutions is of more significance for the overall integrity of the system. For the case investigated, the use of four ring coordinates was found to be sufficient for this purpose.

In other applications (sports equipment is an obvious example) in which the deflections of the impacting structures are comparable to any local contact deformation, many more coordinates would be required. This would be true particularly in cases involving single impacts.

4. CONCLUSIONS

The predicted behaviour of the system during contact has been shown to fall into two regimes, depending on the number of generalised coordinates used to model the elastic body. When only a few ring coordinates are included (up to 12 in the case reported), it has been seen that the behaviour is dominated by the highest coupled mode. This depends on the contact model used; in this work, a linear spring with energy dissipation at the contact neglected. When a larger number of ring coordinates is included, the flexibility of the ring itself (expressed in terms of its higher coupled modes) plays an increasingly important role and when sufficient of these higher modes are present, a converged solution is obtained. If too few ring modes are used, a misleading prediction of the contact duration may be obtained. Further work is required to validate the existing model experimentally and to extend the model for rotor dynamics applications by incorporating shaft rotation, contact friction, dissipation and gyroscopic effects.

REFERENCES

(1) Y-L Chen *Effect of containment flexibility on the impact dynamics of a rotor.* PhD Thesis. University of Nottingham. July 1997

(2) K L Johnson *Contact Mechanics.* Cambridge University Press. 1985

(3) L. Meirovitch *Elements of Vibration Analysis* McGraw-Hill 1975

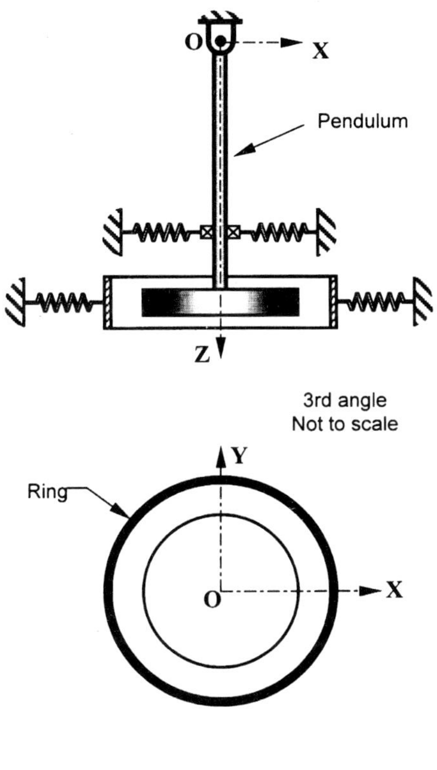

Figure 1 Schematic view of system

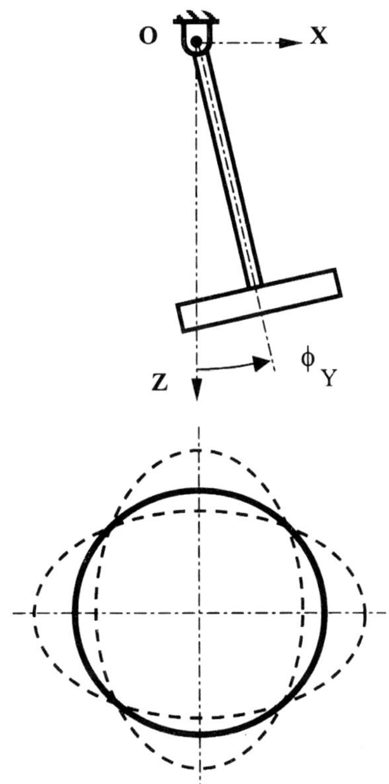

$n = 2$ deformation of ring

Figure 2 Generalised coordinates

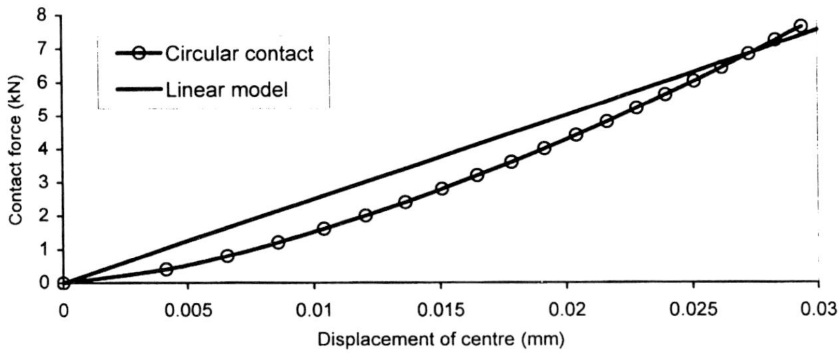

Figure 3 Hertzian contact and the linear model

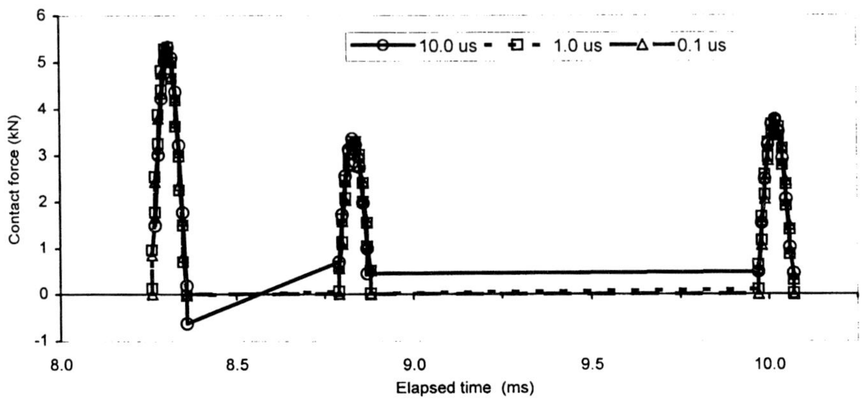

Figure 4 Effect of contact tolerance on predicted force

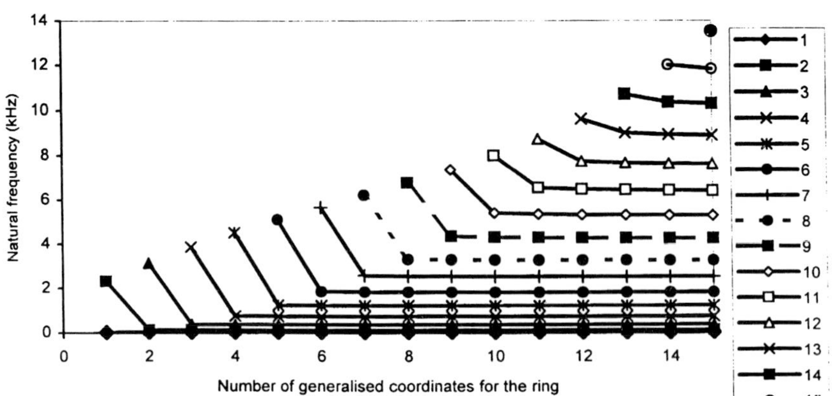

Figure 5 Effect of the number of ring coordinates on the coupled natural frequencies of the system

C553/003 © IMechE 1998

Figure 6 Coupled mode shapes with 11 ring coordinates

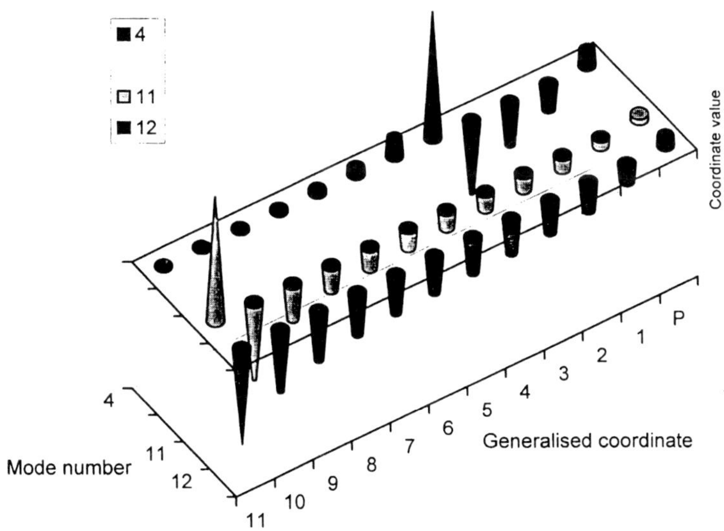

Figure 7 Contribution of each generalised coordinate to coupled modes 4, 11 & 12

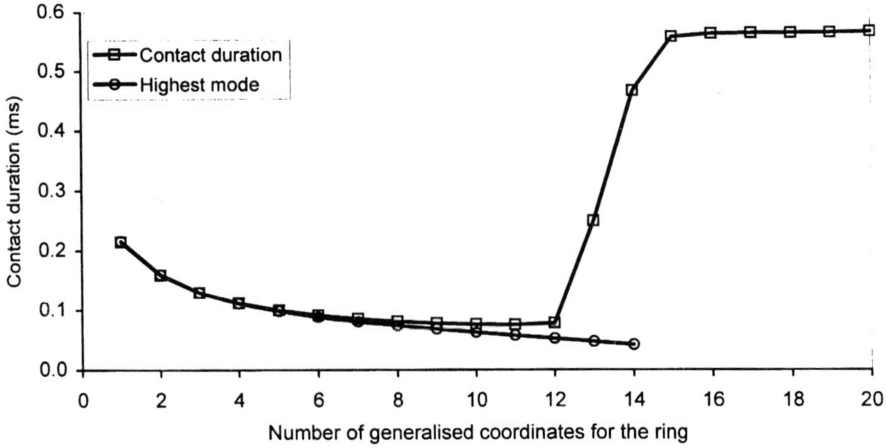

Figure 8 Effect of the number of ring coordinates on the contact duration

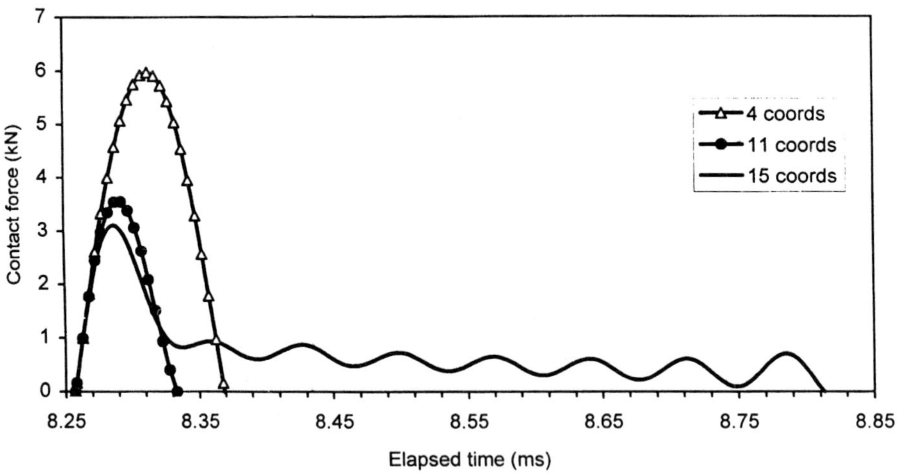

Figure 9 Effect of the number of ring coordinates on the contact force

C553/003 © IMechE 1998

C553/035/98

A novel six degree-of-freedom wind tunnel motion rig

P M ENTWISTLE BSc, AMIMechE
Fraser–Nash Consultancy Limited, Dorking, UK

ABSTRACT

Parallel mechanisms are increasingly being used in applications requiring simultaneous motion about a number of axes. Probably the most widely recognised parallel mechanism, usually called a Stewart Platform, is seen on aircraft flight simulators. However, parallel mechanisms are also used in applications as diverse as multi-axis machine tools and force-feedback joysticks. This paper discusses the process used to design a motion system, based on a modified Stewart Platform, to fly an aircraft model in a wind tunnel. The paper introduces some important considerations for the design of this important class of mechanism.

1. BACKGROUND

Military aircraft are increasingly using flight envelopes that involve non-linear aerodynamics. Predictions of the flight dynamics in these situations are particularly difficult. The most practical approach is to use dynamic wind tunnel testing to provide data to support complex mathematical models. Most dynamic test rigs currently used in wind tunnels only provide small amplitude, single degree-of-freedom oscillatory motions, or continuous rolling motions. These test rigs cannot produce the range of motions necessary to investigate non-linear effects. To overcome this limitation the Defence Evaluation and Research Agency (DERA) at Bedford conceived a novel six degree-of-freedom test rig for installation in their low speed wind tunnel. DERA contracted Frazer-Nash Consultancy Ltd (FNC) to develop the concept and design the test rig.

This paper reviews the requirements for the test rig, explores the limitations of existing multiple degree-of-freedom dynamic test rigs, and outlines the development of the new test rig.

2. DESIGN CONFIGURATION

A survey of existing wind tunnel rig mechanisms revealed that they are limited in their capability to directly separate rotational and translational motions. For small amplitude motions this is not too significant a problem.

Conversely, for the large amplitude high rate motions characteristic of manoeuvring aircraft time-history effects are significant and the aerodynamics may become very non-linear. In this case, it becomes essential to measure the rotational and translation characteristics separately, necessitating a five or six degree-of-freedom mechanism. A number of existing test rig configurations could be modified to generate five or six degrees-of-freedom. However, in the majority of cases these would be serial mechanisms, where the motion elements piggy-back one on another. The end results would be low payload to structure weight ratio and hence high actuator power requirements.

An unconventional (in a wind tunnel application) parallel mechanism for providing six degree-of-freedom motion combined with rigidity, quick response and high payload to structure weight ratio is the Stewart platform, widely used for flight simulators, robot manipulators an machine tools. Stewart's original design aims in 1965 for his platform were:

- Not more than six actuators to avoid redundancy and reduce cost.
- Each actuator reacting on the foundation to avoid interaction between actuators.
- Each actuator operating directly on the load to maximise performance for given power.
- High payload/structure weight ratio to maximise performance for given power.
- Low friction motions to reduce power losses and obtain high response.
- Each actuator identified with one motion to simplify the control.

These aims are equally applicable to the large amplitude aircraft motion rig and are achieved by the Stewart platform, with the exception of the last. However this is not a significant factor in the light of the capabilities of modern computer based control systems.

The most common application of a Stewart platform, the flight simulator, is shown in Figure 1. Here, a triangular top platform is mounted upon six actuators arranged in three pairs. Each actuator pair is attached to one corner of the platform by a common three-axis joint. The bottom end of the actuator is connected to the base by a two-axis joint. The floor joints for each pair of actuators have a common axis, called the "free axis". Thus the position of each attachment point on the platform can be controlled within the plate containing the actuator pair. The "uncontrolled" rotation of this plane about the common axis is determined by the "controlled" positions of the other two attachment points.

C553/035 © IMechE 1998

The Stewart platform can be adapted for use as a wind tunnel model support system by mounting a model in front of the platform. However, the 'six-post' configuration used for flight simulators is too bulky for installation in a wind tunnel. In addition, the motion envelope of the platform is limited by the stroke of the actuators, which in turn will limit the angle through which a model mounted ahead of the platform can be rotated.

These difficulties can be overcome by replacing each symmetric actuator pair by a primary actuator attached to the platform corner. This actuator is controlled in a side-to-side motion by a secondary actuator attached to the lower end of the primary actuator . The cylinder of the primary actuator and the entire secondary actuator are now below the floor of the working section, and the platform is supported by three rather than six struts. The blockage levels of the structure are considerably reduced and the motion envelope significantly increased.

The modified Stewart platform arrangement, shown in Figure 2, has a number of advantages which make it ideal for a large-amplitude motion rig:

- Full six degree-of-freedom motion.
- High payload/structure weight ratio.
- Low support interference - no rig structure behind the model, coupled with reducing blockage levels with increasing angle of attack.
- High stiffness - triangulated structure, with stiffness increasing with angle of attack.
- Versatility - a wide range of applications.
- Ease of installation - the mechanism can be mounted from the existing tunnel structure with minimum modification.

However, in this configuration the primary actuators have to carry significant side loads as well as axial loads. It was recognised that this could limit the performance. Following comparison of the potential advantages and limitations of the new design, it was decided that the modified Stewart platform had sufficient advantages to outweigh this potential limitation.

3. DESIGN REQUIREMENTS

The design requirements for the test rig were based on potential full-scale manoeuvre capabilities and the need for frequency domain aerodynamic model identification. Target model motion parameters were as follows:

Static Incidence Angles		Motion Amplitudes	
angle of attack	= 0° - 90°	pitch	= ±30°
sideslip angle	= ±30°	yaw	= ±15°
cone angle	= ±60°	roll	= ±60°
		x, y, z	= ±0.5 m

For a typical lightweight dynamic model sized for the 13ft x 9ft low-speed wind tunnel, angular velocities of the order of 500deg/s and motion frequencies of the order of 2.5Hz would be needed.

Ensuring that the design of a parallel mechanism meets all the target motion parameters is a complex task, especially as any one motion will generally involve all six actuators. It was decided therefore to use extremes of the required motion to define two design points. The design points would be used to generate the layout and define the actuator requirements. The performance of the resulting design could then be assessed against the overall requirements. The two design points selected were:

1. Sinusoidal pitching motion at a frequency of 2.5Hz from 0° to 60° angle of attack.
2. Sinusoidal coning motion at a frequency of 1.25Hz with an amplitude of ±60° at a mean angle of attack of 60°.

4. ANALYSIS OF RIG MECHANISM

Kinematic analysis of the Stewart platform mechanism as a robot manipulator has received considerable attention in recent years. The general Stewart platform analysis is complex; fortunately the configurations discussed here enable considerable simplification of the governing equations.

The analysis of the mechanism can be divided into two parts - the inverse kinematics and the forward kinematics.

4.1 Inverse Kinematics

The inverse problem is that of determining the six actuator lengths for a given model position and attitude. For the modified Stewart platform this is a straightforward operation, involving a sequence of vector additions and rotations. These equations have been programmed for the large amplitude motion rig on a PC using VisualBasic. Figure 3 shows an example of the screen output from the program.

The solution of the inverse equations enabled the actuator motions corresponding to the two preliminary dynamic design cases to be determined. Typical actuator motion trajectories are shown in Figure 4.

4.2 Forward Kinematics

The inverse solution is sufficient for basic sizing of the rig mechanism. However, analysis of the dynamic behaviour of the mechanism requires the solution of the forward kinematic equations - i.e. the determination of the model position and attitude for a given set of actuator lengths. Unfortunately, it is a characteristic of parallel mechanisms that the forward kinematic solution involves coupled non-linear equations with multiple roots, generally requiring a numerical rather than analytical solution.

Two features of the rig mechanism emerge form this analysis:

a) Multiple assembly modes - for any given set of actuator lengths there are up to 16 different ways of assembling the mechanism.

 C553/035 © IMechE 1998

b) Singular positions - for certain model positions and attitudes the forward kinematic equations become singular. In mathematical terms the mechanism gains a degree of freedom, in practical terms the mechanism would be unstable.

In this case the multiple assembly modes are not in themselves a problem, since they are generally widely separated and most are precluded by practical limits on the mechanism. Singular positions on the other hand result in a loss of control and are clearly undesirable! A limited number of simple singular positions for the Stewart platform have been identified in the literature, all lying outside the usual mechanism workspace and hence not causing a problem. However, during the design of the mechanism it was found that the expanded workspace of the large amplitude rig mechanism contained a large number of previously unknown and more complex singular positions. This meant that the range of motions that could be generated by the mechanism were restricted, and the design points were revised.

The application of forward and inverse kinematics to the design of the mechanism is discussed on the following sections.

5. DESIGN OF THE RIG MECHANISM

The motion system is designed to be mounted from the existing working section floor turntable in the 13ft x 9ft low-speed wind tunnel at DERA Bedford. The general layout of the motion system consists of four sections, these are:

a) Actuators - actuators that pass through the tunnel floor are referred to as 'primary actuators' and actuators below the tunnel floor as 'secondary actuators'.

b) Base Triangle - the size and position of the base triangle of the motion system was controlled by a number of conflicting parameters. These included:

- the size, layout and orientation of the existing turntable
- the vertical height at which the actuators are pivoted
- the size of the primary actuators
- the distance between the primary and secondary attachment points
- the motion envelope and the resulting angles of the primary and secondary actuators
- the position of the model relative to the turntable

c) Top Platform - the top platform is based on a 500mm equilateral triangle between the joint pivot points. This was thought to be large enough to allow a reasonable degree of control and small enough to maximise the motions that could be produced within the limits of actuator performance.

d) Model - since it is intended to use existing and models, the test rig was designed to provide the necessary interface.

6. DESIGN APPROACH

Important features of the basic rig specification were:

- Model to be kept a minimum of 0.5m from wind tunnel roof, walls and floor.
- Hardware in tunnel to be kept to a minimum.
- Rig to be attached to existing half-model balance mounting points on the working section floor turntable.
- Minimal modifications to floor turntable.
- Standard parts used where possible.
- Maximum load on turntable to be less than 3000 kg.

The design of the motion system was approached from two different aspects. These were defined by the issues concerning performance limits and geometric constraints. Arriving at a design solution involved satisfying a number of conflicting requirements. This was achieved through the combined use of a 3D model constructed in I-Deas Master Series and the bespoke inverse kinematics program (Rigsim).

The Rigsim software allows the plane to be positioned in a particular orientation, from which it outputs the primary actuator lengths and angles. The I-Deas Model works in a reverse manner. The actuator lengths are fed into the 3D model geometry, from which the I-Deas software solves for the position of the plane. Figure 5 shows a general view of the I-Deas model. It can be considered that Rigsim works in a similar manner to the control system, and the I-Deas model responds in a similar manner to the motion system. Using these two simulation tools together allowed the design constraints to be explored in detail.

The original design point motion limits of the plane were introduced into the Rigsim program from which actuator lengths for the given rig configuration were obtained. The actuator lengths were then fed back into the I-Deas model.

Use of the I-Deas model highlighted that portions of the motion defined by the original design points were achievable. It was found that these motions (as opposed to positions) were trying to pass through a singularity. New design points were defined, restricting the stroke limits of the actuators so that the geometry did not pass through or get too close to these types of positions.

6.1 Performance Limits

The original dynamic motion requirements gave a maximum velocity of 8m/s for the primary actuators. If this was the only design requirement for the actuators it would have been achievable. However, combined with the constraints imposed by other requirements in the design (in particular the high bending loads imposed on the primary actuators) this level of performance was found to be unachievable. It was decided therefore to limit the actuators to a maximum velocity of 5m/s whilst maintaining the required motion frequency capability, and accepting the resultant reduction in motion amplitude. Since the design points were based on sinusoidal motion, at 2.5Hz with a maximum velocity of 5m/s the corresponding maximum

C553/035 © IMechE 1998

dynamic stroke for the primary actuators is 0.64m and the maximum acceleration is 78 m/s^2. These parameters were used as the basis for the final design.

In order to minimise the inertial loads on the system without compromising the dynamic envelope, a design decision was made to restrict the maximum total working stroke of the primary actuators to slightly more than the dynamic stroke, i.e. around 0.7m. However, this would have severely limited the static model positioning range. To extend the static workspace the primary actuator rods were designed to have interchangeable extension sections. Different sets of rod extensions allow the actuator stroke to be used to cover a number of different static position ranges.

To support the high bending loads imposed on the primary actuators, double ended actuator rods were specified. The separation of the bearing at each end of the cylinder helps to minimise the bearing loads.

6.2 Geometric Constraints

The geometric constraints fall into three categories:

a) Mechanism kinematics - unstable (singular) system conditions.
b) Hardware clashes - clashes between the actuators and the surrounding rig and wind tunnel structure.
c) Platform joint limitations - required angular range of motion for the top platform joints of about 130°.

Arriving at a practical design solution involved satisfying these conflicting requirements. At the end of a lengthy trade-off study, three new dynamic design points were chosen within the bounds of geometric and performance constraints as described above:

1. Sinusoidal pitching motion at a frequency of 2.5Hz from 15° to 60° angle of attack, (with short actuator extensions).
2. Sinusoidal pitching motion at a frequency of 2.5Hz from 0° to 30° angle of attack, (with long actuator extensions).
3. Sinusoidal coning motion at a frequency of 1.25Hz with an amplitude of ±45° at a mean angle of attack of 38°.

These motions assume that the model remains on the wind tunnel centreline and over the centre of the turntable. It is possible that different range of motions could be achieved without exceeding the actuator performance limits, or approaching geometric constraints by moving the model away from this centre position.

7. MECHANICAL DESIGN

Figure 5 shows an I-Deas model of the complete rig mechanism mounted in the 13ft x 9ft wind tunnel working section.

Three degree-of-freedom joints are required between the primary actuator rods and the top platform. These joints have been designed specifically for this application to accommodate up to ±85° actuator rod offset and to be as compact as possible. An extra degree-of-freedom is provided on the platform side of the joint, to allow the elements of the joint to align themselves with the direction of motion when the joint angle exceeds 45°. Without this extra degree-of-freedom the joint may not transmit the motion at large angles. To reduce the articulation of the joints, the attachments on the top platform have been angled to align with the mid points of the motion envelope.

The bottom platform assembly provides the attachment points for the lower end of the primary actuators and the secondary actuators. The actuators are mounted in pairs of gimbal blocks that lie on the free axis of the actuator pair. The assembly attaches to the four existing half-model balance mounting points on the wind tunnel floor turntable.

A finite element model of the support structure was constructed and subjected to representative load cases. Maximum stresses in the support structure arc well within the structural strength of the material and sufficiently low to give an infinite fatigue life. Assuming the rest of the structure is rigid, the maximum deflection at the model is 2mm.

At the initial stages of the design a variety of types of actuator were evaluated. Hydraulic actuators were chosen because they offered the necessary level of control ability while satisfying the power requirements; however, it was established that actuators that could meet the performance criteria would have to be designed specifically for this application. Of particular concern were the large side loads that had to be carried by the primary actuators and the high flow rates necessary to achieve the velocities.

Design features of the actuators and associated hydraulic system are:

- Maximum velocity limited to 5m/s.
- Working stroke of the primary actuator limited to 0.7m.
- Primary actuators have equal area hydraulic cylinders.
- Secondary actuators have unequal area cylinders with regenerative supply.
- Peak flow accumulators mounted close to each actuator.
- Servo valves mounted on the actuators.

8. SUMMARY

The configuration selection and mechanical design of a novel configuration for a six degree-of-freedom large-amplitude motion rig for dynamic wind tunnel testing has been described. The rig, based on the Stewart platform mechanism widely used for flight simulators, offers:

- A full six degree-of-freedom movement.
- Large amplitude, high rate motion capability.
- High payload/structure weight ratio.
- Low support interference.
- High structural stiffness.

- Wide range of applications.

Mechanical design of the rig and fabrication of the major structural components has been completed. Manufacture and testing of a prototype primary actuator is under way. When commissioned, the large amplitude motion rig will provide a unique and highly versatile capability for both static and dynamic wind tunnel testing.

Frazer-Nash Consultancy Ltd would like to acknowledge the support of DERA in carrying out this work.

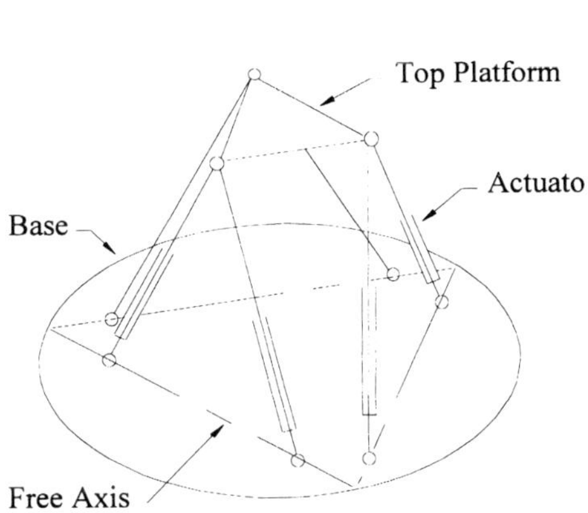

Figure 1 Conventional Stewart Platform

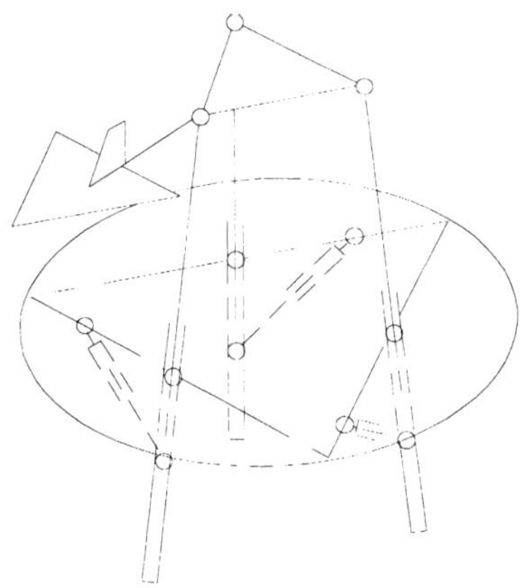

Figure 2 Modified Geometry

© IMechE 1998 C553/035

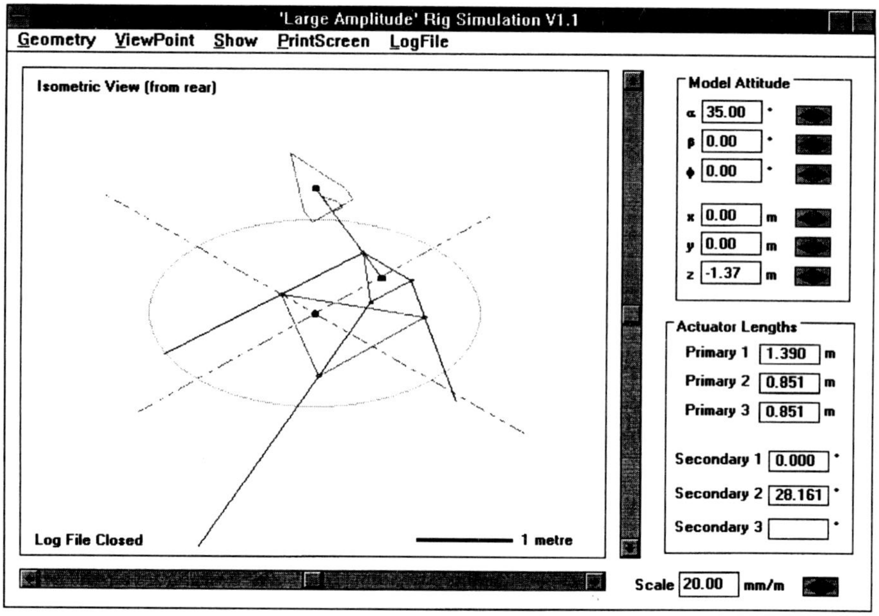

Figure 3 -Inverse Kinematics Program

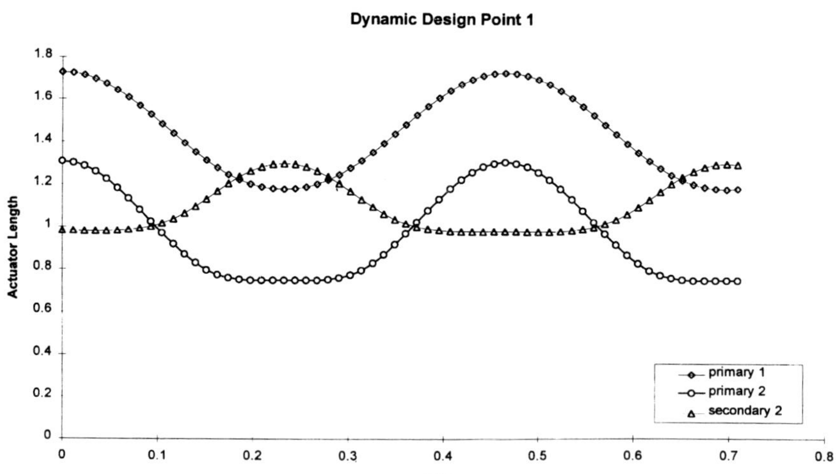

Figure 4 - Typical Actuator Trajectories

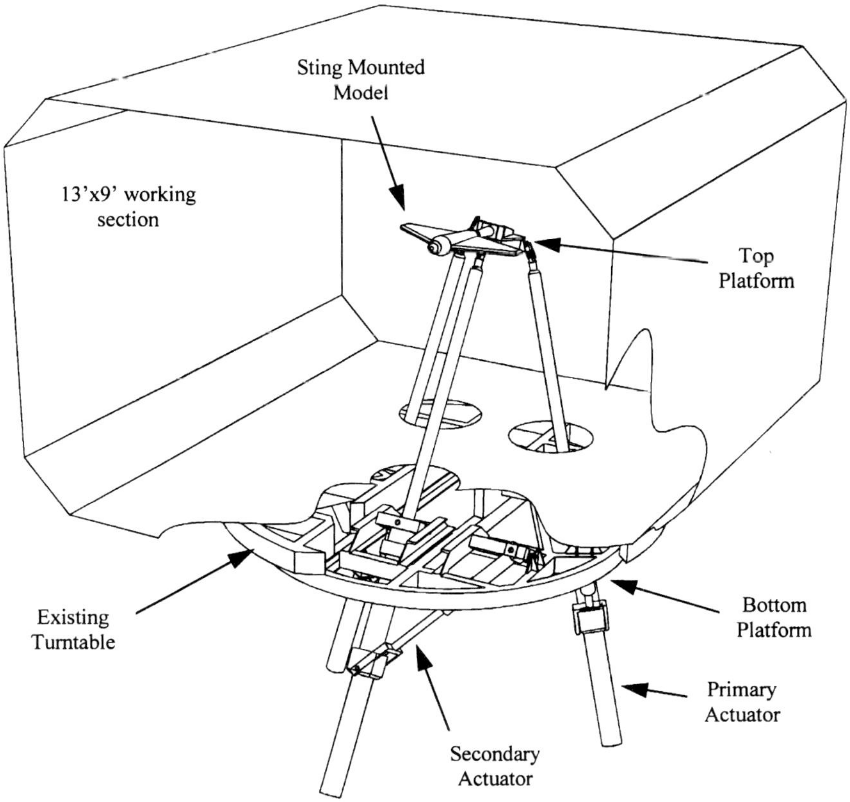

Figure 5 - Rig Installed in Wind Tunnel

Structural and Mechanism Dynamics

C553/045/98

Multi-body dynamics applications in vehicle engineering

R S SHARP BSc, MSc, FIMechE
School of Mechanical Engineering, Cranfield University, Bedford, UK

Synopsis
The paper includes discussion of the implications of modern multibody systems analysis methods and computer software for vehicle dynamics. The different points of view and interests of users are considered, together with appropriate strategies for organising relevant activities. The main dynamical foundations for commercially available software systems are explained briefly and their implications are mentioned. The immense differences in speed of simulation between systems with different fundamental multibody strategies are exposed. Other customer requirements from software suppliers are noted.

1 INTRODUCTION

Discussions of multibody dynamics applications tend to be complex, in spite of intentions to separate issues and simplify. The problem derives from the multitude of issues involved and the strong coupling between them. The issues concern history; people and purposes; strategies for use of multibody methods and the achievement of purposes; multibody formalisms, how they work, what the difficulties are and what each implies; modelling methods; symbolic and numerical computation; and commercial computer software with its interfaces to the user on the one hand and other computer aided engineering packages on the other, its documentation and support etc. The paper will address each of these issues to some extent. The detail will necessarily be limited and the structure imperfect. The focus will be on road vehicle dynamics.

2 A BRIEF HISTORY

In a classic paper in vehicle dynamics, Segel established in 1956 that the motions of a car on a flat surface in response to road wheel steer inputs can be predicted with accuracy in the linear range of operation, extending to about 0.3g lateral acceleration, (1). By 1969, capability to represent the general motions of an automobile, travelling over normal roads and even impacting roadside furniture was demonstrated by McHenry, (2). In the early 1970's, the McHenry model was further developed, becoming the Highway Vehicle Object Simulation

© IMechE 1998

Model (HVOSM) which was used in many US research programmes and some outside the United States. The vehicle model and all the FORTRAN coding of HVOSM were hand derived, by virtue of many man-years of work by high-grade engineering staff. In the model development, the enormous algebraic complexity of the full equations of general motion for an automobile, treated as a multibody system, obliged the analysts to compromise strongly between detailed accuracy and tractability of the problem. Important matters of principle were established, for example that the description of the motion of each road wheel relative to the car body is critical from the point of view of connecting the tyre forces to the vehicle properly but is quite uncritical in relation to the inertial contributions of the wheels. It was established that complex models, especially when embedded in computer programs, can hardly be changed, for fear of upsetting the situation in unknown ways, so that when a model like HVOSM was found not to contain an essential feature of a new task, e.g. suspension compliance, it was necessary to start again.

Also in the early 1970's, the foundations of ADAMS and DADS, today's most used multibody codes, were laid by the doctoral work of Orlandea (3). A few years later, NEWEUL was being developed in Stuttgart (4), AUTOLEV in California (5), MEDYNA at DLR (6), and SD-EXACT at Symbolic Dynamics (7), while towards the end of the decade, MESA VERDE was started in Karlsruhe (8). Many further multibody systems developments have followed, notable among them being SD-FAST (developed from SD-EXACT) and AutoSim, each influenced by Kane (9), and MADYMO, which has had wide application to crash victim simulation. Recently, automotive companies have become users of SIMPACK (to some extent a successor to MEDYNA) and WORKING MODEL. There are now many other software systems for multibody modelling and simulation. The names of the systems and their originators can be found in review literature such as (10, 11). A much fuller account of historical issues can be found in the excellent review by Schiehlen (12). To some extent, the commercial status of the various codes can be gleaned from the reviews but this changes rapidly, so current information may have to be obtained from advertisements, agents, the Internet etc.

In recent years, the world's automotive industries have embraced predictive engineering to a serious extent. Modelling and simulation are seen as effective tools for getting high quality products at minimal cost. Vehicle dynamics modelling is just a part of the whole movement towards the greater use of computers in product engineering. With the experience of HVOSM and other such models behind us, and with the present wide availability of very effective alternatives to manual methods, it would be very foolish to spend a lot of time in manual model building. Few people need convincing of this.

3 PEOPLE AND PURPOSES

The people most concerned with the issues under consideration are members of vehicle and sub-system design teams, development engineers with responsibilities for curing faults, accident investigators and vehicle dynamics researchers. Most students, as undergraduates, deal with simple systems, for which multibody analysis is unnecessary, so they typically know little of the matter when they graduate. Higher degree students may well have had some exposure to the issues and may even have some skill with a particular software system. These people naturally have different perspectives on multibody modelling and simulation matters and may have quite different levels of proficiency and experience in relation to abstraction and simplification of real systems.

C553/045 © IMechE 1998

There are several possible objectives behind multibody analysis. At the simpler end, there may be mechanism kinematics aims, the determination of equilibrium states or linear system operations like stability and modal calculations, while more generally, system simulation is likely to be the purpose. Note that the simplicity notion used here is in connection with computational burden at solution time - describing the rather intricate geometry associated with an accurate determination of suspension and steering kinematics or linearising an equation set to describe small perturbations from a general equilibrium state may be far from simple. Kinematic and equilibrium solutions and linear system results of all kinds tend to be obtained very quickly, whereas simulation of large systems is likely to be slower than the analyst would like, so that simulation speed is an important consideration in some instances. Speed may be vital to the business of the analyst remaining connected to the problem in parametric studies; the simulation may be the core of a parameter optimisation scheme; it may take parameter values from files which represent production tolerances on parts, for later statistical analysis of simulation results; or hardware or people may be "in the loop", making it imperative that real time simulation is possible. Thus, speed of solution varies from being of little consequence to being of paramount importance.

A model building job may be a long-term activity, following the course of the design of a new vehicle or sub-system. Such a mission will necessarily proceed rather slowly and the model building can advance under little time pressure. On the other hand, if the model becomes necessary as a result of a fault condition being discovered in vehicles already with customers, fast model building is essential. Otherwise, decisions on actions to be taken will have to be made before results of simulations are available and the whole point of the modelling will be lost. In research sections, model building is normally a relatively short preliminary to much longer studies of behaviour, so it is advantageous but not essential to build models very quickly. The properties when built are likely to be of greater significance.

4 STRATEGIES FOR MODELLING AND SIMULATION

A simple way of thinking about vehicle dynamics modelling is to imagine a single model so comprehensive in its detailed representation of a vehicle that it is a virtual vehicle for motion simulation. This manner of thinking will appeal particularly to companies that design and manufacture vehicles, in each of which they clearly have a deep and special interest. Such companies are well known to be concerned with competition from other makers and they aim for "best in class" performance levels. Such concerns for competitive products imply the need, according to this thinking, for very detailed models of competitors' vehicles also. According to this thinking, the appropriate virtual vehicle would be used for any purpose that arose.

There is a strong danger of the work getting out of hand with this approach, since the model building itself will be long and costly, with many difficult judgements having to be made. The analyst would need a plethora of decisions about where to include compliance, hysteresis, friction, geometric details, cross-coupling effects etc. Without knowing the purpose of the analysis, it is almost impossible to decide such things. On the basis that features may be needed, including everything that can be thought of implies that the model will never be complete and it will be totally infeasible to determine by properly controlled experiments a full set of parameters representing a real vehicle. Even if such a comprehensive model were put together, it would simulate extremely slowly and, depending on the application, there are implied financial costs and losses of engineering benefit from this.

Alternatively we may think of each model in relation to its purpose, not embarking on model creation until the purpose is very clear. Then good modelling decisions depend on seeing dependencies, or the lack of them, between vehicle design features and behavioural characteristics. The model purpose determines the behaviour of interest and the connections between the behaviour and the design indicate what the model should contain. With the passage of time, a modelling group will be involved with a large number of different models and the extent to which existing models can provide a convenient basis for new models, without the original work having to be re-done, is an important feature of any methodology. Remember that this is one of the big drawbacks of manual methods.

For researchers and development engineers, the nature of the next problem cannot be anticipated, but for designers, each cycle is a refinement of the previous process and the course of events is more predictable. The model based predictive activity can be seen as the theoretical parallel to practical prototype vehicle testing, in which particular tests are used to determine specific aspects of behavioural quality, over which certain parts of the vehicle design will be known to have especially strong influence. To put flesh on these bones, let us consider a few examples out of the many which could be used.

"Boulevard Jerk" is a car body vertical vibration phenomenon that occurs on smooth roads with friction locking of the main suspension. Then typically, the body will resonate on the tyre vertical compliance with a frequency of around 4 Hz in response to the small excitation present. The essential ingredients in the event are the small but significant excitation and the main suspension friction. In exactly the same way as a prototype vehicle will be tested for the phenomenon, it is suggested that a simulation test will be devised and operated routinely in the vehicle design process. Due to the nature of the problem, the boulevard jerk simulation model would contain the car body with vertical, roll and pitch freedoms, simple suspension kinematic properties with each wheel having a single freedom but including Coulomb friction forces, tyre vertical compliances and a suitable road profile model. Due to the smallness of the excitation relevant to the event, it would not make any difference whether or not suspension limit stops were included in the model. If they were included, parameter values would be needed but this represents work without benefit, so they are better omitted. The integrity and efficiency of the simulation work depends on an understanding in principle of design feature and vehicle behaviour relationships.

To find out about the limit stops, the excitation must be at the vigorous end of the spectrum and, of course, limit stop models must be included. The vehicle loading condition and manoeuvring matter. Bump stops are most involved when the wheel loading is high by virtue of the payload and vigorous manoeuvring occurs. Rebound stops are most involved when the payload is least. Under the high excitations under which it makes sense to examine the limit stops, the suspension friction, so vital to the boulevard jerk, would not be so important. An "equivalent" viscous damping, being much easier to simulate, could represent its influence. For determining frequency response properties, relating to ride or handling or both at moderate levels of excitation, a linear model containing neither friction nor limit stops is required.

Rollover may be possible with certain vehicles under certain running conditions. To test for rollover potential in the theoretical world, a vehicle model that captures the essentials of rollover, and running conditions that are most challenging to it are needed. We may surmise that our "vehicle" must have all six body degrees of freedom, since braking and speed variation may be involved, the suspension may jack up etc.; we will need a detailed model of the suspension and steering geometry, at least to relate the tyre forces to the vehicle body (the inertial contributions of the wheels may not be too significant); we will need an accurate

 C553/045 © IMechE 1998

representation of the shear force saturation behaviour of the tyres on a high friction surface; and we will need to simulate slalom manoeuvres at various pylon separations (to hit worst case conditions involving roll resonance). If it were to transpire that high speed reversing could provide an even greater challenge, we should have to simulate that too. Road profiles and/or braking to encourage suspension jacking may be a significant issue in some cases. It is clear that considerable expertise is needed to enable proper judgements to be made concerning the model and the operating conditions for each trial of many.

The need to follow a slalom course in this example raises another problem. A straightforward vehicle model requires a steering wheel input as one of its controls, but the input to cause the vehicle to follow a course will typically not be known. Finding the necessary input by trial and error is usually not practical, so that an effective driver model, enabling the "vehicle" to follow a defined path, or an inversion capability, so that the input required for a specified path response can be found, will be useful.

In an ideal world, the skill to make the proper judgements will be available; new models will be built quickly with much benefit from previous models; parameter demands of the models will be the minimum for the purpose in hand; and the manner in which the simulation results are to be interpreted, when they become available, will be known ahead of the results. Run times will be minimal within the context. Parameter values will be changeable easily, so that parametric studies can be done efficiently. Repetitive needs, such as for tyre force and moment descriptions and for driver models for path following, will be satisfied by "library" facilities or modules that can be joined to the vehicle models as needed. Simulation results will be observable in a manner that is friendly to the user. In the rollover test, for example, animated output or time histories of the vehicle roll angle would each indicate clearly the main result - did the vehicle roll over or not. More subtle aspects, like the relationship between pitching and rolling in a rollover, may not be so clear without animation. A very high degree of flexibility is apparently necessary. In order to benefit from flexibility, good choices must be made and good choices depend on skill.

5 MULTIBODY METHODS

Multibody software may be numerical only or symbolic/numerical. In the former case, the equations of motion and constraint are effectively a set of rules for the manipulation of matrix arrays of numbers representing successive states of the system through time, while in the latter, the equations themselves are derived symbolically and will normally be manipulated into some desirable form before they are embedded in a simulation program, which deals in numbers representing system states. The equations themselves can be inspected, although this may not be useful due to their complexity, but the manipulations may be very valuable, since the possibility arises of performing operations only once in the symbolic world which, in the numerical only world have to be done possibly several times at each time step. Multibody modellers may be complete with solvers, plotters and animators and they may have interfaces to other software systems. Design data may be importable from a CAD database; models may be exportable to control system design suites, etc. Alternatively, the multibody modeller may be a special module, whose total job is to produce the symbolic equations of motion for a system within a certain class of systems (13).

At the base of a multibody method is a dynamics formalism. The simplest and best known formalism is that requiring the individual bodies of a multibody system to be notionally separated out from each other, making each a free body, and associating with each body the appropriate inter-body forces of action and reaction. Constraint equations describe the inter-relationships between the coordinates and possibly velocities of the bodies when they are joined together to make the real multibody system. A spherical joint between two bodies will give rise to a vector constraint equation (three scalar equations) expressing the condition that the joint centre, as seen from each of the two bodies, has the same absolute displacement from the origin, for example. The equations of motion for the separate bodies contain the forces of interaction. Conceptually, the simplest approach to solving the problem is to give the whole set of differential and algebraic equations (DAE's) to a powerful and appropriate solver, which will find the motion and the interaction forces concurrently. It transpires that the equations are ill conditioned in this form (12) and, in vehicle problems using the obligatory fixed reference axis system, large global displacements combine with small local displacements to cause numerical difficulties, so that the burden on the solver is heavy. Simulations tend to be painfully slow (14, 15, 16), although research into solvers for DAE's continues (17).

There are many more equations than there are degrees of freedom for the whole system, so that equations can be manipulated to eliminate variables (dependent coordinates and inter-body forces) from the analysis until a minimal set of equations in the corresponding minimal coordinate set (generalised coordinates) is established. This is possible once and for all if the constraint equations can be solved analytically to give dependent coordinates and/or velocities (to be eliminated) in terms of independent coordinates and/or velocities (to be retained). A simple illustrative example, the double pendulum, is described in an Appendix. However, if the constraint equations cannot be solved analytically in this way, the above procedure can only be followed numerically, at each time step. A corresponding computational burden has to be accepted. This happens with closed kinematic loops. Then the displacement constraint relations are non-linear and they are not solvable analytically. However, they are linear in velocities when differentiated (15), so that the velocities can be separated into independent and dependent ones and the latter can be eliminated from the equations, see Fig. 1. If the equations in this form are simply integrated, the displacement level constraint equations will fail to be satisfied by virtue of drift, so the displacement equations need to be retained in some way to prevent this occurrence – see below.

Non-holonomic constraints involve non-integrable relationships between velocities. In vehicle dynamics they arise typically if wheels are assumed to roll without slip in problems of more than one dimension. Suppose a car is parked in an open, flat, high-friction area and radial line marks are appended to the tyre sidewalls and to the points on the ground nearest to them. The car is then driven slowly, without tyre slip, round the area, and eventually returned to the precise location from which it started. Although the car body can be re-positioned precisely, the tyre marks will not, in general, align now. It is fundamental to the problem that this is so and, of course, the lack of position gearing between the wheel spin and the car position is reflected in the mathematical description of the situation. An example is provided in Fig. 2. In vehicle dynamic analysis, it is usually necessary to treat the tyre/road contact as giving rise to shear forces that are functions of longitudinal slip, lateral slip, load and camber angle, as opposed to constraints on the motion, so non-holonomic constraints are a somewhat academic issue. It is often discussed at some length, however, because it has been very important in the development of multibody methods. In devising a general-purpose multibody analysis system, it is important to employ an unrestricted formalism, so that, whatever reasonable mechanical system is of interest, it can be treated properly. In such a case, the analyst does not need to be concerned with whether or not his/her system is holonomic.

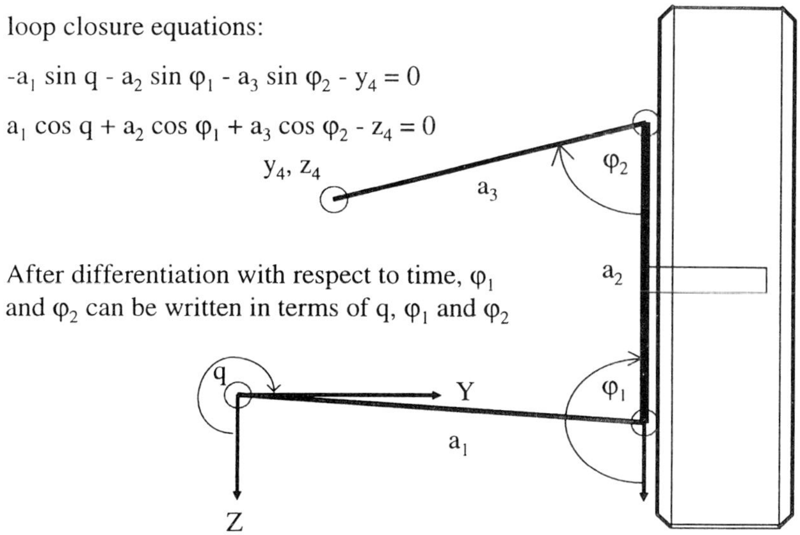

Fig. 1 A closed kinematic loop and the loop closure equations.

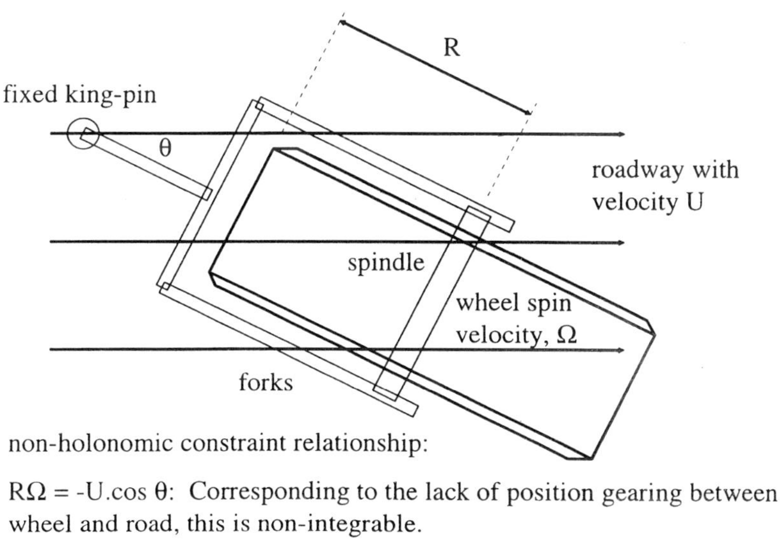

Fig. 2 Plan view of castoring wheel, illustrating a non-holonomic constraint due to rolling without longitudinal slip.

An alternative approach to the free body one is to aim directly at equations of lowest number in a minimal coordinate set before numerical integration is done. Rather than eliminating coordinates by using constraint equations, the possibility arises that coordinates will be introduced only if they are needed by virtue of the degrees of freedom existing. Also, a formulation of the equation of motion problem in which the internal forces do not have to be considered can be found. Such coordinates are generalised coordinates and the key to non-involvement with the internal forces is some form of energy approach. Best known in this class is Lagrange's second method, in which "inertia forces" are deduced by differentiation of kinetic energy expressions and generalised forces are derived from "virtual work" calculations. Internal forces of constraint need not be considered since they are non-working effects. However, this method is not suitable as a general basis for multibody analysis, since it works only for holonomic systems and the algebra (or numerical) implications are very unfavourable. The use of "virtual work" is a key step in Lagrange's method but it is this principle which prevents the method dealing with non-holonomic systems. If "virtual power" is used instead, the difficulty is removed, since virtual velocities remain independent when virtual displacements are not in non-holonomic cases.

Therefore, the most common basis for automated modellers is "virtual power", known also as Jourdain's principle (12) and embedded in Kane's equations (9). The last form has the advantage that the equations are derived through a highly structured process that lends itself to computer programming. With "virtual power", generalised speeds corresponding to the degrees of freedom are used (the minimal set), coordinates are independent of speeds, constraint forces are not considered and all types of system are accommodated. Moving reference axes can be employed, if the problem nature does not demand an inertial reference system. If it is required to calculate constraint forces, they have to be computed, after the motion solutions have been found, by a back substitution process (12). If friction forces dependent on constraint forces are included in the system description, special computational methods have to be used to deal with such circular dependency. Taking the constraint forces from the previous time step at the numerical integration stage is a simple option. The equations of motion come in first order form and they can be decoupled by inversion of the "mass" matrix, to yield the minimum complexity with respect to numerical integration (15).

6 THE SYMBOLIC ROUTE TO FAST SIMULATION

The symbolic approach to equation generation is more difficult than the numerical one and is very demanding of computer memory. Strategies for containing the equation swell problem as system size increases, together with astonishing advances in memory costs have controlled this aspect and a contemporary PC can be thought of as an appropriate device for symbolic multibody analysis. Symbolic preparation of the problem for numerical solution implies that the symbol manipulations, which are performed only once ahead of numerical processing, can be devoted to making the repeated, later numerical operations minimal. It also implies independence of the symbolic equation building and the numerical solution. To contain memory needs, expressions can be kept in factored form, trigonometric simplifications can be applied, intermediate variables and constants can be introduced and small quantities can be eliminated (15).

If closed kinematic loops are involved, the velocity form of the loop equations can be integrated to give an approximate updating of the state. This approximate solution can be used as the starting point for an iterative solution of the displacement level equations. The

 C553/045 © IMechE 1998

standard solution method is the Newton-Raphson process (19) but satisfactory convergence may be achieved more economically by only using the main diagonal elements of the Jacobian matrix, rather than the full Jacobian. The analyst may be able to choose which level of computation to employ. Which option is best will depend on the length of the simulation runs being done (20).

However, closed kinematic loops are most likely to arise in vehicle dynamics from suspension and steering systems and these are confined by design to move through only small geometrical regions. Suspension and steering limit stops define the limits of possible movements and singular configurations cannot be reached. Thus, in this context, there is no need to consider singular situations and there is virtually no loss of accuracy involved if the kinematics problems are solved off-line, the geometric solutions are expressed in analytic or table function form and the functions are used as constraint functions in the multibody model building, to avoid dealing with closed loops (18). Exemplary bump steer and camber change results from a kinematic suspension analysis are shown in Fig. 3, together with the coefficients of the polynomials that fit the results closely. Elastic effects due to suspension bush compliances can be dealt with in a similar way, in this case the off-line calculations being equilibrium ones rather than purely kinematic ones (18).

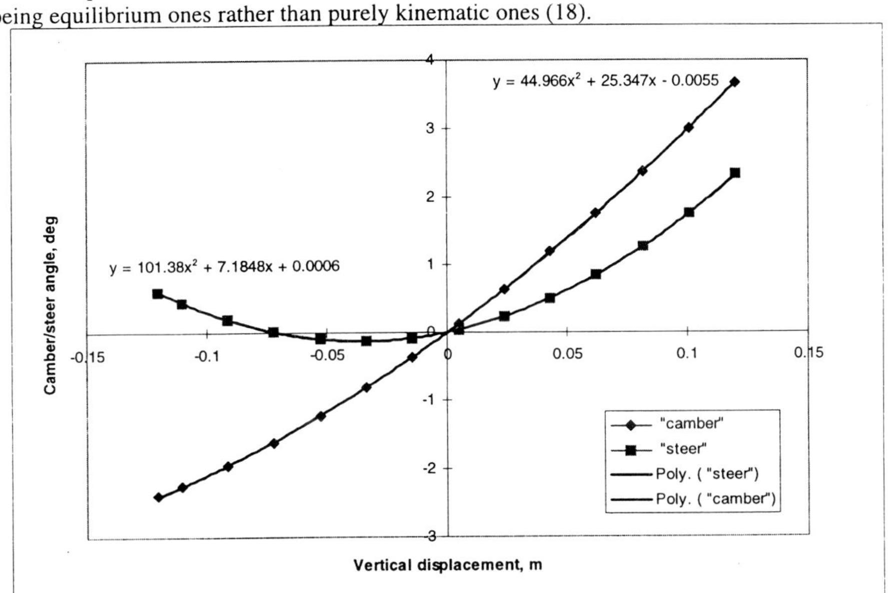

Fig. 3 Camber and steer angles as functions of vertical wheel travel from kinematic analysis; with fitted polynomials; (21)

Other sources of economy of simulation time are; re-distributing the mass of the real system, such that the total mass is correct and tyre forces are computed properly but small parts are considered massless, the system order is contained and extensive and intricate computations of terms which are too small to matter are avoided; small angles are treated using small angle theory; and small displacements of a powertrain, say, on its mounting system are declared small and treated by appropriate approximations; expressions formed but not used in the simulation stage are eliminated; and implicit integration of the equations for fast modes with only weak coupling to the remainder of the system, like wheel spin equations (22). Evidence of the enormous differences that can be made, through attention to these details, is provided in (15). The best systems will offer choices in these matters but the user will require knowledge, skill and deep thinking to take the best options.

7 COMMERCIAL SOFTWARE SYSTEMS

Important considerations with commercial systems are; the user interface – what is the modeller like to use; can one see the model appearance as it is built up (graphical user interface); how much learning is needed before one can achieve anything worthwhile; are the results convenient, understandable and attractive; can kinematics, equilibrium, linearisation and inverse solutions be obtained; can results be transported to other software for post-processing; can parametric data be imported from other software; - the reliability and the failure modes – is the system very sensitive to errors of detail; are error messages sensible or hieroglyphic; is the documentation good; is personal support available; is the software being used enough to become thoroughly debugged and to be continuously improving. A fuller treatment of these factors is given in (13).

8 CONCLUSIONS

Contemporary multibody dynamics software systems contain enormously impressive aids to productive model building and system simulation. Possession of such a facility, together with the skill to use it properly, transform the capability of an analyst in terms of productivity and cost effectiveness. In the early stages of the transformation process, of course, there are the costs of buying or writing the software and the by no means negligible costs of acquiring skill in its use. The gains are made afterwards, the more so as usage increases and skills are developed further by practice. For an organisation, there are considerable potential benefits in having a well-considered strategy for the development, storage and retrieval of models and there are benefits to the commercial world from students being exposed to the technology during their training. Nevertheless, there is a strong possibility that the majority of users of the software will have no real idea of what happens inside the "black box" and that their effective use of the tools will be prejudiced by this lack of base understanding.

Building efficient models seems to depend on appreciating the purpose, accurately associating design features with behavioural properties in a general sense and recognising to some extent how the software works – what makes for easy model building, easily solved equations, robust numerical processes etc. Those parts of vehicle dynamics analysis which are repetitive, like including tyre shear force descriptions, are best done with a modular strategy, so that old work does not have to be redone any more than necessary. However, analysts do have to keep refreshing their memories of the contents and workings of the modules. Otherwise, they lose track of what problem it is which is being solved.

The market dominant systems are first generation numerical modellers. They are completely self-contained as far as simulation is concerned and offer plotting and animation output options. Prepared modules are also available for tyre forces, common suspension types, driver steering control etc. and interfaces to other software to ease operations like parametric data entry, statistical variation and simulation schedule design, control system design and output signal processing exist. These systems are expensive and quite difficult to learn but, in the automotive industry, they provide a commonality of thought and data, making

C553/045 © IMechE 1998

for relative ease of communication and collaboration between suppliers and assemblers. The multibody formalism relied on makes the construction of the system equations relatively simple and places the main burden in the whole problem on the equation solvers. Numerical problems in the solvers are severe and much research has been stimulated in recent years to improve speed and robustness. Notwithstanding this research, it remains the case that these simulations are usually very much slower than those from the "hand-written" era and real time simulation is possible for only the very simplest of systems, for which multibody methods do not have much advantage anyway.

Second generation multibody modellers generally employ a formalism directed towards a minimal equation set in a minimum number of generalised coordinates. They are symbolic, moving to numerics only when the problem becomes specific through the entry of parametric data. The equations of motion are better conditioned in this case and symbolic operation allows preparation of the equations for solution in the optimum form by often intensive symbolic pre-processing, which is justified totally because it only has to be done once per analysis. The best of these codes have shown that they can yield simulations that are significantly faster than manual methods give and surprisingly elaborate (full) vehicle models can be run in real time. These systems are less commercially developed than the first generation systems and the multibody modeller usually will prepare the problem for some independent simulation system, like FORTRAN, C, ACSL or SIMULINK to do the solving. The quality of the whole operation depends on the synergistic and trouble free matching of the separate systems to each other. With the standardisation of data formats in place today, interfacing problems which used to be common are not expected to recur much in the future.

REFERENCES

(1) Segel, L., Theoretical prediction and experimental substantiation of the responses of the automobile to steering control, in Research in Automobile Stability and Control and in Tyre Performance, Proc. I. Mech. E. (A. D.) 1956-7, Number 7, 310-330.

(2) McHenry, R. R., An analysis of the dynamics of automobiles during simultaneous cornering and ride motions, in Handling of Vehicles under Emergency Conditions, I. Mech. E., London, 1969, 28-48.

(3) Orlandea, N., Chace, M. A. and Calahan, D. A., A sparsity oriented approach to the dynamic analysis and design of mechanical systems, Parts 1 and 11, Journal of Engineering for Industry, **99**, 1977, 773-784.

(4) Schiehlen, W. and Kreuzer, E., Rechnergestützus Aufstellen der Bewegungsgleichungen gewöhnlicher Mehrkörpersysteme, Ingenieur-Archiv, **46**, 1977, 185-194.

(5) Levinson, D. A., Equations of motion for multi-rigid-body systems via symbolic manipulation, Journal of Spacecraft and Rockets, **14**, 1977, 479-487.

(6) Duffek, W., Führer, C., Schwartz, W. and Wallrapp, O., Analysis and simulation of rail and road vehicles with the program MEDYNA, in The Dynamics of Vehicles on roads and on tracks (O. Nordström ed.), Swets and Zeitlinger, Lisse, 1986, 71-85.

(7) Rosenthal, D. E. and Sherman, M. A., Symbolic multibody equations via Kane's method, AAS-Paper 83-303, AAS Publ. Office, San Diego, 1983.

(8) Wittenburg, J. and Wolz, U., MESA VERDE: A symbolic program for non-linear articulated-rigid-body dynamics, Proc. 10th ASME Design and Engineering Conf. Mechanical Vibration and Noise, Cincinnati, 1985.

(9) Kane, T. R., and Levinson, D. A., Dynamics: Theory and Applications, McGraw-Hill,

New York, 1985.
(10) Schiehlen, W. (ed.), Multibody System Handbook, Springer-Verlag, Berlin, 1990.
(11) Kortüm, W. and Sharp, R. S. (eds), Multibody Computer Codes in Vehicle System Dynamics, Swets and Zeitlinger, Amsterdam, 1993.
(12) Schiehlen, W., Multibody System Dynamics: Roots and Perspectives, Multibody System Dynamics, **1**(2), 1997, 149-188.
(13) Kortüm, W., Schiehlen, W. and Hoffmann, M., Progress in integrated system analysis and design software for controlled vehicles, in The Dynamics of Vehicles on roads and on tracks (Zhiyun Shen ed.), Swets and Zeitlinger, Lisse, 1994, 274-296.
(14) Sayers, M. W., Symbolic computer language for multibody systems, Journal of Guidance and Control, **14**(6), 1991, 1153-1250.
(15) Sayers, M. W., Automated formulation of efficient vehicle simulation codes by symbolic computation (AUTOSIM), in The Dynamics of Vehicles on roads and on tracks (R. Anderson ed.), Swets and Zeitlinger, Amsterdam / Lisse, 1990, 474-487.
(16) Sayers, M. W., Computer modeling and simulation of vehicle dynamics, UMTRI Research Review, **23**(1), 1992, 1-15.
(17) Hairer, E. and Wanner, G., Solving Ordinary Differential Equations *II*. Stiff and Differential-Algebraic Problems, Springer-Verlag, Berlin, 1991.
(18) Mousseau, C. W., Sayers, M. W. and Fagan, D. J., Symbolic quasi-static and dynamic analyses of automobile models, in The Dynamics of Vehicles on roads and on tracks (G. Sauvage ed.), Swets and Zeitlinger, Amsterdam / Lisse, 1992, 446-459.
(19) Paul, B., Kinematics and Dynamics of Planar Machinery, Prentice-Hall, Englewood Cliffs (NJ), 1979.
(20) AutoSim™ Reference Manual, Version 2.0, 1993.
(21) Sharp, R. S., Use of the symbolic modelling code AutoSim for vehicle dynamics, in Automotive Vehicle Technologies, AUTOTECH '97, Mech. Eng. Publ., Bury St Edmunds, 1997-7, 189-197.
(22) Rill, G., Vehicle dynamics in real-time simulation, in The Dynamics of Vchicles on roads and on tracks (M. Apetaur ed.), Swets and Zeitlinger, Lisse, 1988, 337-347.

C553/045 © IMechE 1998

APPENDIX

A multibody system analysis is illustrated by one of the simplest possible cases - the planar double pendulum, Fig. A1

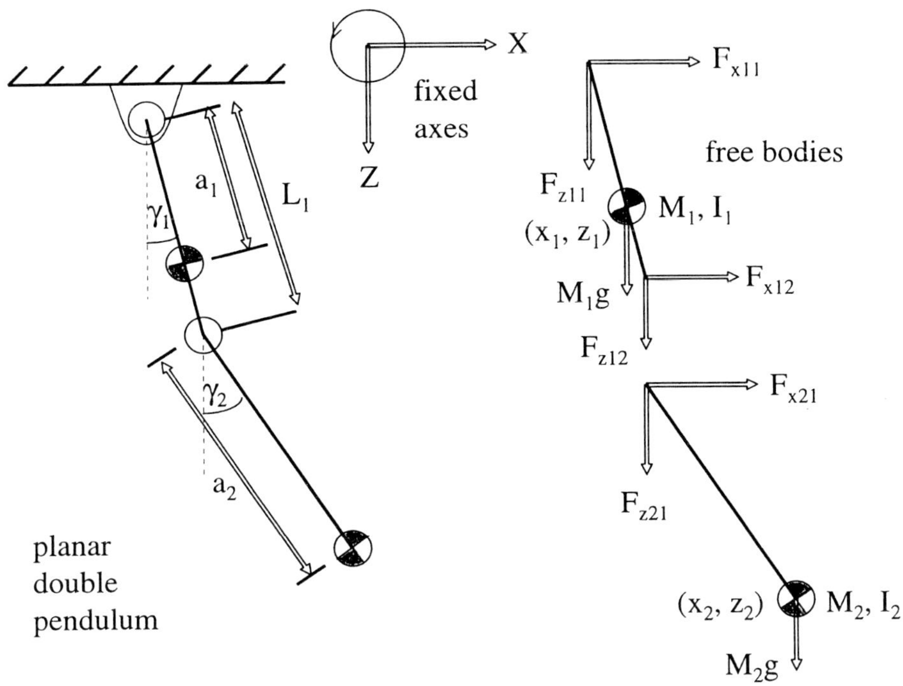

Fig. A1 The planar double pendulum.

Newton-Euler equations:

$M_1\ddot{x}_1 = F_{x11} + F_{x12};\qquad M_1\ddot{z}_1 = F_{z11} + F_{z12} + M_1 g;$

$I_1\ddot{\gamma}_1 = a_1(-F_{x11}\cos\gamma_1 + F_{z11}\sin\gamma_1) + ((L_1 - a_1)(-F_{z12}\sin\gamma_1 + F_{x12}\cos\gamma_1)$

$M_2\ddot{x}_2 = F_{x21};\qquad M_2\ddot{z}_2 = F_{z21} + M_2 g;\qquad I_2\ddot{\gamma}_2 = a_2(-F_{x21}\cos\gamma_2 + F_{z21}\sin\gamma_2)$

action/reaction:

$F_{x12} + F_{x21} = 0;\qquad F_{z12} + F_{z21} = 0;$

position constraints:

$x_1 - a_1\sin\gamma_1 = 0;\qquad z_1 + a_1(1-\cos\gamma_1) = 0;\qquad x_2 - a_2\sin\gamma_2 = x_1 + (L_1 - a_1)\sin\gamma_1;$

$z_2 + a_2(1-\cos\gamma_2) = z_1 + (a_1 - L_1)(1-\cos\gamma_1)$

making 6 second order differential equations and 6 equations of constraint in the 12 variables; x_1, z_1, γ_1, x_2, z_2, γ_2, F_{z11}, F_{z11}, F_{x12}, F_{x21}, F_{z12} and F_{z21}.

The DAE approach to the solution of these equations is to simply give them to the solver and let it find the values of all the variables in the problem through time. Alternatively, the internal forces and the redundant coordinates can be eliminated by combining equations, to yield - using the shorthand notation S_1=sin γ_1; C_1=cos γ_1; S_2=sin γ_2; C_2=cos γ_2 and reducing the equations to state space form:

$$\{I_2 + I_1 + M_1 a_1^2 + M_2((a_2 + L_1 C_2)^2 + (L_1 S_2)^2)\}\dot{\Gamma}_1 + \{I_2 + a_2 M_2(a_2 + L_1 C_2)\}\dot{\Gamma}_2 =$$
$$-(g a_1 M_1 C_1 + M_2(g C_2(a_2 + L_1 C_2) S_1 + S_2(g a_2 C_1 + L_1(-a_2 \Gamma_2(2\Gamma_1 + \Gamma_2) + g S_1 S_2))))$$
$$\dot{\gamma}_1 = \Gamma_1$$
$$\{I_2 + a_2 M_2(a_2 + L_1 C_2)\}\dot{\Gamma}_1 + \{I_2 + M_2 a_2^2\}\dot{\Gamma}_2 = -a_2 M_2(g C_2 S_1 + (L_1 \Gamma_1^2 + g C_1) S_2)$$
$$\dot{\gamma}_2 = \Gamma_2$$

This is the form in which Lagrange's second method, virtual work, virtual power or Kane's equations yield the results. They are still not optimal for simulation, since the "mass" matrix above would need inverting at each time step - it can be seen to contain the time varying sines and cosines - and this will be extravagant. Symbolic inversion and pre-multiplication by the result gives the preferred form:

$$\dot{\Gamma}_2 = -(a_2 M_2(g C_2 S_1 + (L_1 \Gamma_1^2 + g C_1) S_2) - (I_2 + a_2 M_2(a_2 + L_1 C_2))(g a_1 M_1 C_1 + M_2(g C_2(a_2 + L_1 C_2) S_1$$
$$+ S_2(g a_2 C_1 + L_1(-a_2 \Gamma_2(2\Gamma_1 + \Gamma_2) + g S_1 S_2))))/(I_2 + I_1 + M_1 a_1^2 + M_2((a_2 + L_1 C_2)^2 + (L_1 S_2)^2)))/$$
$$(I_2 + M_2 a_2^2 - (I_2 + a_2 M_2(a_2 + L_1 C_2))^2/(I_2 + I_1 + M_1 a_1^2 + M_2((a_2 + L_1 C_2)^2 + (L_1 S_2)^2)))$$
$$\dot{\gamma}_2 = \Gamma_2$$
$$\dot{\Gamma}_1 = -(g a_1 M_1 C_1 + \dot{\Gamma}_2(I_2 + a_2 M_2(a_2 + L_1 C_2)) + M_2(g C_2(a_2 + L_1 C_2) S_1 + S_2(g a_2 C_1 + L_1(-a_2 \Gamma_2$$
$$(2\Gamma_1 + \Gamma_2) + g S_1 S_2)))) / (I_2 + I_1 + M_1 a_1^2 + M_2((a_2 + L_1 C_2)^2 + (L_1 S_2)^2))$$
$$\dot{\gamma}_1 = \Gamma_1$$

It should be easy to appreciate that the algebra involved in manipulating even these simple system equations is formidable and the results shown here have been generated using AutoSim. If they had been derived by hand, one should doubt the wisdom of the analyst and the accuracy of the results too. In the AutoSim program, the precomputed constants and the repeated variables facilities were switched off, in order to remain in the base variables which should be recognisable. To make a fast simulation program, these facilities would be switched back on. With this done, the simulation program evaluates the left hand side rates of change by 31 multiply/divide operations, 18 add/subtracts and 4 function calls, to evaluate the sines and cosines. This should give some idea of the benefit to be obtained from effective use of relatively simple software engineering methods.

C553/045 © IMechE 1998

C553/047/98

Vehicle dynamics and control for improving handling and active safety: from 4WS to DYC

M ABE
Kanagawa Institute of Technology, Japan

Recent trends of vehicle chassis control are reviewed, paying attention to the basic nature of the vehicle dynamics, tire characteristics and the effectiveness and limit of 4WS and DYC. It is pointed out that taking the non-linear characteristics of tire and vehicle dynamics into consideration, introducing the chassis control from 4WS to DYC is inevitable for improving handling performance and active safety in the vehicle motion with larger side-slip angle and/or higher lateral acceleration. Emphasis is placed on the need to propose a control strategy based on thorough observation and understanding of the tire and vehicle dynamics. A side-slip model following DYC by sliding control is introduced and the effects of the control are discussed.

1 INTRODUCTION

The vehicle chassis control is in general to control vehicle lateral, vertical and longitudinal motions in order to improve handling performance, ride comfort and traction/braking performance. The main interest here is the chassis control for improving the vehicle handling performance and active safety.

The most common chassis control for the improvement of handling performance is four wheel steering (4WS). The 4WS control depends on tire lateral force which is proportional to the steer angle, a control command, in a range where the lateral acceleration is small. In this range, the control law can rather easily be introduced by using two degrée of freedom linear vehicle model. However, in higher lateral acceleration, the lateral force is not necessary proportional to the steer angle due to its saturation property to side-slip angle. The lateral force also becomes strongly depend on tire vertical load and longitudinal force as well, so that the control law becomes rather sensitive to the vehicle motion and its environmental conditions.

A direct yaw moment control (DYC) controlling vehicle motion by a yaw moment which is actively generated by the intentional distribution of the tire longitudinal forces is becoming

© IMechE 1998 C553/047

one of the most promising means of chassis control. One of the major advantages of this control method is that the tire longitudinal force has no feed-back from the vehicle lateral motion as long as it is within the limit of the tire capacity due to the vertical load. Therefore, a precise yaw moment, required to control the vehicle lateral motion, can be developed and the vehicle motion with the control becomes more robust to fluctuation in the vehicle running condition and its environment.

It is obvious that any chassis control aiming at improving vehicle handling performance must rely on the tire lateral force and that the tire longitudinal force can also be used for the chassis control. These two forces each rely upon the vertical load and are interdependent. This relationship makes the vehicle dynamic characteristics very complex and a comprehensive study is required to develop more effective chassis control strategies.

2 NOMENCLATURE

m: vehicle mass
I: yaw moment of inertia
$l_{f,r}$: distance between mass center and axle
l: wheel base
$K_{f,r}$: tire cornering power
u, V: vehicle forward speed
v: vehicle side-slip velocity
β: side-slip angle
r: yaw rate
δ: steering angle
$\beta_{f,r}$: tire side-slip angle
$Y_{f,r}$: axle cornering force
$X_{f,r}$: tire longitudinal force
$W_{f,r}$: tire vertical load
M_z: control yaw moment
μ: friction coefficient
s: Laplace transform variable
$_{f,r}$: front and rear

3 LINEAR VEHICLE DYNAMICS AND CLASSICAL 4WS

The vehicle dynamics of an automobile is a typical example of complicated multi-body dynamics. However, the basic nature of the lateral dynamics is described by the following equations of motion for two degree of freedom vehicle plane model:

$$m(\dot{v} + ur) = Y_f + Y_r$$
$$I\dot{r} = l_f Y_f - l_r Y_r \qquad (1)$$

If the tire characteristic is linear, in other words, the tire lateral forces are proportional to side-slip angles, the following are derived as the basic equations for the vehicle lateral motion with

C553/047 © IMechE 1998

4WS:

$$m V \dot{\beta}+2(K_f+K_r)\beta+\left\{m V+\frac{2}{V}(l_f K_f-l_r K_r)\right\} r=2 K_f \delta_f+2 K_r \delta_r$$

$$2(l_f K_f-l_r K_r)\beta+I \dot{r}+\frac{2(l_f{}^2 K_f+l_r{}^2 K_r)}{V}=2 l_f K_f \delta_f-2 l_r K_r \delta_r \tag{2}$$

From the beginning of the 1980s, 4WS has been regarded as a promising tool to improve the vehicle handling and has been installed in some passenger cars in the market. Even so, to date there has been no general consensus on what type of control law is suitable for the 4WS. However, both zero side-slip controls by feed-forward control and combined feed-forward and feed-back control have been commonly adopted as a control law for the initial stage of 4WS.

It is very easy to introduce from Eq.(2) the following proportional relation of rear wheel steer angle to front wheel one as a control law to get the steady state side slip angle equal to zero.

$$\delta_r=-\frac{l_r-\dfrac{m l_f}{2 l K_r} V^2}{l_f+\dfrac{m l_r}{2 l K_f} V^2} \delta_f \tag{3}$$

Putting a transfer function of rear wheel steer angle to front wheel angle into Eq.(2) and letting the side-slip angle constantly equal to zero, the following transfer function arises, which gives not only steady state side-slip angle but also transient response of side-slip angle equal to zero:

$$\frac{\delta_r}{\delta_f}(s)=-\frac{l_r-\dfrac{m l_f}{2 l K_r} V^2+\dfrac{I V}{2 l K_r} s}{l_f+\dfrac{m l_r}{2 l K_f} V^2+\dfrac{I V}{2 l K_f} s} \tag{4}$$

This is a feed-forward type of zero side-slip 4WS control.

Steering a rear wheel proportionally to both front wheel steer angle and yaw rate can bring about the constant zero side-slip response. It is easy to introduce the control law as follows by putting this proportional relation into Eq.(2):

$$\delta_r=-\frac{K_f}{K_r} \delta_f+\frac{m V^2+2(l_f K_f-l_r K_r)}{2 K_r V} r \tag{5}$$

This is a combined feed-forward and feed-back type of zero side-slip 4WS control.

It is found from Eq.(2) that if not only rear wheel steer angle but also additional front wheel one is actively controlled, it is possible to obtain the arbitrary response of vehicle side-slip and yaw rate by the active steering controls. This is because there are two manipulated variables, while the vehicle motion determined by the two state variables. The objective of the active control algorithm is to get a desirable vehicle response to steering wheel input. Therefore a model response to steering wheel input is often given and the control to follow the model response is introduced for the active steer control by using a pole assignment method, LQ control, etc.

4 SOPHISTICATED 4WS CONTROL

As shown in the above discussion, the control law for the 4WS can easily be introduced by using two degree of freedom vehicle plane model within a range of linear dynamics. However, it is likely that the intended control performance may not be obtained under cornering conditions where vehicle lateral acceleration is high. In this case the tire characteristics enter the non-linear region and thus the linear control theory become inappropriate. The tire lateral force is also strongly dependent on the tire vertical load as well as the longitudinal force, thus the vehicle dynamic parameters will be subject to change from the nominal value due to such changes as running conditions.

In order to negotiate with these parameter variations and/or non-linear aspects on the dynamics of the controlled system, in general, there is an issue as to which is appropriate; robust control or adaptive control? In view of these variable dynamic parameters researchers on 4WS control are now trying to further study the control law considering such theories as adaptive control and robust control.

In the adaptive control system, it is necessary to have a process for identifying a real time vehicle response parameters but there is a problem that where steering input from the driver is very small, the accuracy of the identification will be deteriorated. Also, the theoretical treatment of a stability condition for designing an adaptive control system is complicated and thus it has a disadvantage in that it is rather difficult to design a control system. Based on this view, there are many studies being conducted on techniques to design a robust control system assuming from the beginning that there will be changes in the parameters, instead of trying strictly to identify the errors in the vehicle dynamic parameters.

Horiuchi, et al.[1] made 4WS control configuration of two degree of freedom control system basing on H-infinity control theory. They designed a control system intended to have the lateral acceleration and yaw rate response independently and also to improve the system robustness against the changes of vehicle dynamics and external disturbances. They reported that with computer simulation and by testing an experimental vehicle, they were able to obtain the lateral acceleration and yaw rate response which they have sought , as shown in **Fig.1**. Hirano, et al.[2] proposed an application of a robust control theory to 4WS control. This theory is an adaptation of H-infinity control theory and mu-synthesis in relation to the changes in cornering stiffness caused by the tire non-linear characteristics. Simulated results of the effects of the control are shown in **Fig.2**.

A sliding control is regarded as one of robust control. There are research works applying sliding controls to 4WS. Ro, et al.[3] reported that they have obtained a good robust performance by adopting a sliding mode control for 4WS. Kaminaga[4] proposed a method of sliding mode control with some adaptive concept by called "intelligent sliding surface" to improve the critical cornering property when the tire side-slip angle becomes large while maintaining 4WS vehicle stable by restricting the tire side-slip.

Although these robust control systems are useful when the changes in the tire cornering power is not significant, it is conceivable that it may face a running condition in which robust

C553/047 © IMechE 1998

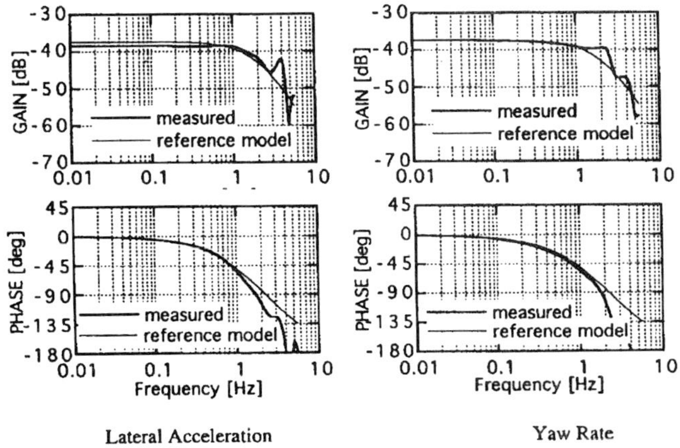

Fig.1 Results of two degree of freedom/H infinity controller synthesis[1]

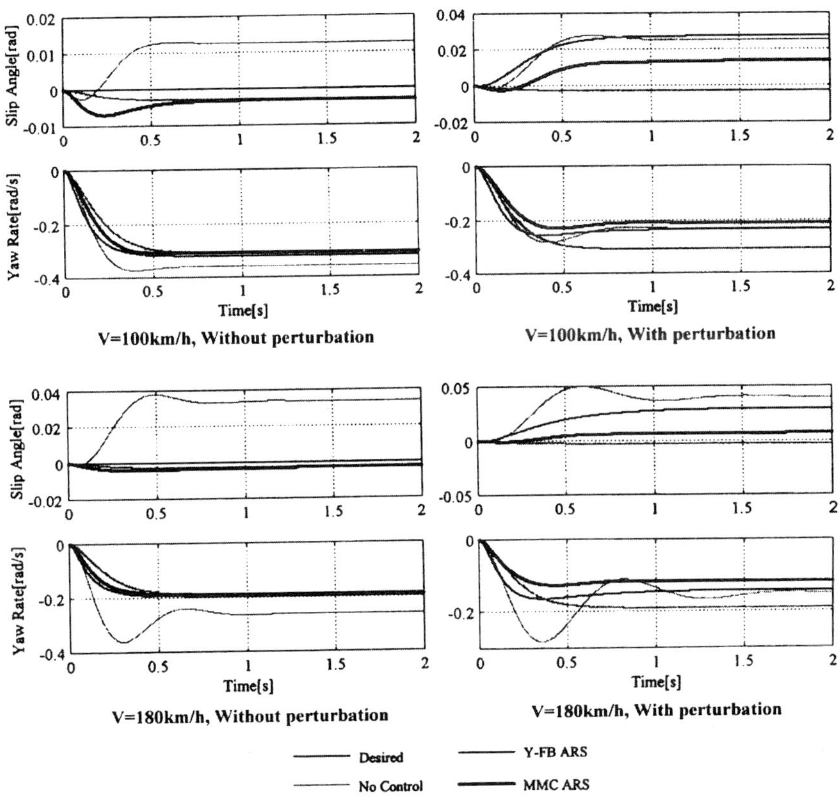

Fig.2 Effects of robust active rear steering control[2]

control theory can not be sufficient to absorb conditions encountered when the tire cornering power is extremely reduced. Especially when the tire force required exceeds the maximum frictional force and the cornering power indicates a negative value, it is conceivable then that the control system will aggravate the instability.

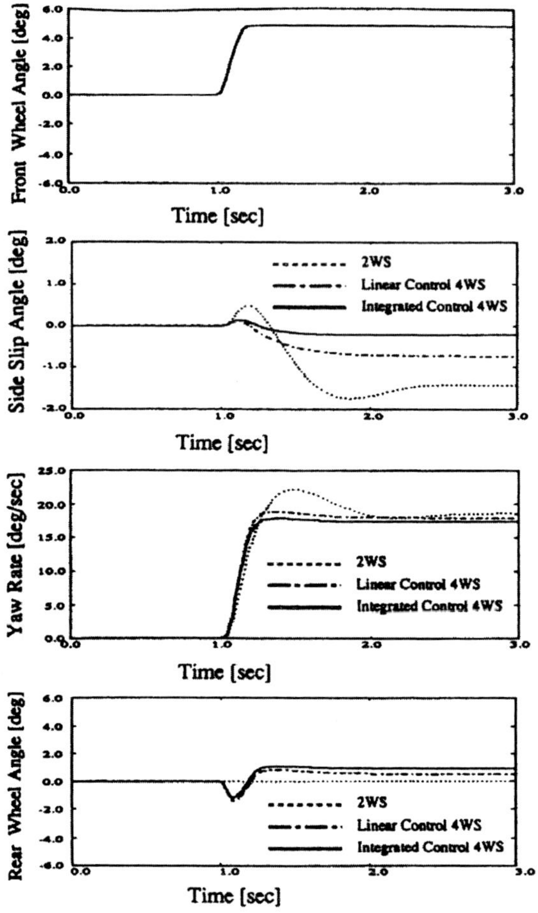

Fig.3 Effects of adaptive 4WS using neural network[5]

There is a concept of adaptation of 4WS control to non-linear characteristics of vehicle dynamics in which neural network theory is used to identify the non-linear vehicle dynamics [5]. The effect of the adaptive 4WS is shown in **Fig.3**. The method is used to compose the non-linear 4WS control to compensate the linear control law for the side-slip error due to non-

C553/047 © IMechE 1998

linearity. A primary advantage of using a neural network technique for the adaptive control is that basically it does not need any theoretical stability condition for adaptive identification and control. The effect of the self tuning adaptive 4WS control to adapt to the parameter variations are shown in [6].

The saturation property of a tire lateral force to side-slip angle, which is a main cause of the non-linear characteristics of a vehicle motion, depends primarily on the friction coefficient between tire and road surface. Wakamatsu, et al.[7] proposed the least square estimation of the friction coefficient and a 4WS control method of yaw rate feed-back by an internal model control structure with two degree of freedom. The estimation technique of the friction coefficient is to use the least square estimation of the actual yaw rate ratio to the yaw rate calculated by a linear model, depending on the fact that yaw rate step response is changed by the friction coefficient. Good yaw rate and body side-slip response of the vehicle with this control method was obtained as shown in **Fig.4**. As any 4WS control depends upon tire lateral force, it has some limit due to the tire saturation property of lateral force to side-slip-angle, even though some sophisticated adaptation to it is taken into consideration in the control law.

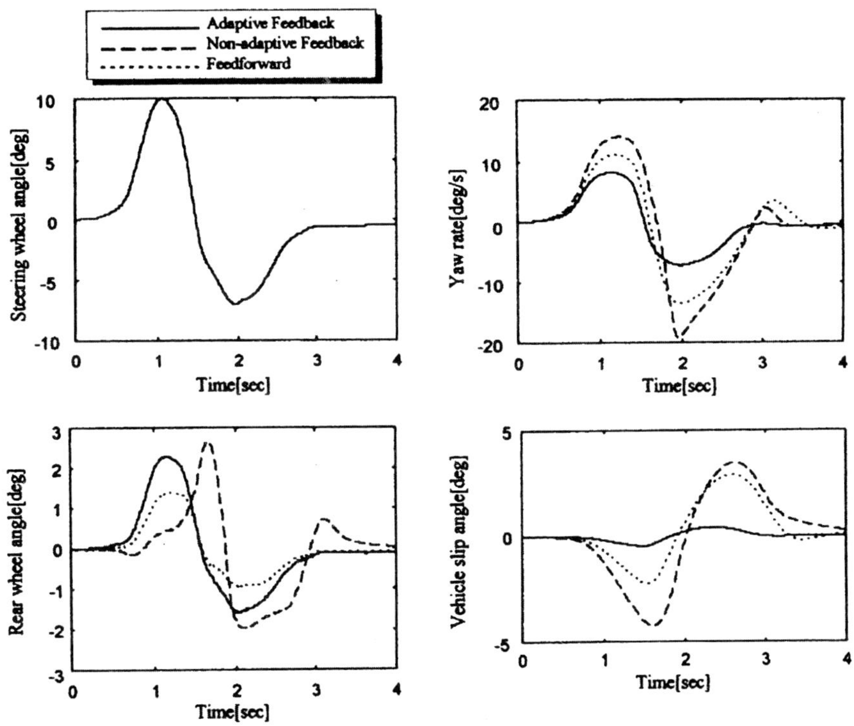

Simulation results of time responses
(v=30m/s, μ=0.4)

Fig.4 Effects of adaptive yaw rate feedback 4WS[7]

5 NON-LINEAR CHARACTERISTICS OF VEHICLE LATERAL DYNAMICS

From Eq.(2) the term $(l_rK_r\text{-}l_fK_f)$ plays an important part of a cross coupling of side-slip and yaw motions. This is a yaw moment caused by unit vehicle side-slip angle. When this value is positive, side-slip motion yields a restoring yaw moment and reduces the side-slip angle, on the other hand, when it is negative, the yaw moment results in increase of the side-slip angle. The former is termed Understeer and the latter is Oversteer. In order to extend this idea to non-linear region, Shibahata, et al.[8] proposed to calculate $(l_rY_r\text{-}l_fY_f)$ against vehicle side-slip angle and discussed how the yaw moment generated by the lateral force of vehicle front and rear wheels changes in response to the vehicle side-slip angle. **Figure 5** illustrates an example of the phenomenon.

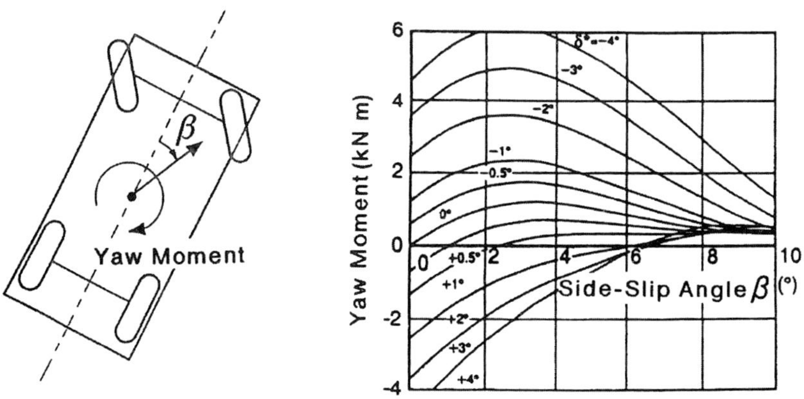

Fig.5 Yaw moment generated by tire lateral forces against vehicle side-slip angle[8]

One of the curved lines in this figure describes a relationship of yaw moment generated by tire lateral force against the vehicle side-slip angle when the difference in the side-slip angles between the front and rear wheels is constant. The difference of the side-slip angles between the front and rear wheels can be determined by the front wheel steer angle and yaw rate. From this curved line, it is found that when the side-slip angle is small, the yaw moment is proportional to the side-slip angle; that is, the yaw moment to reduce the vehicle side-slip is generated but the yaw moment will be saturated and starts to decrease at a certain point. This is caused by the fact that the rear tire lateral force will be saturated due to the increase of the side-slip angle, regardless of a similar characteristics of the front tires. Alternatively, increase in the vehicle side-slip angle causes a reduction of the cornering stiffness of the rear tire and restoring yaw moment, which recursively increases the side-slip angle. Because of this characteristic, a vehicle might execute an unstable motion, called "spin motion", when the vehicle is in a large side-slip motion.

A more precise discussion of the vehicle stability in non-linear region is possible in the following manner. Equation(1) can be rewritten for a small perturbation around an equilibrium point as follows:

 C553/047 © IMechE 1998

$$mV\dot{\beta} + \left(\frac{\partial Y_f}{\partial \beta_f} + \frac{\partial Y_r}{\partial \beta_r}\right)\beta + \left(mV + \frac{l_f \frac{\partial Y_f}{\partial \beta_f} - l_r \frac{\partial Y_r}{\partial \beta_r}}{V}\right) r = \frac{\partial Y_f}{\partial \beta_f}\delta_f$$

$$\left(l_f \frac{\partial Y_f}{\partial \beta_f} - l_r \frac{\partial Y_r}{\partial \beta_r}\right)\beta + I\dot{r} + \frac{\left(l_f^{\,2} \frac{\partial Y_f}{\partial \beta_f} + l_r^{\,2} \frac{\partial Y_r}{\partial \beta_r}\right)}{V} r = l_f \frac{\partial Y_f}{\partial \beta_f}\delta_f \qquad (6)$$

According to the stability theorem of non-linear systems, stability is determined by the roots of the characteristic equation with respect to the singular point. The characteristic equation of the system described by Eq(6) is:

$$s^2 + Bs + A = 0 \qquad (7)$$

with

$$A = \frac{\frac{\partial Y_f}{\partial \beta_f}\frac{\partial Y_r}{\partial \beta_r} l^2}{mIV^2} - \frac{2\left(l_f \frac{\partial Y_f}{\partial \beta_f} - l_r \frac{\partial Y_r}{\partial \beta_r}\right)}{I}$$

$$B = \frac{m\left(l_f^{\,2} \frac{\partial Y_f}{\partial \beta_f} + l_r^{\,2} \frac{\partial Y_r}{\partial \beta_r}\right) + I\left(\frac{\partial Y_f}{\partial \beta_f} + \frac{\partial Y_r}{\partial \beta_r}\right)}{mIV} \qquad (8)$$

where $\partial Y_f / \partial \beta_f$, $\partial Y_r / \partial \beta_r$ represent the slope of the tire lateral force against side-slip angle at the equilibrium point considered. These coefficients represent the equivalent cornering power at the front and rear respectively.

For the motion to be stable, A and B must be positive. For $A < 0$ the following instability condition arises:

$$\frac{\partial Y_r}{\partial \beta_r} < \frac{m l_f V^2 \frac{\partial Y_f}{\partial \beta_f}}{m l_r V^2 + l^2 \frac{\partial Y_f}{\partial \beta_f}} \qquad (9)$$

It is interesting to find from Eq.(9) that the non-linear vehicle motion tends to unstable motion predominantly by the decrease of the equivalent cornering power of the rear tire with increase of the side-slip angle. This corresponds with the reduction of restoring yaw moment due to non-linear characteristics of rear tire with increase of the vehicle side-slip angle which is pointed out by Shibahata[8].

6 SIDE-SLIP CONTROL BY DYC

From the above discussion, controlling vehicle side-slip with additional active yaw moment results in more stable vehicle motion under near limit maneuvering. A yaw moment which is actively generated by the intentional distribution of the tire longitudinal forces has been regarded as a promising yaw moment for the control.

6.1 Recent DYC controlling side-slip angle

Ono el al.[9] proposed a method for designing a control law of DYC with H-infinity control to stabilize the motion by allowing such a variation of the equivalent cornering power, even within the range of the limitation where the vehicle equilibrium point is unstable. The effect of such a control for improving the vehicle performance is shown in **Fig.6**.

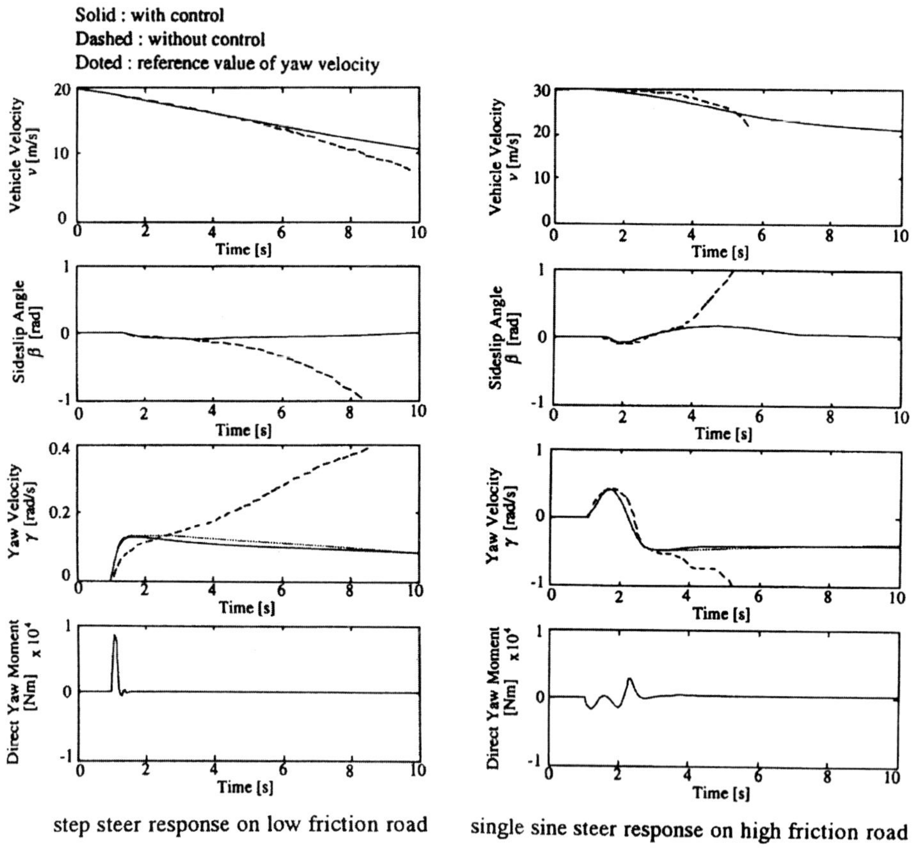

Fig.6 Effects of DYC with H-infinity control[9]

 C553/047 © IMechE 1998

Zanten, et al.[10] reported by citing Shibahata's paper that since the instability of vehicle motion at its limit is caused by the decrease of restoring yaw moment at a large side-slip angle, it is necessary to control the side-slip along with the control of yaw motion for maintaining the stability and thus they introduced a control law for controlling the right and left brake forces. Alberti, et al.[11] also proposed a control strategy for the controlling of side-slip rate in addition to the yaw rate. That is to say, if the product of side-slip angle and the side-slip

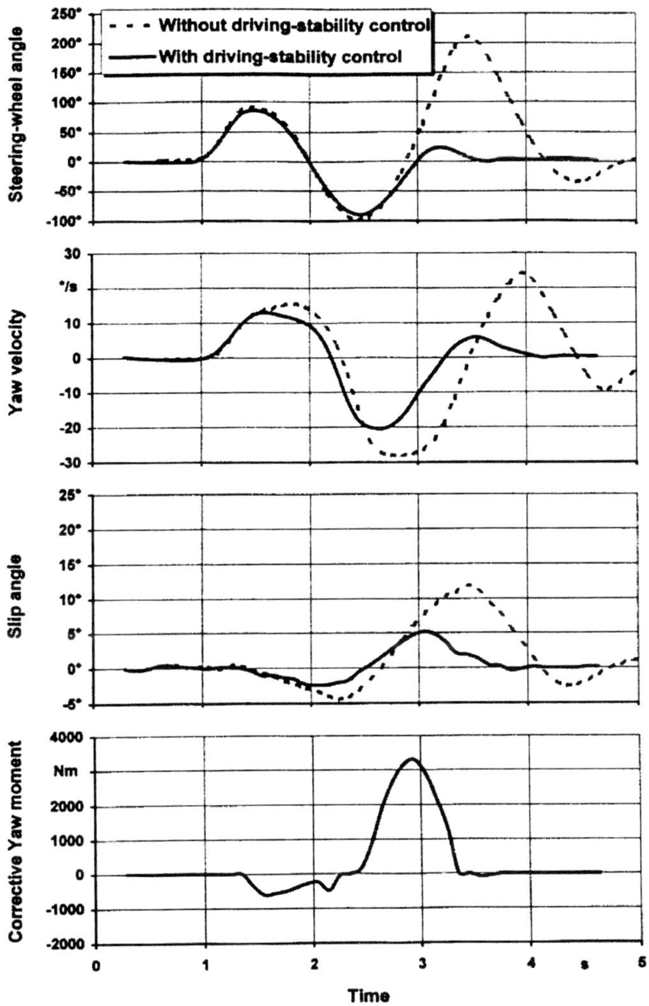

Lane-change tests with steering correction ($v \approx 20$ m/s, $\mu \approx 0.4$)

Fig.7 Effect of the driving stability control[11]

rate is negative, there will be no need for the control but if it is positive, then a yaw moment, generated by the brake forces, will be needed to control the vehicle motion until it becomes negative. Thus control of the side-slip rate is employed to minimize the absolute value of side-slip. The effect of this control is shown in **Fig.7**. Inagaki et al.[12],[13] further analyzed the stability of vehicle motion by phase-plane of side-slip angle and successfully proposed a control law of vehicle motion depending on yaw moment by brake force so that the state of vehicle motion will be within a following condition:

$$|C_1\dot{\beta} + C_2\beta| < 1 \qquad (C_1 > 0, \quad C_2 > 0) \tag{10}$$

The effect of this control is shown in **Fig.8**.

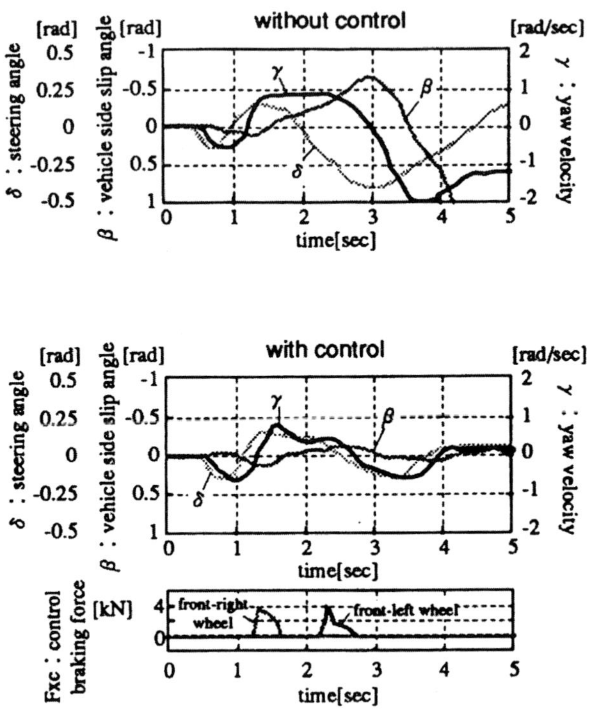

Effect of the control in the pyron slalom task

Fig.8 Effect of the vehicle-stability-control[12]

6.2 Side-slip model following DYC

In these DYC proposed in the above mentioned studies, it is inevitably necessary to detect the side-slip angle, however, there is no requirement to strictly follow a given side-slip motion in any control law of the DYC. A model following DYC by a sliding control is now proposed. A

 C553/047 © IMechE 1998

side-slip response of the linear two degree of freedom vehicle plane model with constant speed is set as the reference model response of side-slip angle for the control. Then there will be no control requirement so long as the vehicle motion remains within a linear characteristic region. The response error of the vehicle side-slip angle from the model response primarily caused by the tire nonlinear characteristics and acceleration or braking would require the control yaw moment.

From the equation of motion, Eq.(6), yields the following description for the ratio of steady state side-slip gain to yaw rate gain:

$$\frac{G_\delta^\beta(0)}{G_\delta^r(0)} = \frac{l_r}{V}\left(1 - \frac{ml_f V^2}{2l_r l}\frac{1}{\frac{\partial Y_r}{\partial \beta_r}}\right) \tag{11}$$

This shows that if the equivalent cornering power of the rear tire is reduced due to low friction coefficient between tire and road surface, large side-slip angle, load transfer from rear to front during braking, etc., the negative value of side-slip angle to generate some definite yaw rate increases. This means that adopting a yaw rate response as a model response for the control yields a larger side-slip response when the vehicle subjected to the deterioration of rear tire characteristics. On the other hand, if the side-slip motion is controlled to follow a model response, the vehicle response gain is autonomously restricted with the deterioration of rear tire characteristics. This is a reason why, from the stability point of view, it is better to adopt the side-slip response as a model response for the sliding control rather than to adopt the yaw rate response.

Based upon the above view, the model response of the side-slip angle is given as follows:

$$\frac{\beta}{\delta}(s) = G_B \frac{1 + T_B s}{1 + \frac{Q}{P}s + \frac{1}{P}s^2} \tag{12}$$

This equation is rewritten as a form of sliding surface for the sliding control:

$$S = \ddot{\beta} + Q\dot{\beta} + P\beta - PG_B T_B \dot{\delta} - PG_B \delta = 0 \tag{13}$$

where P, Q, G_B, T_B are expressed by the vehicle parameters mentioned in Section 2.

The sliding condition is:

$$\dot{S} + kS = 0 \tag{14}$$

and equations for the vehicle motion are:

$$mV(\dot{\beta} + r) = Y_f + Y_r$$

$$I\dot{r} = l_f Y_f - l_r Y_r + M_z \tag{15}$$

From these equations, the control yaw moment for the DYC is derived as:

$$\frac{M_Z}{I}=-\frac{(l_f Y_f - l_r Y_r)}{I}+c_1\left(\frac{\partial Y_f}{\partial \beta_f}+\frac{\partial Y_r}{\partial \beta_r}\right)\left(\frac{Y_f+Y_r}{mV}-r\right)+c_2 s\delta$$
$$+\frac{1}{s+c_3}\left[c_4\left(\frac{Y_f+Y_r}{mV}-r\right)+c_5\beta+(c_6 s+c_7)\delta\right] \tag{16}$$

where, c_i $(i=1-7)$ are expressed by P, Q, G_B, T_B and k.

The following tire model, which is called here the on-board-tire-model[14], is to be used for the calculation of lateral forces and their slops - equivalent cornering powers, in the Eq.(16):

$$Y_T=-\left[K_T\beta_T-sign(\beta_T)\frac{K_T^{\,2}}{4\mu W_T}\beta_T^{\,2}\right]\sqrt{1-\left(\frac{X_T}{\mu W_T}\right)^2} \qquad (\text{when: } |\beta_T|<\frac{2\mu W_T}{K_T})$$
$$=-sign(\dot{\beta}_T)\mu W_T\sqrt{1-\left(\frac{X_T}{\mu W_T}\right)^2} \qquad (\text{when: } |\beta_T|\geq\frac{2\mu W_T}{K_T}) \tag{17}$$

where the subscript $_T$ is $_f$ and $_r$, representing front and rear respectively.

6.3 Estimation of side-slip angle

It is essential to detect the side-slip angle to be used in the control. Since directly sensing side-slip angle is not necessarily easy, a number of estimation methods have been proposed. The most common approach to obtain the side-slip angle is to use pseudo-integration of lateral acceleration and yaw rate, though this will cause a significant integration error.

The practical reason why the integration is used and not necessarily the state observer, is that a real vehicle on road frequently has such an initial condition as steering wheel angle, yaw rate and side-slip angle are simultaneously equal to zero. In other words it is straight running and its state can easily be detected. Another reason is that a Luenberger type of observer can not simply be applied because of non-linear characteristics of vehicle dynamics. Therefore some adaptive observer is needed especially considering the saturation property of the tire depending on the friction coefficient between tire and road surface, however, due to this nonlinear characteristics, a general form of stability condition for adaptive identification can not easily be introduced into this case.

From the above point of view it is intended to use the lateral forces calculated by the on-board-tire-model instead of directly sensed lateral acceleration in the integration in order to estimate the side-slip angle. The side-slip motion is described by the following equation:

$$\dot{v}=\frac{Y_f+Y_r}{m}-ur \tag{18}$$

and side-slip angle is defined as:

$$\beta=\tan^{-1}\left(\frac{v}{u}\right) \tag{19}$$

C553/047 © IMechE 1998

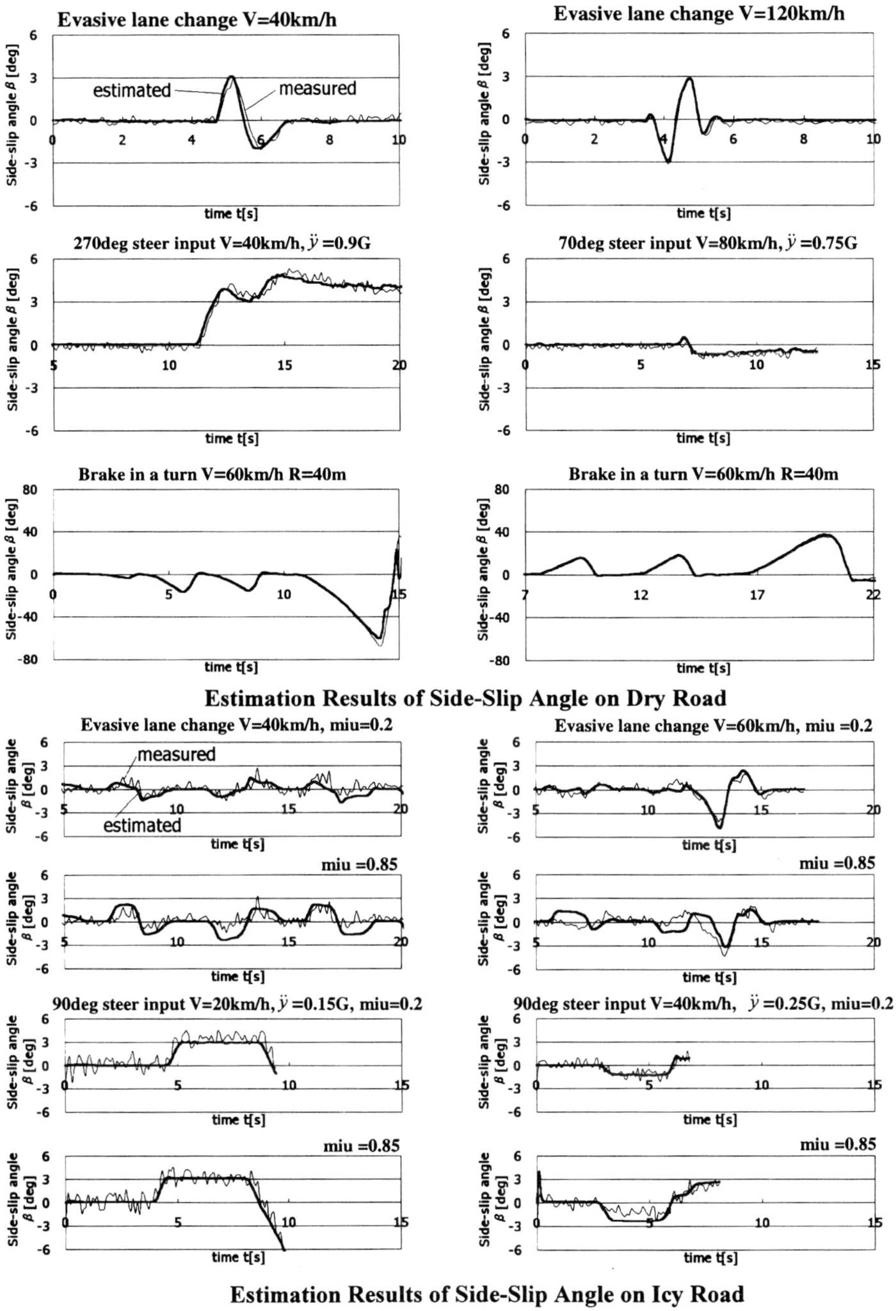

Fig.9 Experimental results of side-slip angle estimation[14]

These two equations are valid no matter how the vehicle running changes with acceleration or braking. In Eq.(18), Y_f and Y_r are given by using Eq.(17).

By on-line real time integration of Eq.(18), the side-slip angle defined by Eq.(19) can be found. Directly sensed steering angle and yaw rate is used in this integration, thus this estimation procedure may be called a hardware-in-the-loop on-line simulation. That is to say a real yaw motion of a vehicle coupled with side-slip motion is regarded as being linked in the on-line simulation loop. The side-slip angle estimated in this process is reflexively used in the calculation of tire forces using the on-board-tire-model, therefore the side-slip angle should be found with less significant integration error. Tire vertical loads and longitudinal forces in Eq.(17) are to be estimated using directly detected longitudinal acceleration.

The experimental results of the estimation are shown in **Fig.9**. It is found that the side-slip angle can be estimated precisely even under such severe running conditions as evasive lane change maneuvering with quick and large steering angle, turning with near limit lateral acceleration with large step steer input and brake-in-a-turn with steering correction. The experimental results of the estimation on a low friction road prove that the estimation method introduced is effective during running on a low friction road as well. Some estimation errors are recognized in the estimation during maneuvering with near limit lateral acceleration when the high friction coefficient is used in the on-board-tire-model, even if so however, it is found that no catastrophic failure can be seen in the estimation. Furthermore almost the same accuracy of the estimation is possible with the friction coefficient set as on dry road during running with normal maneuvering on a low friction road[14],[15].

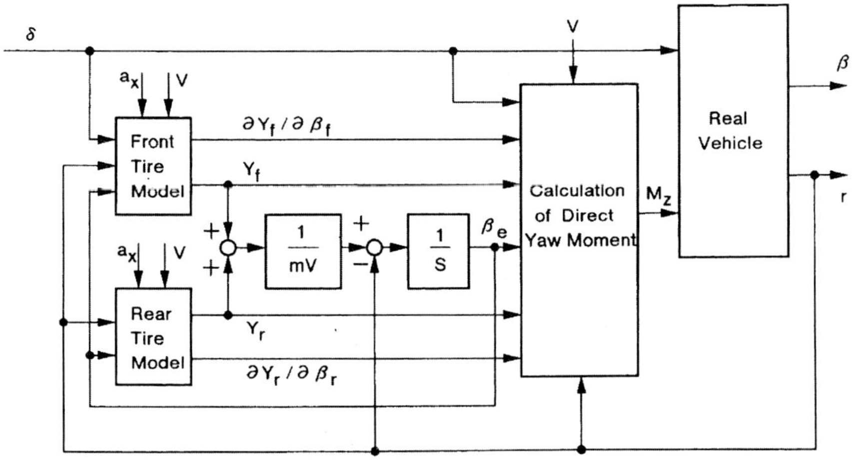

Fig.10 Block diagram of the control system

6.4 Effects of the DYC

The **figure 10** shows a block diagram of the control system with the estimation of side-slip angle, in which the on-board-tire-model is used for the calculation of the control yaw moment as well as the estimation. It is easy to expect that the proposed DYC is effective in non-linear maneuvering range of vehicle motion.

 C553/047 © IMechE 1998

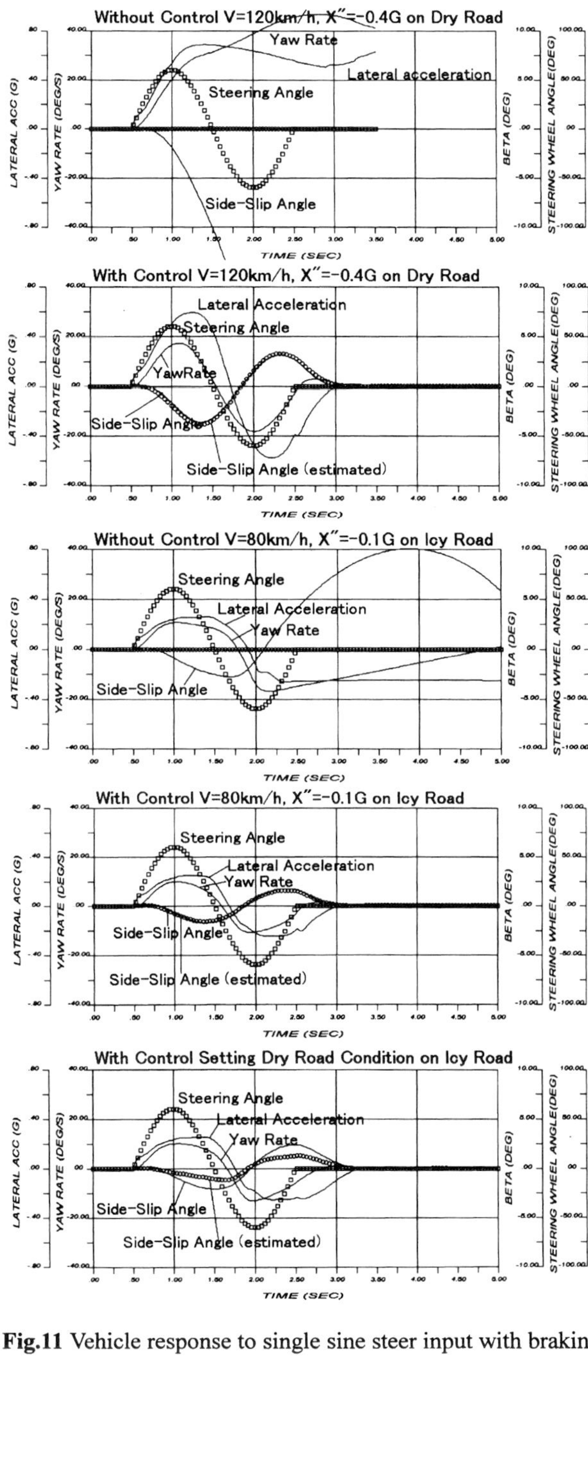

Fig.11 Vehicle response to single sine steer input with braking

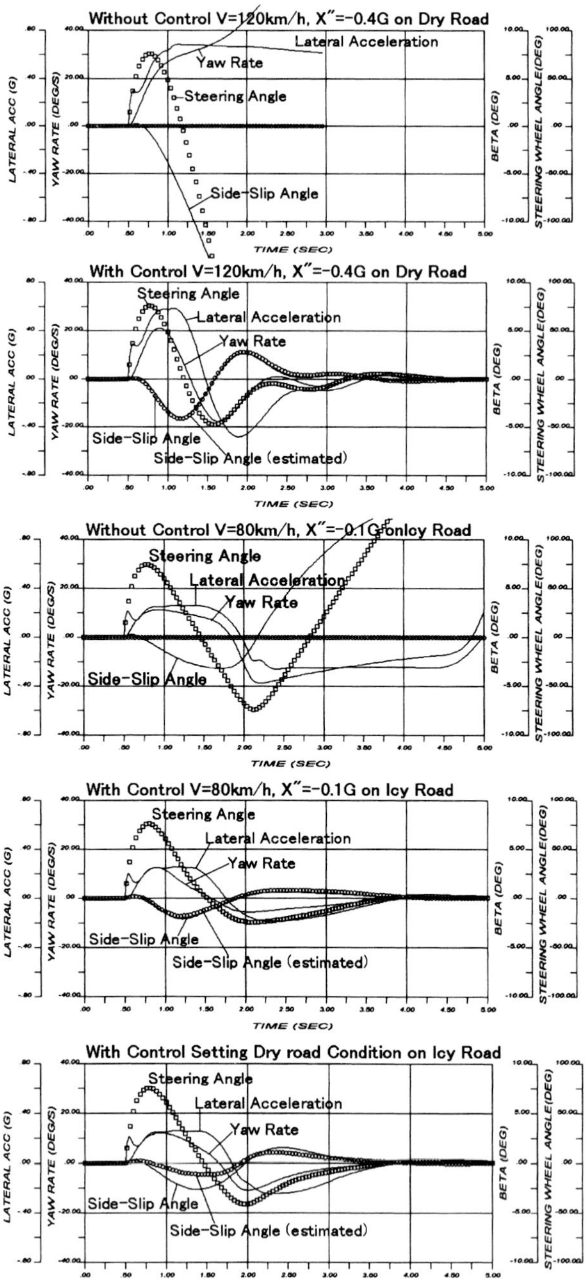

Fig.12 Driver-vehicle system response subjected to lane change with braking

In order to prove the effects of the DYC on improving vehicle handling performance during severe maneuvering, the computer simulation of the vehicle handling response to severe input of single sine wave steering with braking, and the driver-vehicle system response subjected to the severe lane change which may correspond to an evasive maneuver for a real vehicle on a road, was carried out. The computational results are shown in **Figs.11, 12**. Though the vehicle responses without the control are unstable, the vehicle with the DYC shows stable responses to a single sine wave steer input even on a low friction road surface. The closed-loop driver-vehicle system with the control shows stable lane-change responses in the simulation as well.

It seems essential to know the friction coefficient between tire and road surface for the use in the on-board-tire-model, however, the computational results indicate that the friction coefficient is not necessarily needed to be known so precisely. This is because the vehicle performance even on a low friction road is significantly improved by the proposed DYC with the friction coefficient in the on-board-tire-model fixed as on dry road one, though the side-slip angle itself can not necessarily be estimated so precisely[14],[16].

7 CONCLUDING REMARKS

Recent trends of 4WS and DYC have been reviewed., Their effectiveness and limit has been qualified by paying attention to the basic nature of the vehicle dynamics and tire characteristics. From the view point of non-linear characteristics of tire and vehicle dynamics, changing the chassis control from 4WS to DYC seems inevitable for improving handling performance and active safety in a vehicle motion with larger side-slip angle and higher lateral acceleration.

A deep understanding of vehicle non-linear characteristics gave the view that a side-slip model following DYC is better rather than a yaw rate model following one for stabilizing the vehicle motion in nonlinear region. Based on this view, a side-slip model following DYC by sliding control was introduced, in which the on-board-tire-model is used for the estimation of side-slip angle as well as the calculation of the control yaw moment. An investigation with computer simulation proved the significant effects of the method of control. The results suggested that an adaptation of the tire-road friction coefficient in the on-board-tire-model to the variation of the friction coefficient of a real road surface is not necessarily needed.

DYC is one approach which complements the limit of 4WS control. It has been emphasized that DYC is more effective on the vehicle motion control in non-linear range of vehicle dynamics and tire characteristics, and the control law for DYC should be introduced on the assumption of this non-linear dynamics. Furthermore the integration/coordination control of 4WS and DYC should be a promising control to bring about a higher limit performance of vehicle handling. A reasonable control law for the integration/coordination control is expected to be proposed[17].

ACKNOWLEDGEMENTS

The author would like to acknowledge Dr. P. C. Brook, University of Leeds for his review of this paper.

REFERENCES

[1] Horiuchi, S. et al. : Two degree of freedom/H infinity controller synthesis for active four wheel steering vehicles. Vehicle System Dynamics Supplement 25, 1996

[2] Hirano, Y. et al. : Development of robust active rear steering control. Proc. of AVEC'96, Aachen, 1996

[3] Ro, P. et al. : Four wheel steering system for vehicle handling improvement: a robust model reference control using the sliding mode control. Proc. of IMechE Vol.210, 1996

[4] Kaminaga, M. : Vehicle control using intelligent sliding surface. Proc. of AVEC'96, Aachen, 1996

[5] Nagai, M. et al. : Integration of linear systems and neural networks for the design of nonlinear four-wheel-steering system. Proc. of AVEC94 Symposium JSAE October, 1994.

[6] Yuhara, N. et al. : A robust adaptive rear wheel steering control system for handling improvement of four-wheel steering vehicles. Proc. of 12th IAVSD Symposium, August, 1991

[7] Wakamatsu, K. et al. : Adaptive yaw rate feedback 4WS with friction coefficient estimator between tire and road surface. Proc. of AVEC'96, Aachen, 1996

[8] Shibahata, Y. et al. : The improvement of vehicle maneuverability by direct yaw moment control. Proc. of AVEC92, Yokohama, 1992.

[9] Ono, E. et al. : Analysis and control for vehicle stability. Proc. of the 3rd International Symposium on Motion and Vibration Control. Chiba, Japan 1996

[10] van Zanten, A.T. et al. : Control aspect of the BOSCH-VDC. Proc. of AVEC96, Aachen, 1996.

[11] Alverti, V. et al. : Improved driving stability by active braking of the individual wheels. Proc. of AVEC96, Aachen, 1996.

[12] Inagaki, S. et al. : Analysis on vehicle stability in critical cornering using phase-plane method. Proc. of AVEC94, Tsukuba, 1994

[13] Koibuchi, K. et al. : Vehicle stability control in limit cornering by active brake. SAE Paper 960487.

[14] Abe, M. et al. : Direct yaw moment control by sliding control using on-board-tire-model. Proc. of MOVIC98, Zurich, 1998

[15] Abe, M. et al. : Estimation of vehicle side-slip angle for DYC by using on-board-tire-model. Proc. of AVEC98, Nagoya, 1998

[16] Furukawa, Y. et al. : Direct yaw moment control with estimating side-slip angle by using on-board-tire-model. Proc. of AVEC98, Nagoya, 1998

[17] Furukawa, Y. et al. : Advanced chassis control systems for vehicle handling and active safety. Vehicle System Dynamics Vol.28, 2-3, 1997

 C553/047 © IMechE 1998

C553/039/98

Vehicle modular simulation in system dynamics

R KÜBLER MASME, VDI and **W SCHIEHLEN**
Institute B of Mechanics, University of Stuttgart, Germany

Synopsis. Modeling and simulation of complex vehicle systems are often relieved by a modular approach in which the global system is divided into subsystems. Advantages raise from independent and parallel modeling of subsystems, over easy exchange of the resulting modules, to the option to use different software for each module. Each subsystem may be simulated independently and in parallel with adapted numerical integration methods and step sizes. Data exchange between subsystems only takes place at discrete times with variable communication intervals. Different methods of simulator coupling are investigated and it will be shown that stability is guaranteed only for iterative simulator coupling schemes. A wheel suspension serves as example.

1 INTRODUCTION

Modeling and simulation of complex engineering systems, such as automobiles in vehicle dynamics, require the modeling of many components from different engineering fields, e.g. mechanics, control and electronics. In general, the global system has to be divided into subsystems due to different physical disciplines or to make it treatable for a team of engineers. Many advantages raise from this approach, once the interfaces of the subsystems have been defined: Subsystems can be modeled independently from each other and in parallel from experts in the corresponding engineering field, exchange and modification of a subsystem is independent of any other components, different and independent software tools can be used for every physical discipline, and it is even possible to hide the internal dynamics of a subsystem during simulation of the global system, see e.g. Rükgauer (1).

To carry out the overall simulation, all subsystems have to be coupled to obtain the global system. The coupling of models can be achieved on three different levels of model description, as illustrated in Figure 1 for mechatronic systems.

In the physical model description the engineering system is represented by a physical model with physical parameters, such as masses or dimensions in the case of multibody

© IMechE 1998 C553/039

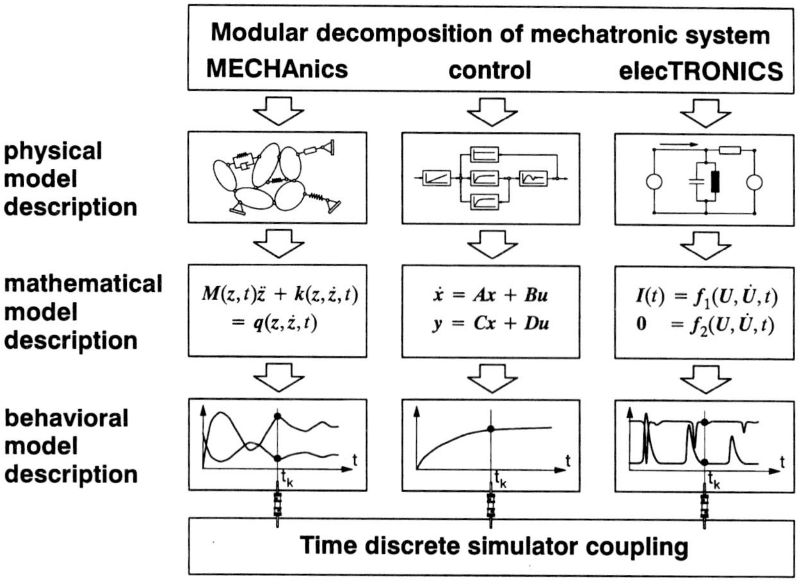

Figure 1: Modular modeling of mechatronic systems

systems. The mathematical model description is an abstract representation of the engineering system by mathematical equations, e.g. the equations of motion of multibody systems. The results of simulation of the mathematical model, for instance positions and velocities of bodies, is considered finall yas the behavioral model description.

Standardization of model description and coupling of models on all three description levels are very lively research subjects. Standardization of mechatronic systems on the physical model description level including the standardization of multibody system data within STEP is under development, Dürr et al. (2), Dürr and Schiehlen (3). Standardization on mathematical model description may be achieved with the hardware description language VHDL–AMS (4). A textual system description language based on the nonlinear state space representation has been proposed by Junker et al. (5). The coupling on behavioral description level has successfully been tested by coupling of different simulators, see for example Wünsche et al. (6).

Coupling of models in behavioral model description is also referred to as *modular simulation* or *simulator coupling*, respectively. Simulation of the global system is realized by a time discrete linker and scheduler which combines the inputs and outputs of the corresponding subsystems and establishes communication between the subsystems to discrete times, Figure 1. The main advantage of coupling on the behavioral description level is the possibility to use integrated software packages for each subsystem, where preprocessor, solver and postprocessor are linked together resulting in a very user–friendly solution.

This paper is devoted to the theoretical aspects of simulator coupling. Section 2 describes the mathematical formulation of subsystems chosen as multibody systems and defines the input and output variables of the general state space equations. In Section 3 simulators are introduced as discrete dynamical systems and the requirements simulators have to meet in order to qualify for coupling are discussed. Section 4 classifies the methods

C553/039 © IMechE 1998

of simulator coupling and gives conditions of convergence. In Section 5 first results are presented for the modular simulation of a wheel suspension model.

2 MATHEMATICAL DESCRIPTION OF MULTIBODY SYSTEMS

In this Section, the mathematical description of one module or subsystem chosen as multibody system is described. A multibody system with p rigid bodies and q holonomic, rheonomic constraints, resulting in $f = 6p - q$ degrees of freedom, is considered, Schiehlen (7–10). Inputs to this subsystem are purely time–varying forces and prescribed motions, the resulting absolute motions and constraint forces are considered as output quantities, Figure 2.

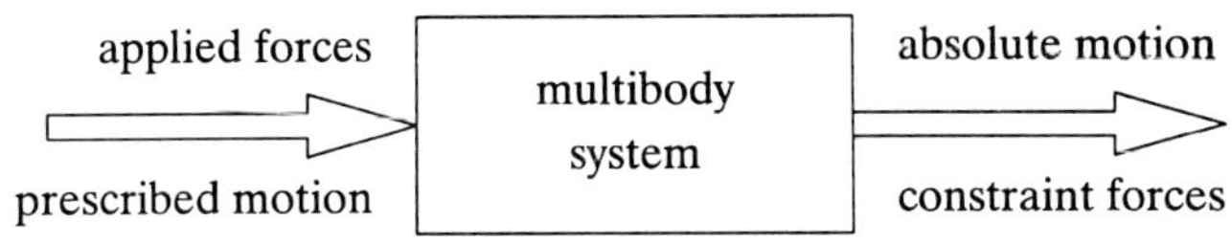

Figure 2: Input and output of multibody systems

The applied forces $\boldsymbol{u}_f(t)$ together with the absolute positions $\boldsymbol{u}_m(t)$, velocities $\dot{\boldsymbol{u}}_m(t)$ and accelerations $\ddot{\boldsymbol{u}}_m(t)$ of the prescribed motions represented in the inertial frame are components of the input vector

$$\boldsymbol{u}(t) = \begin{bmatrix}\boldsymbol{u}_f^T & \boldsymbol{u}_m^T & \dot{\boldsymbol{u}}_m^T & \ddot{\boldsymbol{u}}_m^T\end{bmatrix}^T . \tag{1}$$

The state vector

$$\boldsymbol{x}(t) = \begin{bmatrix}\boldsymbol{z}^T & \dot{\boldsymbol{z}}^T\end{bmatrix}^T, \qquad \boldsymbol{x} \in \mathbb{R}^{2f} \tag{2}$$

is composed of generalized coordinates and velocities.

The equations of motion in minimal formulation with the input vector $\boldsymbol{u}$ and the state vector $\boldsymbol{x}$ can be represented as the nonlinear state equations, Schiehlen (7), Kübler and Schiehlen (11)

$$\dot{\boldsymbol{x}} = \boldsymbol{f}(\boldsymbol{x}, \boldsymbol{u}, t). \tag{3}$$

The positions, velocities and accelerations of the absolute motions together with the external constraint forces represented in the inertial frame result in an output vector $\boldsymbol{y}(t)$ which is determined using the state and input vectors $\boldsymbol{x}$ and $\boldsymbol{u}$ by the nonlinear output equations

$$\boldsymbol{y} = \boldsymbol{g}(\boldsymbol{x}, \boldsymbol{u}, t). \tag{4}$$

If the output equation is explicitly dependend on the input, the corresponding system is called a system with *feed–through.* This is an important property for the following investigation of simulator coupling schemes. If two simulators are coupled both of which simulate subsystems with feed–through, so called *algebraic loops* can be appear. It turns out that algebraic loops may destabilize the modular simulation.

3 DESCRIPTION OF SIMULATORS

In this Section simulators are introduced as discrete dynamical systems, and the requirements simulators have to meet in order to qualify for coupling are discussed.

A global time step of a simulator from time t_k to t_{k+1} with global step size $H = t_{k+1} - t_k$ is shown in Figure 3.

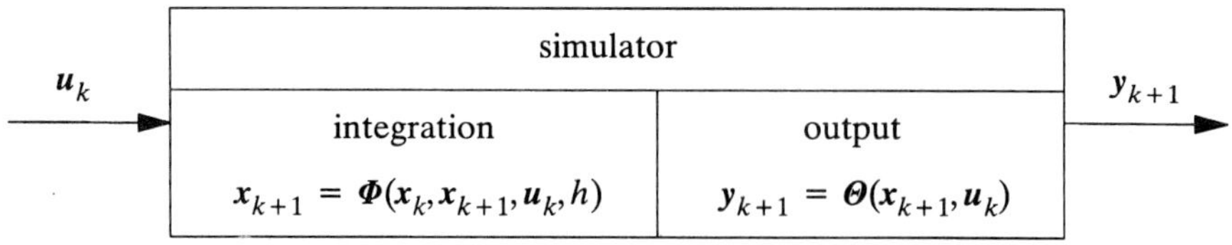

Figure 3: Mathematical description of simulators

The variables are separated in dynamic state variables $\boldsymbol{x}$ and algebraic output variables $\boldsymbol{y}$. After receiving the input variables $\boldsymbol{u}_k$, the integrator $\boldsymbol{\Phi}$ is calculating the states $\boldsymbol{x}_{k+1}$ with variable or constant local step size $h \leq H$. For intermediate times the inputs $\boldsymbol{u}$ are calculated by extrapolation, if only $\boldsymbol{u}_k$ is available, or interpolation, if $\boldsymbol{u}_{k+1}$ was also given to the simulator as input. Extrapolation and interpolation is necessary for Runge–Kutta methods with more than one stage, for implicit integration schemes, and for any integration method, if the local step size is smaller than the global step size.

After integration, the nonlinear function $\boldsymbol{\Theta}$ is evaluated at time t_{k+1} to obtain the outputs $\boldsymbol{y}_{k+1}$ of the simulator. The algebraic output equations depend on the already calculated state variables $\boldsymbol{x}_{k+1}$ and the input variables $\boldsymbol{u}_{k+1}$ which are either given to the simulator or have to be calculated again by extrapolation.

Finally, the requirements a simulator has to meet for coupling are summarized:

- time incremental simulation is supported,
- some mechanism for data exchange to other software is supported,
- numerical integration to a prescribed point in time by means of constant or variable step size is possible,
- the global time step can be reset from outside and recalculated with new inputs.

4 METHODS OF SIMULATOR COUPLING

On the basis of the previous description of simulators our method of simulator coupling is presented which leads to the solution of nonlinear algebraic equations for each global time step. It will be shown, that stability is not guaranteed for coupling schemes not involving the solution of equations.

Figure 4 shows the coupling of two simulators with iteration of the global integration step for a given global step size H. The input of simulator I is the output of simulator II and vice versa. After one global integration step with extrapolation of the inputs, the output of the other simulator is used as new input to recalculate the global step with interpolation of the inputs. This process is repeated until convergence is reached.

 C553/039 © IMechE 1998

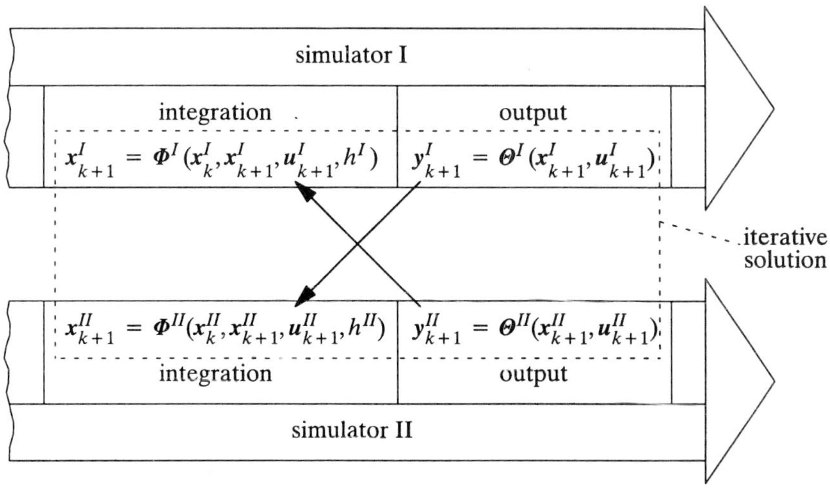

Figure 4: Simulator coupling with iteration of global integration step

4.1 Simulator coupling without iteration

A straightforward solution to the problem of simulator coupling is to calculate the outputs to given inputs once with each simulator and to use this result with the other simulator for the next global step. This approach corresponds to the iterative scheme, Figure 4, if the iteration is stopped after the first step, and is consequently refered to as *simulator coupling without iteration.*

For integration of the state variables and evaluation of the output equations extrapolation has to be used for the inputs. Therefore, the outputs of simulator I can be expressed as a function of the inputs of the previous step:

$$\boldsymbol{y}^I_{k+1} = \Theta^I(\boldsymbol{x}^I_{k+1}, \boldsymbol{u}^I_k) = \Psi^I(\boldsymbol{u}^I_k). \tag{5}$$

The equivalent formulation is obtained for simulator II:

$$\boldsymbol{y}^{II}_{k+1} = \Psi^{II}(\boldsymbol{u}^{II}_k). \tag{6}$$

Introducing the global output vector

$$\boldsymbol{y}_k = \begin{bmatrix} \boldsymbol{y}^I_k \\ \boldsymbol{y}^{II}_k \end{bmatrix} \tag{7}$$

and using the coupling equations

$$\boldsymbol{u}^I_k = \boldsymbol{y}^{II}_k, \tag{8a}$$
$$\boldsymbol{u}^{II}_k = \boldsymbol{y}^I_k, \tag{8b}$$

equations (5) and (6) can be written as

$$\boldsymbol{y}_{k+1} = \Psi(\boldsymbol{y}_k). \tag{9}$$

Equation (9) defines a discrete dynamical system which is the basis of the following investigation on stability of the modular simulation in addition to the dynamics of the numerical integrators inside the simulators used.

For any numerical integration scheme necessary conditions for convergence are consistency and zero–stability, Hairer et al. (12). Consistency means, that the discrete system which is used to approximate the continuous system approaches to the continous system, if the step size goes to zero. Obviously, this is true for the introduced simulator coupling, if well designed simulators are used.

Zero–stability means, that the discrete system is stable, if the step size approaches zero. It can be shown, that zero–stability of any integration method is independent of the inputs (12):

$$\boldsymbol{x}_{k+1} = \boldsymbol{\Phi}(\boldsymbol{x}_k, \boldsymbol{x}_{k+1}, \boldsymbol{u}, h \to 0) = \boldsymbol{\Phi}(\boldsymbol{x}_k, \boldsymbol{x}_{k+1}). \tag{10}$$

This is due to the fact, that all inputs are multiplied by the step size h in any integration formula. Consequently, for reliable simulators the used integrators (10) are zero–stable.

The remaining question is, if the additional dynamics introduced by the output equations (9) is zero–stable. To get an answer, (9) is linearized for the current time step:

$$\boldsymbol{y}^I_{k+1} = \boldsymbol{\Psi}^I_0 + \boldsymbol{D}^I \boldsymbol{y}^{II}_k, \tag{11a}$$
$$\boldsymbol{y}^{II}_{k+1} = \boldsymbol{\Psi}^{II}_0 + \boldsymbol{D}^{II} \boldsymbol{y}^I_k. \tag{11b}$$

Using the global output vector (7), equations (11) result in the linear discrete system

$$\boldsymbol{y}_{k+1} = \boldsymbol{\Psi}_0 + \boldsymbol{D}\boldsymbol{y}_k \tag{12a}$$

with the system matrix

$$\boldsymbol{D} = \begin{bmatrix} \boldsymbol{0} & \boldsymbol{D}^I \\ \boldsymbol{D}^{II} & \boldsymbol{0} \end{bmatrix}. \tag{12b}$$

Stability of the linear system is independent of the state vector $\boldsymbol{x}_{k+1}$ which can be interpreted as input in this context. Consequently, stability is defined by the autonomous system

$$\boldsymbol{y}_{k+1} = \boldsymbol{D}\boldsymbol{y}_k, \quad \rho(\boldsymbol{D}) \overset{!}{\leq} 1, \tag{13}$$

where ρ denotes the spectral radius. It is obvious, that condition (13) is dependent on the models to be simulated and therefore, no general statement on stability for the coupling scheme without iteration can be made.

However, it can be observed from (12), that for $\boldsymbol{D}^I = \boldsymbol{0}$ or $\boldsymbol{D}^{II} = \boldsymbol{0}$ the spectral radius of $\boldsymbol{D}$ is zero and the stability condition (13) is fulfilled. This condition corresponds to systems where the output equations of one simulator has no feed–through, or no algebraic loop is existent between the two simulators, respectively.

4.2 Iterative simulator coupling

The iterative simulator coupling scheme illustrated in Figure 4 is now investigated. Considering the repeated integration only as a function of the changing inputs, the outputs of simulator I can be expressed as

$$\boldsymbol{y}^I_{k+1} = \boldsymbol{\Theta}^I(\boldsymbol{x}^I_{k+1}, \boldsymbol{u}^I_{k+1}) = \boldsymbol{\Theta}^I(\boldsymbol{\Phi}^I(\boldsymbol{u}^I_{k+1}), \boldsymbol{u}^I_{k+1}) = \boldsymbol{\Psi}^I(\boldsymbol{u}^I_{k+1}). \tag{14}$$

C553/039 © IMechE 1998

Analogous formulation of simulator II leads to

$$\boldsymbol{y}_{k+1}^{II} = \boldsymbol{\Psi}^{II}(\boldsymbol{u}_{k+1}^{II}). \tag{15}$$

Using the coupling equations (8) and the global output vector (7), equations (14) and (15) can be written as the set of nonlinear algebraic equations

$$\boldsymbol{y}_{k+1} = \boldsymbol{\Psi}(\boldsymbol{y}_{k+1}). \tag{16}$$

In contrast to the the discrete dynamical system (9), equation (16) is purely algebraic and consequently does not introduce additional dynamics which could destabilize the modular simulation.

The system of nonlinear algebraic equations (16) have to be solved for each global time step at time t_{k+1} by an iterative method, see Dennis and Schnabel (13), Ortega and Rheinboldt (14). For Gauß–Jacobi and Gauß–Seidel method local convergence is not guaranteed. Newton's method is locally convergent, however, the Jacobian in not available in modular simulation, in general. Consequently, a quasi–Newton method which uses an approximation to the Jacobian has to be used. One of the most reliable quasi–Newton methods is Broyden's method, see Broyden (15), Dennis and Schnabel (13), Press et al. (16), which uses a secant approximation to the Jacobian and is also locally convergent.

Summarizing the analysis of this Section, the following conclusions can be drawn for the different methods of simulator coupling for any number of simulators:

- If no algebraic loops are existent between any simulators, the simulator coupling without iteration is always convergent.
- If there exists at least on algebraic loop between any simulators, convergence is only guaranteed for the iterative coupling schemes.

5 EXAMPLE

A wheel suspension is modeled as the multibody system illustrated in Figure 5. The system has $f = 1$ degree of freedom and the corresponding parameters are also given in Figure 5. Except of the mass m^{II} all parameters are fixed. The introduced simulator coupling schemes will be investigated for variation of the mass ratio $\beta = m^{II}/m^{I}$.

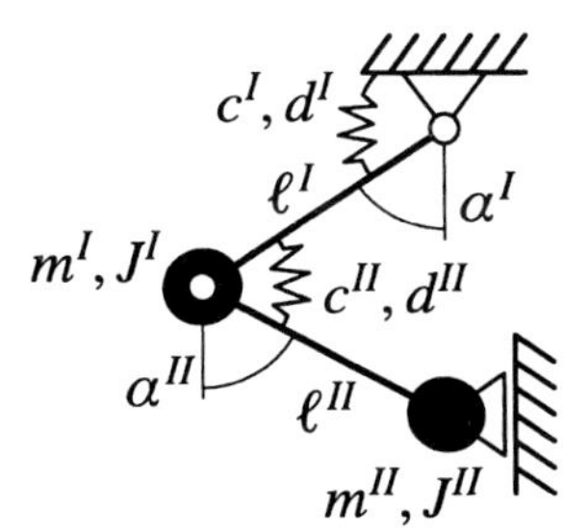

mass	$m^I = 1$
mass	$m^{II} = \beta$
inertia	$J^I = J^{II} = 1$
length	$\ell^I = \ell^{II} = 1$
stiffness	$c^I = c^{II} = 1$
damping	$d^I = d^{II} = 0.1$

Figure 5: Multibody system with parameters

The multibody system is decomposed into two subsystems with corresponding input and output quantities as illustrated in Figure 6.

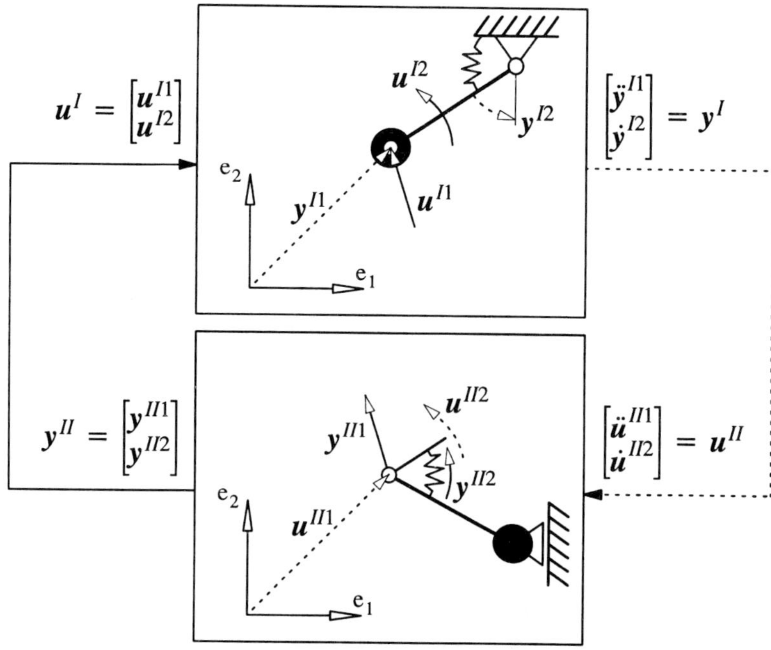

Figure 6: Decomposed multibody system

Subsystem I has $f^I = 1$ degree of freedom and the applied force $\boldsymbol{u}^{I1}$ and the applied torque $\boldsymbol{u}^{I2}$ are components of the input vector $\boldsymbol{u}^I$. The state space formulation (3) of the equations of motion with the state vector $\boldsymbol{x}^I = [\alpha^I \; \dot{\alpha}^I]^T$ has the form

$$\dot{\boldsymbol{x}}^I = \boldsymbol{f}^I(\boldsymbol{x}^I, \boldsymbol{u}^I). \tag{17}$$

The output vector $\boldsymbol{y}^I$ is composed of the absolute acceleration $\ddot{\boldsymbol{y}}^{I1}$ and the related angular velocity $\dot{\boldsymbol{y}}^{I2}$ of body I and is computed from the nonlinear output equations (4):

$$\boldsymbol{y}^I = \boldsymbol{g}^I(\boldsymbol{x}^I, \boldsymbol{u}^I). \tag{18}$$

Subsystem II has no degree of freedom and consequently there is no dynamic equation for this subsystem. The input vector $\boldsymbol{u}^{II}$ of subsystem II is composed of the prescribed acceleration $\ddot{\boldsymbol{u}}^{II1}$ of the joint and the prescribed angular velocity $\dot{\boldsymbol{u}}^{II2}$ of the rotational spring. The output vector $\boldsymbol{y}^{II}$ is composed of the constraint force $\boldsymbol{y}^{II1}$ and the torque $\boldsymbol{y}^{II2}$ generated by the rotational spring and is calculated by the nonlinear output equation

$$\boldsymbol{y}^{II} = \boldsymbol{g}^{II}(\boldsymbol{x}^{II}, \boldsymbol{u}^{II}). \tag{19}$$

Subsystem I and II are coupled by the coupling equations (8):

$$\boldsymbol{u}^I = \boldsymbol{y}^{II}, \tag{20a}$$
$$\boldsymbol{u}^{II} = \boldsymbol{y}^I. \tag{20b}$$

Stability behavior and accuracy are compared for simulator coupling without iteration and the iterative simulator coupling scheme subject to variation of the mass ratio β,

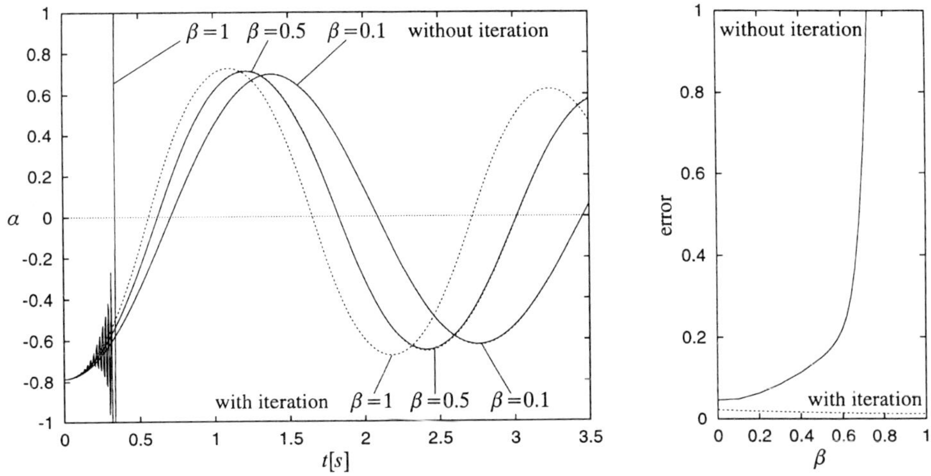

Figure 7: Simulator coupling without iteration and iterative simulator coupling scheme

Figure 7. The trajectory of angle α^I is plotted on the left of Figure 6. The results for the two methods look similar for small mass ratios β, whereas for larger β the simulator coupling without iteration becomes unstable. These results verify the stability analysis of Section 4.1, where the condition for stability of simulator coupling without iteration was found to be dependent on the model to be simulated.

In addition, the right side of Figure 7 shows the comparison of the accuracy of the two simulator coupling schemes. For the method without iteration, the integration error is highly dependend on the mass ratio β and approaches infinity for the unstable configuration. On the other hand, the iterative simulator coupling scheme results in a small, nearly constant integration error for all parameter variations.

6 CONCLUSIONS

A multibody system module is formulated as a subsystem with applied forces and prescribed motions as input and the resulting absolute motions and constraint forces as output. The general state space representation of the multibody system module is given.

Different methods of simulator coupling are investigated. If algebraic loops are not existing between any simulators, the simulator coupling without iteration is always convergent. If there exists at least one algebraic loop between any of the simulators, convergence is only guaranteed for iterative coupling schemes. The presented iterative coupling scheme with iteration of the global integration step leads to a system of nonlinear algebraic equations for each global integration step. It is shown that the nonlinear algebraic equations have to be solved by quasi–Newton methods, in general.

Reliability and accuracy of the new method in modular simulation is demonstrated with a wheel suspension modeled as a multibody system.

REFERENCES

[1] A. Rükgauer. *Modulare Simulation mechatronischer Systeme mit Anwendung in der Fahrzeugdynamik.* Fortschritt-Berichte, Reihe 20, Nr. 248, VDI-Verlag, Düsseldorf, (1997).

[2] R. Dürr; U. Neerpasch; W. Schiehlen; L. Witte. Mechatronik und STEP, Standardisierung eines neutralen Datenformats in STEP für die Simulation mechatronischer Systeme. *Produktdaten Journal*, **2**, 12–19 (1995).

[3] R. Dürr; W. Schiehlen. Exchange of design data for mechatronic systems. In *Booklet of Abstracts, EUROMECH Colloquium 370 'Synthesis of Mechatronic Systems'*, pages 9–11, (1997).

[4] IEEE standard 1076.1. draft.

[5] F. Junker; M. Hahn; J. Lückel. Systematic modelling of mechanical parts in mechatronic systems. *Systems Analysis Modelling Simulation*, **18–19**, 619–624b (1995).

[6] S. Wünsche; C. Clauß; P. Schwarz; F. Winkler. Microsystem design using simulator coupling. In *Proceedings of the European Design and Test Conference 1997*, pages 113–118, (1997).

[7] W. Schiehlen. *Technische Dynamik.* Stuttgart: Teubner, (1986).

[8] W. Schiehlen, editor. *Multibody Systems Handbook.* Berlin: Springer, (1990).

[9] W. Schiehlen, editor. *Advanced Multibody Dynamics.* Dordrecht: Kluwer Academic Publishers, (1993).

[10] W. Schiehlen. Multibody system dynamics: Roots and perspectives. *Multibody System Dynamics*, **1**(2), 149–188 (1997).

[11] R. Kübler; W. Schiehlen. Modular simulation in multibody system dynamics. In S. Idelsohn; E. Onate; E. Dvorkin, editor, *Fourth World Congress on Computational Mechanics*, Buenos Aires, Argentina, (1998).

[12] E. Hairer; S.P. Nørsett; G. Wanner. *Solving Ordinary Differential Equations I. Nonstiff Problems.* Berlin: Springer, (1987).

[13] J.E. Dennis and R.B. Schnabel. *Numerical Methods for Unconstrained Optimization and Nonlinear Equations.* Englewood Cliffs, New Jersey: Prentice Hall, Inc., (1983).

[14] J.M. Ortega; W.C. Rheinboldt. *Iterative Solution of Nonlinear Equations in Several Variables.* New York: Academic Press, (1970).

[15] C.G. Broyden. Solving nonlinear simultaneous equations. *Math. Comp.*, **19**, 577–593 (1965).

[16] Press; B.P. Flannery; S.A. Teukolsky; W.T. Vettering W.H. *Numerical Recipes in C.* New York: Cambridge University Press, (1988).

C553/039 © IMechE 1998

C553/026/98

Multi-body dynamics analysis of the ride quality of an off-road articulated truck

R J ANDERSON BSc, MSc, PhD, PEng, MSCME, MSAE
Department of Mechanical Engineering, Queen's University, Kingston, Canada

SYNOPSIS

This paper describes a multibody model of an off-road utility truck and concentrates on the prediction of ride quality. In particular, the paper presents two approaches to simulating ride quality – 1. the traditional frequency domain analysis using a linear model and 2. a more complete time domain analysis using a non-linear vehicle model. It is shown that the differences between the predictions of the two methods are not significant enough to require the use of non-linear models during, for instance, parametric studies at the design stage of a vehicle.

1. INTRODUCTION

As part of a project to develop a novel off-road utility vehicle for use in support of the Canadian oil industry, a full multibody model of the vehicle was developed and used to provide specifications for the suspension. The A'GEM Multibody Dynamics software package (1,2) developed at Queen's University was used to undertake a number of analyses in support of the truck development program. This paper concentrates on the ride quality analysis for the vehicle and, in particular, on the differences between a traditional frequency domain analysis using a linear model and a more complete time domain analysis using a non-linear vehicle model.

© IMechE 1998 C553/026

The most straightforward prediction of ride quality is made from linear models working in the frequency domain. There are two steps to the process.

As a first step, the natural frequencies and corresponding mode shapes of the vehicle are calculated. These results show at what frequencies the vehicle is likely to experience significant motions regardless of the disturbance. The designer is free at this stage to make changes which will move the natural frequencies and damping ratios to values which promote good ride quality. Basically, the primary rigid body modes of the vehicle should have frequencies in the order of 1 to 2 Hz with damping ratios between 15 and 40%.

In the next step, the vehicle accelerations in response to different road profiles are calculated and compared to standard accelerations in order to give a quantitative estimate of ride quality. The Power Spectral Density (PSD) of a road profile is used as an input to the vehicle and the RMS accelerations in the cab are compared to standards from the International Standard ISO-2631-1978(E) entitled a "Guide for the Evaluation of Human Exposure to Whole-Body Vibration" (3). The comparison yields quantitative ride quality predictions.

The difficulty with this analysis is, as is usually the case in modelling, the fact that the vehicle contains many non-linear elements which must be approximated in order to undertake the linear analysis. The accuracy of the resulting ride quality predictions is therefore questionable.

This paper discusses the inaccuracies resulting from these approximations by repeating the ride quality analysis in the time-domain with non-linearities included. While this is a more complicated and time consuming process, it is more accurate and a comparison of results with those from a linear analysis indicates the level of confidence which should be placed in the traditional linear techniques.

2. QUANTIFYING RIDE QUALITY

A quantitative measure of vertical ride quality can be made by considering the vehicle response to a vertical road profile. The RMS acceleration of the cab of the vehicle is plotted versus frequency in 1/3 octave bands and compared to standards taken from International Standard ISO-2631-1978(E) entitled a "Guide for the Evaluation of Human Exposure to Whole-Body Vibration" (3).

ISO-2631 gives three thresholds for use in estimating the effects of vibrations on humans. They are:

1. ***Reduced Comfort Boundary*** - Specifies the limit at which one encounters difficulties in carrying out such operations as eating, reading and writing. It is specified as the length of time before these activities are affected by the vibration. This limit is in common use for mass transport vehicles such as passenger trains.

2. ***Fatigue-Decreased Proficiency Boundary*** - Specifies the limit above which exposure to vibration will impair working efficiency in many types of time-dependent tasks. This is the limit which is commonly used for workers (e.g. train engineers as opposed to passengers)

C553/026 © IMechE 1998

3. ***Exposure Limit*** - Specifies the limit approximately at the threshold of pain. These limits are a factor of two higher than the Fatigue Boundaries.

3. THE ARTICULATED TRUCK

Figure 1 is a drawing of the articulated off-road truck used as an example in this paper. The truck is supported by leaf springs mounted on front and rear beam axles. The large tires are designed to operate at low pressures to minimize the contact stresses on off-road terrain. Above the suspension the vehicle has three main components – 1. the rear sprung mass which carries the payload; 2. the front sprung mass including the engine and transmission; 3. the cab which has a separate level of suspension connecting it to the front sprung mass.

3.1 The multibody model

The multibody model of the truck is based on nine rigid bodies interconnected by force producing elements. The force producing elements are the tires connecting the axles to the ground, the leaf springs and shock absorbers connecting the axles to the front and rear sprung masses, and the pneumatic suspension elements connecting the cab to the front sprung mass. In addition, there is a constraint element which permits articulation between the front and rear sprung masses.

Figure 1 shows the truck without any payload and is presented simply to give an impression of the actual vehicle. The results presented in this paper are for a vehicle which includes a deck and a crane for which no drawings are available. The total mass of the truck is 10,110 kg and it has a wheelbase of 4.37 m.

If linear properties are assumed for all of the force producing elements, a linear mathematical model results and analysis can proceed in the frequency domain. If the characteristics of the force-producing elements are described by non-linear, piece-wise linear or discontinuous functions, the solution can only be generated in the time domain. The A'GEM software package gives the user both options.

4. THE OFF-ROAD PROFILE

Figure 2 shows the off-road profile used to calculate the vehicle response in order to predict the ride quality of the vehicle. The presentation in Figure 2 is in the form that is used as input to the non-linear, time-domain vehicle model. Calculations of this type use a forcing function that arises from the vertical motions of the tire contact patches as they traverse the profile at a specified, usually constant, speed.

Frequency domain calculations, on the other hand, require a forcing input in the frequency domain. This is found by calculating the Power Spectral Density of the known terrain profile as a function of position. PSD calculations are more typically made in the time domain where a signal is expressed as a function of time. In the time domain case, the calculated PSD is a

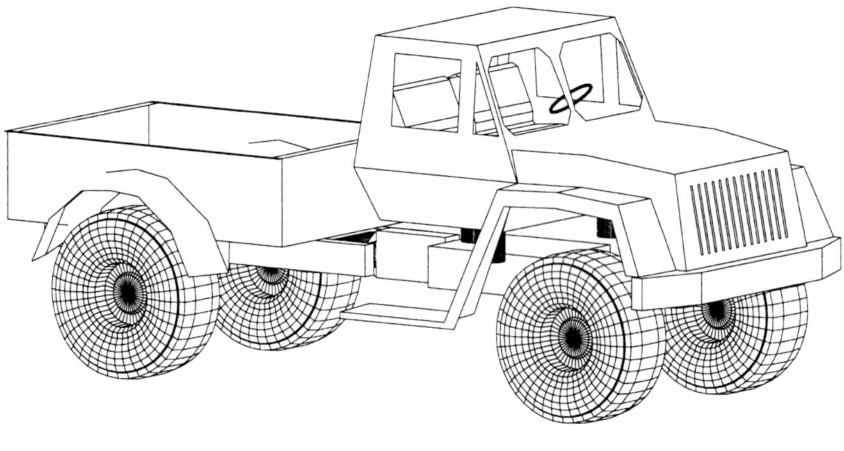

Figure 1: The articulated off-road truck.

function of frequency in cycles/second and the area under the PSD is the mean square value of the signal. As a result, the PSD has units of "amplitude squared per cycle/second" where "amplitude" refers to the units used to measure the signal. Analogously, the terrain PSD has units of metres squared per cycle/metre since the signal amplitude is measured in metres and the time base has been replaced by a position base where metres replace seconds. The resulting profile PSD is presented in Figure 3.

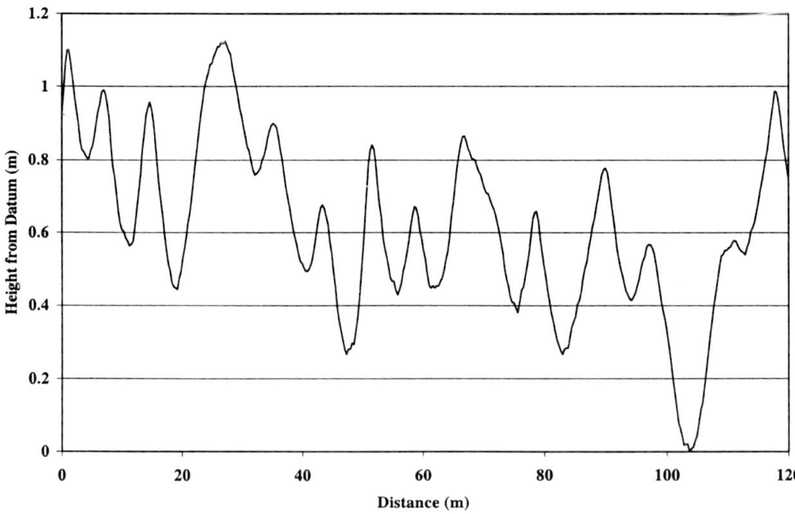

Figure 2. The off-road terrain profile.

 C553/026 © IMechE 1998

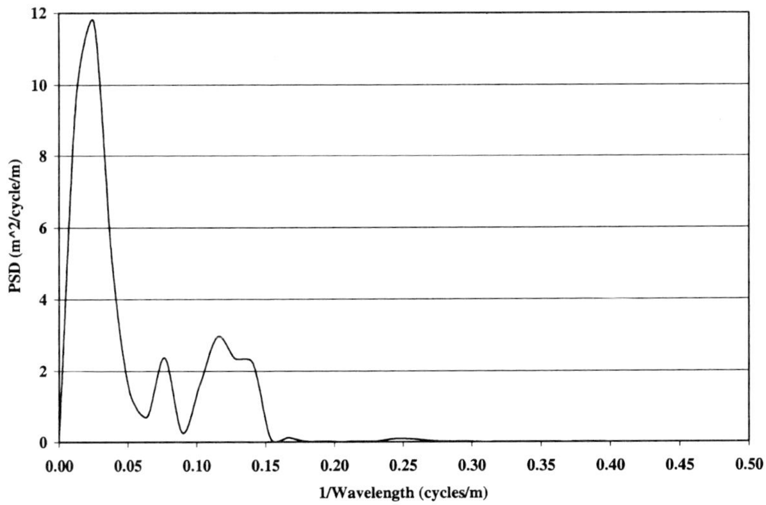

Figure 3: Power spectral density of the off-road profile.

The forcing function entering the equations of motion for the frequency domain model must be a harmonic function of time. The assumption of a constant vehicle speed over the profile allows a simple transformation whereby the vehicle travels one wavelength of the profile in a period which is derived from the vehicle speed and then easily transformed to frequency.

5. RIDE QUALITY PREDICTIONS FOR THE LINEAR MODEL

As a first step in the study, it is necessary to establish the degree of similarity between ride quality predictions made for the linear vehicle model in the frequency and time domains. This section presents a comparison of results for that purpose.

5.1 Linear, frequency domain predictions

The results of the analysis in the frequency domain are presented in Figure 4. It is clear that most of the activity in the vehicle is at the lower end of the frequency range. The indication is that the ride quality is governed by the peaks at centre frequencies of 1.25 and 1.60 Hz and corresponds to a ride quality measure which can be stated as lying between 1 hour and 2.5 hours on the Fatigue-Decreased Proficiency Boundaries (3). For a utility vehicle such as this, it is appropriate to use the fatigue boundaries rather than the reduced comfort boundaries used for passenger comfort.

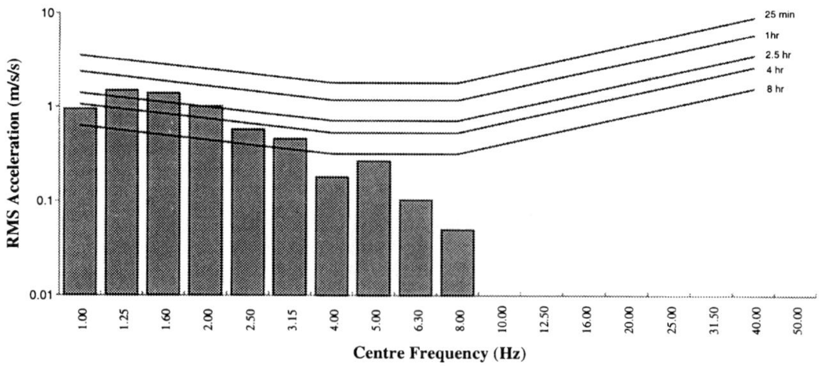

Figure 4. Frequency domain results.

5.2 Linear, time domain predictions

Figure 5 presents the ride quality results obtained by performing time domain simulations of the truck traversing the off-road profile at a speed of 5 m/s. The vertical acceleration in the cab is predicted and saved as a function of time during the simulation. These data are then processed in the same way that one would handle experimentally generated acceleration data when estimating ride quality. The Power Spectral Density is calculated and, from this, RMS accelerations in the third-octave bands specified in the ISO standard are derived. These RMS values are plotted in Figure 5 and compared to the Fatigue-Decreased Proficiency Boundaries.

Ideally, Figures 4 and 5 should be identical since they correspond to the same model of the vehicle, road profile, and operating parameters. The only difference between them is that they have gone through different numerical processes. To get the frequency domain results, a PSD of the terrain was generated and used as a forcing input to the linear vehicle model. The calculated PSD of the vertical accelerations in the cab was then transformed into RMS accelerations in the specified third-octave bands and the result is shown in Figure 4. The time domain results in Figure 5 were generated by a time integration of the equations-of-motion of the vehicle containing only linear components and traversing the terrain profile shown in Figure 2. The predicted vertical cab accelerations were then processed by software which is commonly used to analyze ride quality from experimentally measured accelerations. This software produces the Power Spectral Density of the acceleration signal and manipulates it into the required form for comparison with the ISO standard.

The slight differences seen between Figures 4 and 5, essentially differences in magnitude, are due to the different numerical processes used to generate them. Nevertheless, the agreement between the two is sufficiently good to allow further conclusions to be made between the linear model and models containing non-linear elements.

C553/026 © IMechE 1998

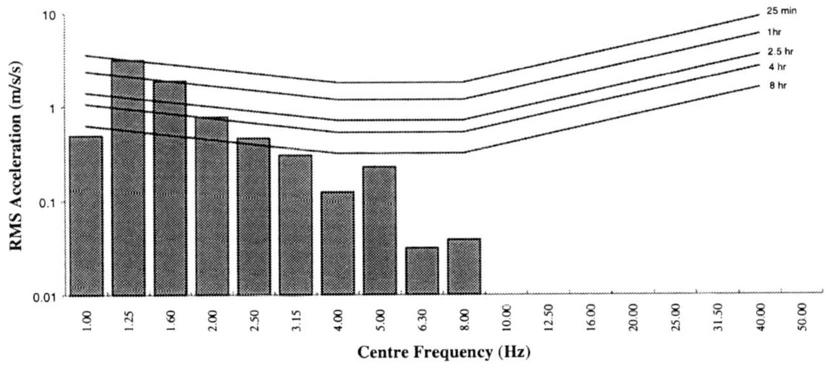

Figure 5. Ride Quality for the Linear Model in the Time Domain.

6. EFFECTS OF NON-LINEAR ELEMENTS

The effects on ride quality of two common types of non-linear elements are considered in this section. These are bump-stops and non-linear shock absorbers.

Bump-stops are used to limit suspension travel and see a great deal of action in off-road vehicles as they traverse the type of terrain used here. They are typically conical blocks of rubber which are engaged after a specified suspension deflection. The model used here is that of a very stiff spring which comes into effect after a suspension deflection of 100 mm.

Shock absorbers are known to be very non-linear in their behaviour. They typically exhibit different damping rates in jounce and rebound. In this study, a simple bi-linear behaviour for the shock absorber which contains the different damping rates is considered.

6.1 Ride Quality with Bump-Stops

Figure 6 shows how the predicted ride quality is affected by the bump-stops. The results shown are for the same operating conditions as previously used (i.e. a constant speed of 5 m/s over the specified terrain). Since this is a non-linear model, a time integration solution followed by a post-processing step which transforms the results into the proper format is used.

The bump-stops introduce accelerations at higher frequencies than experienced in their absence. The results in Figure 6, while showing bars all the way to 50Hz, are not significantly different from those in Figure 5 at the low frequency end. The quality of the ride is still dominated by the peak at 1.25 Hz.

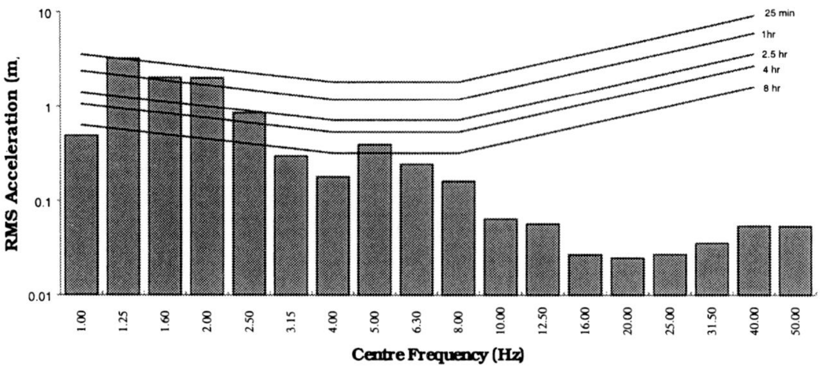

Figure 6. Ride quality with bump-stops.

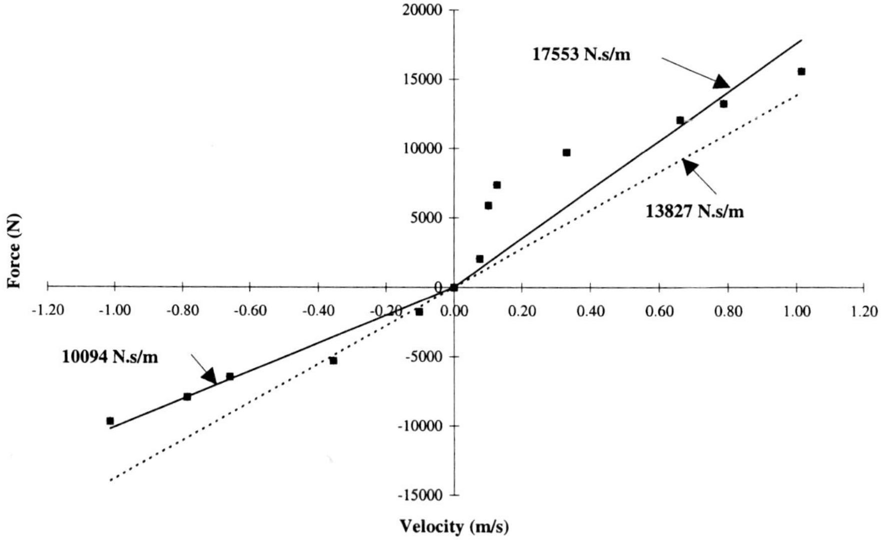

Figure 7. The shock absorber characteristic.

C553/026 © IMechE 1998

6.2 Ride Quality with a Non-Linear Shock Absorber

The shock absorbers used in the truck do not, in reality, exhibit linear behaviour. That is, a plot of force versus velocity does not result in the traditional linear, viscous damping constitutive equation which is known so well. Figure 7 presents both the experimentally measured data for the shock absorber (the points) and curve fits which yield, in the first case, a single linear fit over the range of the data:

$$\mathbf{F} = 13827\mathbf{V}$$

where **F** is the force in the shock absorber, **V** is the velocity across the shock absorber, and **C=13827 Ns/m** is the viscous damping coefficient. Note that the linear curve fit which determines the viscous damping coefficient has been forced to pass through the origin. Another fit which doesn't pass through the origin yields **C=13728 Ns/m** with a force of **2483N** at zero speed. The viscous damping coefficients vary by approximately 0.7% so that either could be chosen but it is imperative that the damping force be zero at zero speed or the linear model will not be in equilibrium at steady state.

The second curve fit is bi-linear with different slopes for jounce and rebound: **17553 Ns/m** and **10094 Ns/m** respectively. The curve again passes through the origin for the same reason discussed above. Neither of the curve fits describes the very non-linear behaviour of the shock absorber accurately but, since the purpose here is simply to estimate the difference between linear and non-linear models, the two approximate damping representations are adequate if not accurate.

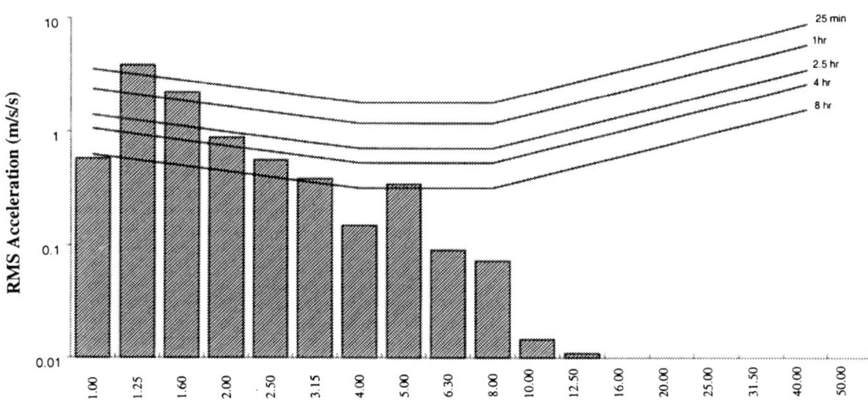

Figure 8. Ride quality with non-linear shock absorbers.

The results for the single linear fit have been presented for the linear simulation (Figures 4 and 5). Figure 8 presents the ride quality for the bi-linear model. Once again it can be observed that the predicted ride quality results are not significantly different when the non-linear shock absorber is used. The dominant peak in the graph is at a centre frequency of 1.25 Hz and corresponds approximately to the 25 minute fatigue-decreased proficiency boundary as was so for all of the cases considered.

7. CONCLUSIONS

The results presented in this paper show that the effect on vertical ride quality prediction of including typical non-linear elements in the multibody model of the vehicle is not large. The differences between the two linear model results shown in Figures 4 and 5 are on the same order of magnitude as the differences introduced by the non-linear elements.

Given the ease of performing the frequency domain predictions and the fact that off-road and, particularly on-road, vertical profiles are most often available as PSD's, it is reassuring that ride quality estimates made in the frequency domain appear to be adequate for design purposes. The ability to obtain useful design information from parameter studies carried out in the frequency domain is reassuring.

8. REFERENCES

1. Anderson, R.J., “The A’GEM Multibody Dynamics Package”, Multibody Computer Codes in Vehicle System Dynamics, edited by Kortüm, W. and Sharp, R.S., Supplement to Vehicle Systems Dynamics, Volume 22, Swets and Zeitlinger, Amsterdam, 1993, pp. 41-44.

2. Anderson, R.J., Cosme, C.A., and Minato, R., “Suspension Design using Multibody Dynamics with CAD Pre- and Post-Processors”, Proc. I.Mech.E. International Conference on Vehicle Ride and Handling, Birmingham, November, 1993, Paper No. C466/015, pp.311-318.

3. International Standard ISO-2631-1978(E), "Guide for the Evaluation of Human Exposure to Whole-Body Vibration".

 C553/026 © IMechE 1998

C553/014/98

Application of multi-body dynamics to derailment safety

J R EVANS BSc, MIMechE and **J CLEMENTSON** MA, MIMechE
AEA Technology Rail, Derby, UK

SYNOPSIS

Derailment of railway vehicles is a rare but potentially dangerous event. New designs of vehicles must be carefully verified to ensure safety from derailment, and when existing vehicles derail, full and careful investigation is required. This paper describes the use of the VAMPIRE® multi-body railway vehicle dynamics simulation package in the design and investigation process. Verification tests for derailment predictions are described, and case studies of actual derailment incidents are given.

1. INTRODUCTION

AEA Technology Rail, formerly British Rail Research, has for many years been in the forefront of the science of railway vehicle dynamics. This extensive experience has driven the development of our world-leading multi-body railway vehicle dynamics simulation package VAMPIRE®.

By comparison with general purpose multi-body dynamics packages VAMPIRE® has two particular strengths. The first is that by concentrating on the railway environment VAMPIRE® is able to make certain well-validated simplifications which result in simulation times up to a hundred times faster than general purpose multi-body dynamics packages. Secondly, VAMPIRE® has evolved robust and efficient techniques for dealing with the special conditions of railway vehicle dynamics such as the severe non-linearities, discontinuities and sometimes very high frequencies inherent in the wheel-rail contact.

The VAMPIRE® package is used by railway vehicle manufacturers and operators in the UK and around the world for the prediction of stability, ride and curving performance of new designs and for the investigation of in-service problems.

© IMechE 1998 C553/014

The ease of use, specialised suspension components such as air springs, and the wide range of post-processing and railway-oriented analysis capabilities make VAMPIRE® an integral part of the design process for railway vehicle manufacturers. Continuing validation, and years of practical experience in railway applications make VAMPIRE® the reliable choice for the investigation of rail vehicle dynamic behaviour.

Confidence and experience of the use of VAMPIRE® in these applications has grown in an environment of significant change in the UK rail industry. Track testing has always been expensive and relatively difficult to organise, but this is particularly the case on the privatised railway. New vehicles must demonstrate compliance to Railtrack Group Standards over a range of vehicle conditions. The railway industry has made increasing use of VAMPIRE® to support this activity, and indeed some of these safety requirements can only be demonstrated by calculation.

In addition, the separation of train operation from infrastructure has increased the importance of independent investigation of real derailment incidents. VAMPIRE® is increasingly used in the investigation of derailments, particularly simulating actual incidents by modelling vehicle behaviour over surveyed track and evaluating the sensitivities to vehicle set-up, speed and specific track features.

2. VALIDATION

Unlike any other dynamics package, VAMPIRE® has been carefully validated for derailment situations by a series of controlled tests using a laboratory coach "Decapod". This vehicle had a fifth central wheelset which could be applied to the track with any desired combination of yaw angle, vertical wheel loads and lateral applied force by means of hydraulic jacks.

The test wheelset was taken to the point of derailment for a wide range of vertical wheel loads and wheelset angles of attack, and the forces required to cause derailment compared with the VAMPIRE® predictions. The agreement was found to be very good, giving confidence in the results of the software even in such extreme conditions.

3. DESIGN ASSESSMENT

3.1 Vehicle Acceptance Criteria

In the UK, Railtrack PLC Group Standards specify acceptance criteria and tests for new and modified vehicles. Typical acceptance tests include :-

- Wheel unloading on twisted track
- Bogie rotational resistance
- Derailment resistance calculation
- On-track ride safety tests
- Vehicle ride performance
- Vehicle kinematic gauge

Simulation of these tests using VAMPIRE® can demonstrate the acceptability of the design prior to final vehicle production. Optimising the design as far as possible using these multi-body dynamics techniques then allows more cost effective track testing. Simulation allows investigation of the full range of vehicle conditions likely to be experienced in service, such as worn wheels or deflated airsprings, to ensure an optimised design.

The derailment resistance calculation involves calculation of the derailment quotient Y/Q (the ratio of lateral to vertical wheel load) for a range of different combinations of curve radii with the maximum permitted cant gradient combined with an imposed track twist to promote maximum wheel unloading on the most derailment prone wheel.
The use of calculations alone has been accepted for this case, as testing on ordinary track would be unlikely to include the worst case combination of conditions, and setting up a special test track to investigate the worst case for a specific vehicle would be very expensive. This calculation is not required for straightforward vehicles which pass both the wheel unloading and bogie rotational resistance tests, but is required in other cases. As a check, the calculations should be submitted to independent scrutiny, and simulations of the wheel unloading and bogie rotation tests should be compared with the measured results to confirm the accuracy of the suspension model.

To investigate the full range of conditions the following variations may need to be studied.

- Range of wheel profiles from new to worn
- Range of rail profiles from new to worn
- Range of wheel to rail friction
- Braking and/or traction may need to be included.

The vehicle model must include realistic representations of all important bodies, suspension components and clearances. Structural flexibilities may be important in calculating wheel unloading. The modelling has to be validated against real laboratory testing of the vehicle as nominal suspension parameters rarely match the as-built condition.

Vehicle acceptance calculations are a particularly challenging application of a dynamics software package. The derailment resistance calculation requires reliable solution of the wheel/rail contact equations in severe flange contact conditions, while kinematic gauging or pantograph sway calculations require the prediction of movements to within a few millimetres several metres above rail level. VAMPIRE® has been carefully formulated to cope with these types of calculation reliably and efficiently.

3.2 Derailment Resistance Calculation Example

This example calculates the response of a vehicle passing over a 150m radius curve with 100mm of cant. The transition curves at the start and end of the curve are of 30m length with the run-off transition including a linear dip of 20mm over 6m in the outer rail, 21.56m after the start of the transition. This ensures that the leading wheelset is at the dip exactly as the trailing wheelset enters the transition curve. A nominal low speed of 10mile/h has been used.

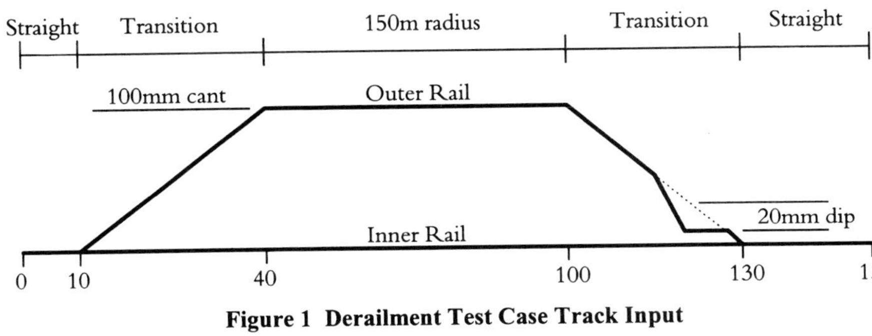

Figure 1 Derailment Test Case Track Input

This calculation is based on the requirements of Railtrack Group Standard GM/RT2141. The cant gradient of 1:300 represents the maximum permitted on the railway, while the 20mm dip in one rail represents a worst-case twist defect. In order to demonstrate acceptable safety margins, the calculation would need to be carried out for a range of combinations of curve radius, cant, dip position and vehicle condition.

The predicted lateral and vertical forces at the leading outer wheel of a bogie passenger vehicle in this case are shown in Figure 2.

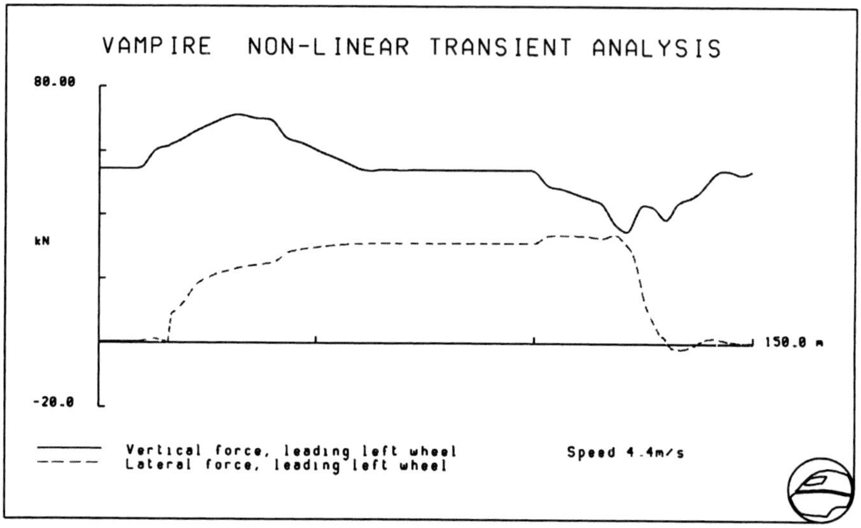

Figure 2 Predicted forces at leading outer wheel

As the vehicle negotiates this track input there is an increase in vertical load on the leading outer wheel in the run-on transition, due to the twist resulting from the cant gradient. Likewise, there is a reduction in the vertical load on the run-off transition, which is significantly augmented by the effect of the additional 20mm dip in the outer rail. In the lateral sense there is a substantial lateral force on the outer wheel through the curve due to the yaw angle of the wheelset relative to the track, which is largely due to the friction force on the inner wheel driving the wheelset into flange contact. This lateral force is lower on the run-on transition, and higher on the run-off transition as a result of the off-loading and on-loading of the inner wheel due to the cant gradients.

Figure 3 shows a plot of the predicted Y/Q ratio corresponding to the forces in Figure 2. As might be expected, the worst case is when the wheel is at the bottom of the dip, with a maximum value of Y/Q of 0.93, which is well within the permitted limit of 1.2.

C553/014 © IMechE 1998

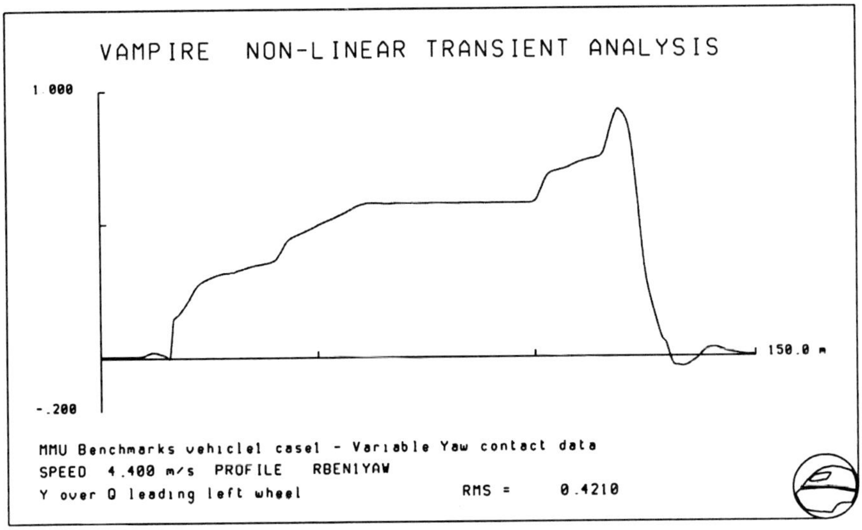

Figure 3 Predicted Y/Q ratio at leading outer wheel

4. DERAILMENT INVESTIGATION

Most derailment incidents are the result of an adverse combination of a number of factors, often under the control of different parties in the privatised railway, and computer simulation is useful in assessing the relative importance of the different causal factors to assist in prioritising recommendations and, if necessary, apportioning responsibility.

The simulation of derailment incidents can be a difficult process, and may require experienced interpretation of conflicting factors. A number of issues must be considered.

4.1 Track Geometry Information

The track geometry will usually be recorded by manual survey. The need for restoration of the line to traffic following a derailment often limits the amount of time available for the survey. This leads to some limitations which need to be understood.

- The track geometry measurement starts with a survey of the unloaded track. A vehicle is then moved onto the track and the deflection of the track under the load is measured at each sleeper in turn, to establish the loaded profile. Despite the use of high quality surveying equipment the accuracy of each individual measurement in the vertical direction is limited to typically ±2mm. This accuracy is not sufficient for high speed situations, and the measurements need to be smoothed. The relatively coarse spacing of the measurements limits the amount of smoothing before detail in the geometry of the track defect being investigated is lost. Sometimes the manual survey may be backed up by comparing with data from the High Speed Track Recording Coach which carries out regular surveys of the track, and can supply measurements at closer spacings.

- The survey is limited in length, usually to approximately 150 metres before the point of derailment and up to 15 metres beyond. Particularly for higher speed incidents, this may not be sufficient to establish the full dynamic behaviour of the vehicle at the point of derailment. Various techniques may be employed to extend the track to give a longer lead-in to the derailment site.
- The track after the point of derailment may have been destroyed. The track data may have to be artificially extended beyond the derailment point, particularly if derailment occurred on a trailing wheelset.

4.2 Modelling Issues

There are a number of factors affecting the vehicle model and the wheel and rail contact conditions which also need to be considered.

- Obtaining adequate parametric information for the modelling of many older types of vehicle can be difficult. Additionally, the condition of an individual vehicle may be an important factor, and the use of generic data for a vehicle type may not be sufficient. Derailed vehicles may be subjected to a torsion test to establish the important effect of structural twisting or suspension set-up on the distribution of wheel loads, but this evidence may be destroyed in a serious incident.
- The tightness of the couplings between adjoining vehicles may be a significant factor in some cases, and may not be easy to establish after the event.
- The wheel and rail profiles may be important in the behaviour of the vehicle, and can be measured, although sometimes damage to the wheels may hinder this process.
- Some important parameters for the simulation, such as the effective coefficient of friction, can not be readily measured, and have to be estimated on the basis of experience.

After consideration of these issues very good simulations of derailments have been demonstrated with vehicles predicted to derail at or close to the actual point of derailment.

However, most derailments result from the simultaneous combination of a number of extreme factors in the condition of the vehicle and its operation. In the absence of complete knowledge of all of these factors it is often the case that actual derailment is not predicted, but a representative vehicle operating in representative conditions is shown to have statistically dangerous levels of the derailment indicators such as Y/Q ratio. The simulation techniques then allow assessment of the sensitivity of the level of risk to different causal factors.

5. DERAILMENT INVESTIGATION EXAMPLES

5.1 Vehicle dynamic response

An important use of VAMPIRE® simulation has been in the investigation of derailments caused by the excitation of vehicle suspensions by cyclic track dips ('Cyclic Top') such as those illustrated in Figure 4.

This type of derailment typically involves a two-axle wagon, and is the result of the excitation of a severe body pitching motion, which is insufficiently damped by the friction suspension. VAMPIRE® can be used to confirm the mechanism of derailment, and to investigate the sensitivity to speed, vehicle loading and suspension type. Certain vehicles are found to be more susceptible to this type of derailment than others. Generally, models of other susceptible vehicles are run over the same track for comparison purposes.

C553/014 © IMechE 1998

Figure 4 'Cyclic Top' Derailment Site

The cyclic motion can be seen initially in the vertical wheel loads and body displacements. As derailment conditions are approached, the wheels start to lift off the rail. Wheel lift in itself does not indicate derailment, as quite large lifts are required to overcome the depth of the flange, and even large wheel lifts must be accompanied by some lateral motion, possibly due to curving or vehicle hunting, before the wheel flange is able to literally jump onto or across the rail head.

Figure 5 shows a typical simulation of a real derailment incident. The real derailment occurred 102.4 metres along the track, which coincides with the largest wheel lift in the predictions. The x-axis relates to the position of the leading wheelset along the track. The peaks of wheel lift occur at 9m intervals, and as the vehicle has a 9m wheelbase, the peaks in the trailing axle correspond to the same track input as the preceding peak for the leading axle.

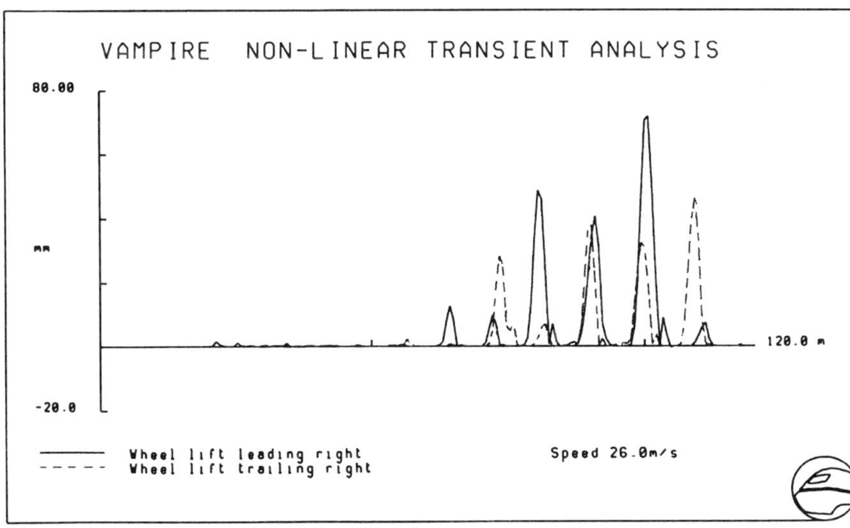

Figure 5 Predicted Wheel Lift Above Rail

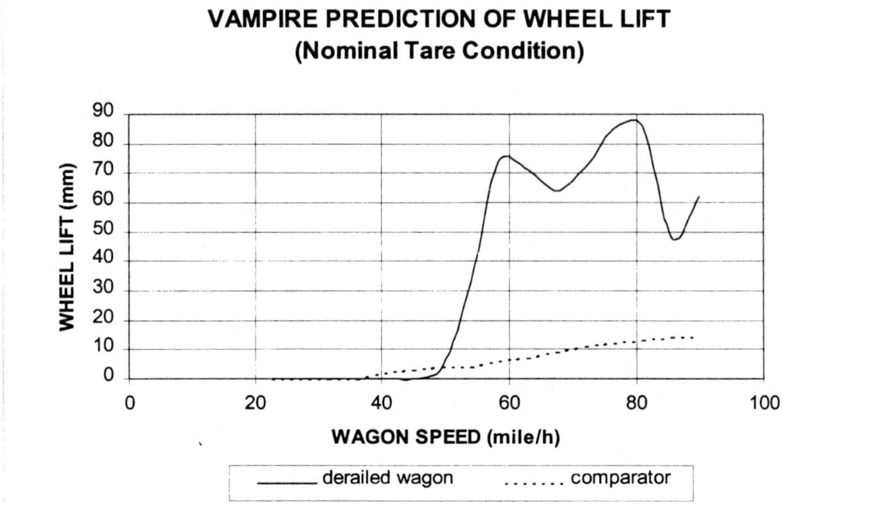

Figure 6 Comparison of wheel lift for different wagons over a range of speeds

The maximum wheel lift can be plotted for a range of speeds to illustrate the speed range for which derailment is most likely. Figure 6 shows the variation of wheel lift with speed for the derailed vehicle and a comparable vehicle with a different suspension type. The derailment occurred at 57 mile/h, and the comparison clearly shows that vehicle speed and the suspension design were each contributory factors.

In another recent example, the cyclic dips were not excessive, but friction surfaces on the wagons had been lubricated. VAMPIRE® simulations were used to demonstrate the level of damping at which the risk of derailment became significant.

5.2 Flange climb on plain line and in switches and crossings

In the assessment of flange climb derailments on plain line the track twist geometry and the results of torsional stiffness tests of the vehicles are compared with the requirements of the relevant standards. Wheel and rail profiles are checked to ensure they are within acceptable limits. Usually one or more of these parameters is found to be outside the normal limit, and VAMPIRE® simulation is only needed if the relative contribution of these factors or others such as wheel-rail friction, vehicle loading and running speed must be established.

There have been several derailments at a particular switch over the past few years involving the same type of electric multiple unit train (Figure 7). The mechanism of each of these derailments has been clearly established. In each case, a sharp corner formed on the switch blade penetrated the flange face of a recently reprofiled wheel causing the wheel to climb up the blade and drop between the switch blade and stock rail. In every case the incident involved derailment of the leading wheelset, leading bogie of the third vehicle in the unit formation (the power car). VAMPIRE® simulations were subsequently used to investigate why this vehicle always derails in preference to other vehicles.

C553/014 © IMechE 1998

Figure 7 Derailed power car bogie

Models of both the power car and trailer car were constructed. The models were run at the speed of 15 mile/h recorded for the most recent derailment, over the actual track geometry surveyed and processed for VAMPIRE® use.

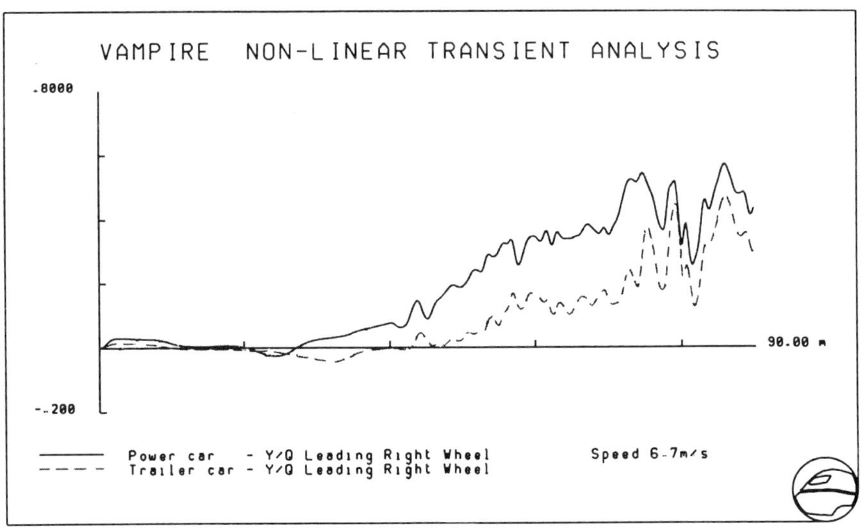

Figure 8 Predicted Values of Y/Q for 0.25 Coefficient of Friction

The Y/Q ratio and lateral force, for the leading right wheel of each vehicle were compared for a range of wheel/rail coefficients of friction. Figure 8 shows the Y/Q ratio, which demonstrates that at the point of derailment, 77m along the track, the flange climb risk is approximately 40% greater for the power vehicle than the trailer vehicle. Much lower Y/Q ratios can cause derailment when a sharp switch blade digs into the flange surface.

5.3 Gauge Spreading Derailment

Derailment is sometimes caused by the wheel-rail forces in curves being sufficient to push the two rails apart, allowing the wheelset to drop between the rails (Figure 9).

Figure 9 Gauge Spreading Derailment

VAMPIRE® can be used to predict the forces induced in a particular track geometry by a particular train. The sensitivity to parameters such as load and train speed and wheel-rail friction level can then be investigated. The forces reacted at the rail can then be compared with practical tests which indicate the forces needed to spread the rails on a section of track.

6. CONCLUSION

Driven by the needs of the rail industry, VAMPIRE® has developed into an important tool for design assessment and derailment investigation. It provides understanding into the behaviour and optimisation of vehicles over the whole of their life and for a wide range of conditions, thus enabling vehicles to be optimised at minimum cost.

Application of VAMPIRE® to derailment investigation is increasing, and careful validation and growing experience in this most difficult and practical of all application areas for rail vehicle dynamic simulation gives increasing confidence in the results. However, the major limitation remains that reliable conclusions depend on the input of realistic parameters and careful interpretation of results. This reliability can only come with years of experience, associated testing and application.

C553/014 © IMechE 1998

C553/038/98

A new non-linear multi-body approach to durability analysis of vehicles

R M V STURT MA, CEng, MIMechE and **R J SNELL** MEng
Ove Arup & Partners, Coventry, UK

ABSTRACT

Multi-body nonlinear finite element analysis is an established technique, widely used in the automotive industry for assessing crashworthiness and occupant protection of new vehicle designs prior to manufacture of prototypes. Similar techniques can be applied to durability. This in turn permits shorter development programmes with fewer phases of prototypes. This paper illustrates the methods with a simulation of a vehicle driving over a kerb at 30mph – a typical "suspension abuse" loadcase. This class of problem is traditionally analysed using a multi-step approach (load calculation, linear stress analysis, life calculation). The new technique presented here is a "virtual prototype test": a vehicle model is driven over a representation of the test track, with tyre deformation and other nonlinear dynamic effects included. The first potential advantage of this technique is that the loading on the vehicle can be quantified with confidence, owing to finite element tyre models. The second is that more realistic stress distributions can be calculated due to the inclusion of dynamic nonlinear effects. Thirdly, any permanent deformation and areas of plasticity can be quantified. In addition, by animating results, the analyst obtains an intuitive feel for the causes and effects of the loading that may help to identify and resolve durability problems.

INTRODUCTION

Durability is one of the principal concerns in the design of vehicles. The term is used here to mean the vehicle's ability to survive the loading encountered during its life, from one-off abuse conditions such as driving over a kerb, to less severe loading that is repeated millions of times during the life of the vehicle such as accelerating and braking. Traditionally, durability performance has been achieved largely by test and development backed up by linear finite element analysis. However, with each new vehicle programme becoming more compressed than its predecessor, and having fewer prototypes, there is less and less opportunity to test. It is becoming essential for designs to be right first time, with testing taking place only as a confirmatory exercise at the end. Test failures at this stage can compromise the timing of the whole programme. Therefore, the accuracy of the analytical techniques is becoming critical.

C553/038 © IMechE 1998

In this paper, we demonstrate the new nonlinear dynamic analysis technique. Vehicle on test track situations can be simulated more realistically than by the more traditional analysis methods, because the test is modelled as it is performed in reality. The modelling techniques have not been correlated to test and require further development, but there is clearly the potential to make an important contribution to durability analysis of vehicles.

The analysis software is LS-DYNA, a code used to solve a wide range of nonlinear dynamic problems. In the automotive industry, it is principally used for crashworthiness simulation.

DISADVANTAGES OF TRADITIONAL DURABILITY ANALYSIS

The three main steps in traditional durability analysis of vehicle structures are:

- Calculate loads to apply to the structure at each input point. This may be done with the aid of a dynamic calculation using a rigid body mechanisms package, often in combination with measured forces at the wheel spindle. Loads are measured or calculated as time-histories. In other cases, loads may simply be peak tyre contact patch forces estimated by experience. The input point forces would then be determined by static mechanisms analysis. The difficulty of making accurate prediction of the loads is recognised as being a key limitation of durability analysis.
- Perform linear static FE analysis to obtain stress distribution, often as a series of unit loadcases. A development of this technique is to use a dynamic modal model (1) to calculate the dynamic stress distribution.
- Combine loadcases according to input load calculation; often a fatigue life calculation is performed and linked to the load time histories to determine number of repeats to crack initiation.

The traditional multi-step approach to durability analysis has several inherent disadvantages, which together reduce the predictive ability of the calculations. This paper focuses on two aspects: determining the loads, and calculation of stresses. The final step, prediction of fatigue life from the stresses, is not covered here. The problems considered in this paper are:

- Difficulty in quantifying loads. For vehicle on test track situations, results are dependent on the quality of the tyre model. As will be shown in this paper, it is essential to include the flexibility of body and suspension components – rigid body calculations may be grossly inaccurate.
- Linear stress analysis techniques ignore yielding. The effects of local yielding on fatigue life at individual elements can be allowed for in a fatigue post-processor, but the change in stress distribution in other elements as a result of the yielding would not be accounted for.
- Linear stress analysis techniques ignore nonlinear geometry effects. For example, forces applied perpendicular to flat panels are assumed to be reacted entirely by local bending. In reality, as the panel bends, the change in geometry allows load to be carried in membrane tension. This effect can be highly significant at the top of a shock tower, for example. It is possible to run most of the commonly used linear codes with geometric and material nonlinearity included, but it then becomes impossible to superpose different loadcases in the normal manner, and many of the advantages of the traditional techniques are lost.

C553/038 © IMechE 1998

- Static stress analysis cannot account for dynamic effects: the various components of the vehicle respond dynamically to the input load time histories and hence the true stress distributions are poorly represented by static loadcases in which equilibrium must be assumed.
- Loads on different points of the vehicle body may be out-of-phase (i.e. may peak at different times). It is not possible to select one combination of loads that represents the worst-case loading condition, because different areas of the structure may reach their peak stresses at different times.

Because of these difficulties, it is unlikely that stress distributions used for durability analysis accurately reflect the real situation. Therefore, no matter how sophisticated the life calculation that forms the final step, results will always be circumspect.

In addition, it is often desirable to quantify the damage to a vehicle from the various "abuse" loadcases (for many of these, localised yielding is tolerated provided that the permanent deflections are within pre-set limits). It is not possible to calculate the permanent deformation using linear stress analysis techniques.

MULTI-BODY NONLINEAR ANALYSIS

Multi-body nonlinear finite element analysis is generally carried out using codes such as LS-DYNA (2). The codes have evolved to treat the most complex classes of problem such as crash simulation. These are characterised by large deformations and displacements, highly nonlinear material behaviour, arbitrary contact conditions and multiple body motion. Most of the codes use explicit time integration; this technique avoids the mathematical problems of inverting large matrices and allows a greater range of capabilities. The principal disadvantage is that the timestep is limited by numerical stability (typically to about 1 microsecond), therefore the number of solution steps needed to simulate an event of 0.1 to 1 second duration can be of the order of 10^5 or 10^6, leading to long computing times. As computers become faster, the computing cost is less of a limitation and it is now possible to use explicit solvers to tackle highly complex problems.

An example of a crash simulation in LS-DYNA is shown in Figure 1. The model consists of 114609 shell elements, 33341 solids and 16897 beams. The simulation time is 120ms. The various components of the vehicle (body shell, engine/gearbox, subframes, suspension members, battery, brake servo, etc) and crash dummy occupants are modelled as a series of bodies. Each is deformable, with nonlinear material properties allowing permanent deformation; generalised contact algorithms are used to prevent the bodies penetrating each other.

The technique can be extended to include simulation of the restraint system and occupant response – see Figure 2. The dummy is a fiftieth percentile male Hybrid III, the most commonly used frontal impact dummy. The model dummy is one of the FT-Arup series (3). The level of detail in the model is indicative of current trends – the geometry of every part is meshed from CAD data, and the material properties of the plastics and foams are generated from extensive testing of samples.

The capabilities required for Virtual Prototype simulations of vehicles on test tracks are a subset of those required for crashworthiness analysis:

- Nonlinear material behaviour in limited areas, particularly for extreme loading and abuse cases
- Large displacements
- Some nonlinear geometry effects
- Multiple bodies

It would, therefore, appear that existing modelling techniques using codes such as LS-DYNA could be adapted for Virtual Prototype test track analysis.

OBJECTIVES AND OUTLINE OF THE NEW ANALYSIS TECHNIQUE

The aim in the work reported here was to overcome as many of the disadvantages of the traditional techniques as possible. Most of the errors occur because each stage of the analysis is to some extent removed from the real conditions of the test. The ideal solution is to make the simulation reflect the test as fully as possible, and to do the analysis in a single step. The approach used here was to model the whole vehicle driving on the test track, including suspension and tyres, using LS-DYNA. In theory, all of the disadvantages listed above are overcome by this modelling method. It should be possible to calculate the loads accurately for any road surface, using a finite element tyre (4) together with a fully deformable model of the vehicle. All the dynamic and nonlinear geometric effects are automatically included; and because yielding is represented, permanent damage to the vehicle can be quantified. There is also the practical advantage of performing the whole calculation in a single analysis.

The principal limitation is CPU time. At present it is not practical to simulate events longer than 1-10 seconds, depending on the degree of detail in the vehicle model. As faster computers become available, longer events will be simulated with more of the vehicle modelled as finite elements.

SUSPENSION ABUSE LOAD MODEL

An example is shown in Figure 1. A small hatchback vehicle weighing 997kg is driven over a 100mm kerb at 13.4m/s (30mph). At this speed some localised yielding might be tolerated but the vehicle should remain driveable.

Suspension components are modelled with shell elements with elastoplastic material properties, mounted to the body and to each other using spring elements that represent the compliance of the bushes and/or bolts. These have individually determined force-deflection characteristics that mimic those of the real components. The tyre model consists of orthotropic shell elements with internal pressure determined by a control volume. This paper presents work in progress, and at present most of the vehicle body is modelled as rigid, with deformable elements only at the front shock towers. However, the results suggest that body flexibility is significant in determining loads and stresses even in the suspension; it is intended to include at least a proportion of the body as deformable elements at a later date.

The model statistics are:

- 34284 shell elements
- 396 beam elements

C553/038 © IMechE 1998

- 108 discrete elements
- Simulation time 0.2 sec
- CPU time 59069 sec (16.4 hours) on a single processor of a Cray J916

RESULTS

Figure 3 shows the vehicle model driving over the kerb. The development of stresses in the front suspension components is illustrated in Figure 4.

Quantifying the loading

The force generated at the top of the spring/damper is shown in Figure 5. The peak force (9.79kN) is equivalent to 3.26g times the front corner weight of the vehicle. Estimates for the loading in this type of event (used for design purposes when no other information is available) would typically be around 3g. Given the extreme deformation of the tyre (Figure 6), it is doubtful whether the look-up table method of calculating tyre forces used by rigid body mechanisms programs would be sufficiently accurate for this situation. The facilities available in LS-DYNA for modelling tyres with finite elements confer a very significant advantage over traditional methods, because of the ability to predict loads.

Out-of-phase loading

Force histories at selected points are shown in Figure 7. The results show some evidence of out-of-phase loading: the forces on the body via the subframe peak at around 120ms, while the force through the shock tower peaks at 128ms. A static analysis of the body in which all peak loads were applied simultaneously would be incorrect. However, the loads on the subframe in this analysis are in phase with each other, all peaking at close to 120ms.

Note also that the longitudinal (X) force on the tyre has a lower and earlier peak that the spindle – the dynamic response of the lower control arm amplifies the loading into the suspension.

Effect of flexibility of suspension and body

To test the effect of the flexibility of the suspension, the LS-DYNA model was rerun with the main member of the front subframe made rigid. The model is otherwise identical to the baseline analysis. Forces at selected points are compared with the baseline analysis in Table 1 below.

	Peak force, deformable subframe	Peak force, rigid subframe (% difference)
Subframe to body, front bolt, X	3.96kN	3.06kN (-23%)
Subframe to body, rear bolt, X	6.59kN	9.81kN (+49%)
Subframe to body, front bolt, Y	2.47kN	1.52kN (-38%)
Subframe to body, rear bolt, Y	12.15kN	1.41kN (-88%)
Wheel spindle, X	11.40kN	13.23kN (+16%)

Table 1 – Peak Forces for Deformable vs Rigid Subframe Models

Subframe flexibility affects the load distribution into the body dramatically. If the suspension members were assumed rigid when determining the loads, any subsequent durability calculation would be grossly inaccurate. It is likely that body flexibility would have a similarly significant effect. The Y-force on the rear bolt is of particular interest. It arises from bending of the main subframe member – as the bolts are offset behind the member, they force the subframe to strain axially as it bends (Figure 8). This effect is not shown with the rigid subframe. The lack of body flexibility in the model exaggerates the difference between the two cases, but the conclusion is clear – assumption of rigid suspension components would lead to gross underprediction of the lateral loads on the body at the rear bolt. Consequently, a durability problem in the vehicle body might be missed.

Ability to predict permanent deformation

Permanent deformations in the shock tower are shown in Figure 9. After the event, the maximum deformation of the shock tower is 1.1mm. If linear stress analysis methods were used, the permanent deformation could not be estimated – the only tool at the analyst's disposal would be a "gut feel" interpretation of the stresses, possibly leading to over-conservative design.

CONCLUSIONS

It is now possible to model certain durability tests as they are carried out in practice, using dynamic nonlinear multibody analysis. The techniques presented in this paper are most usefully applied in extreme load conditions, where nonlinear behaviour and consequent errors in the more traditional techniques are most significant, or where the loads generated by tyre deformation are not easily quantified. These cases also have the advantage of relatively short event times.

The potential advantages are:

- Ability to predict the loads on the structure including those due to tyre response
- Ability to predict the amount of permanent deformation
- More accurate stress predictions because fewer unrealistic assumptions are made.
- The whole event can be modelled in a single analysis
- By animating results, the analyst obtains an intuitive feel for the causes and effects of the loading – this in itself may allow rapid identification and resolution of durability problems.

The models have not been correlated to test. The magnitudes of the stresses and forces quoted should therefore be treated with caution. However, the models clearly demonstrate some important principles; the techniques presented could be developed to form the basis of an accurate predictive method for calculating stresses for durability analysis.

REFERENCES

(1) Huang L, Agrawal H & Kurudiyara P, Dynamic Durability Analysis of Automotive Structures, Paper no. 980695, SAE, 1998

(2) Hallquist, J.O., LS-DYNA 940 manual, LSTC, 1997

(3) Moss S, Huang Y, Keer T, Shah B, Development of an Advanced Finite Element Model Database of the Hybrid III Crash Test Dummy Family, Paper no. 971042, SAE, 1997

(4) Sturt R and Shah B, Tyre Modelling in LS-DYNA, 4th LS-DYNA Japan Users Conference, 1997

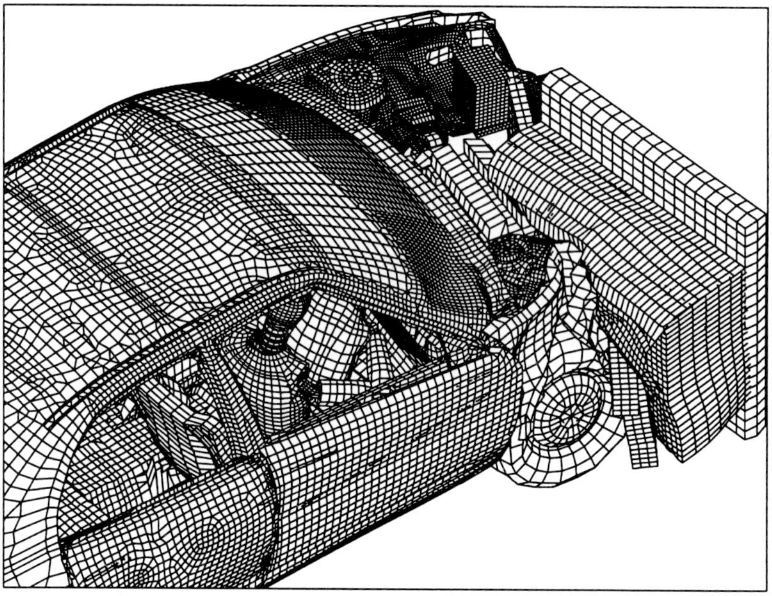

Figure 1 – Simulation of Impact into Offset Deformable Barrier

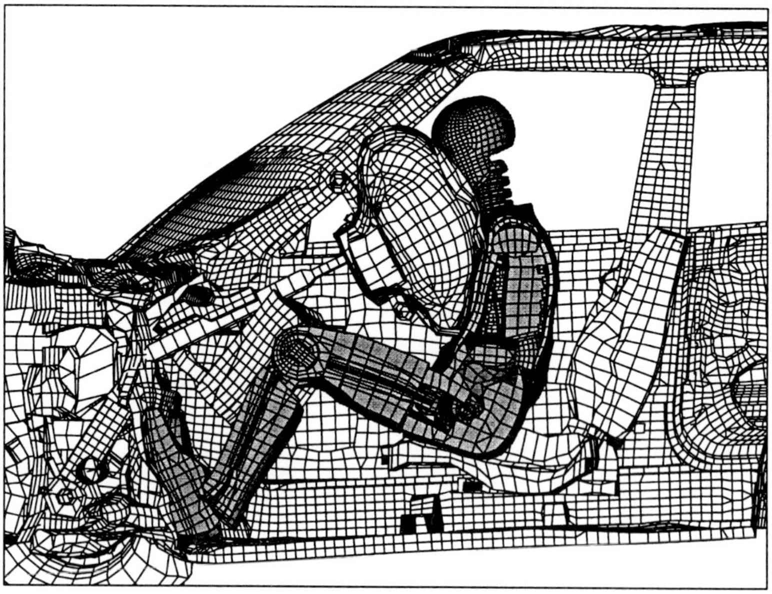

Figure 2 – Model of Hybrid III Crash Dummy in Vehicle

C553/038 © IMechE 1998

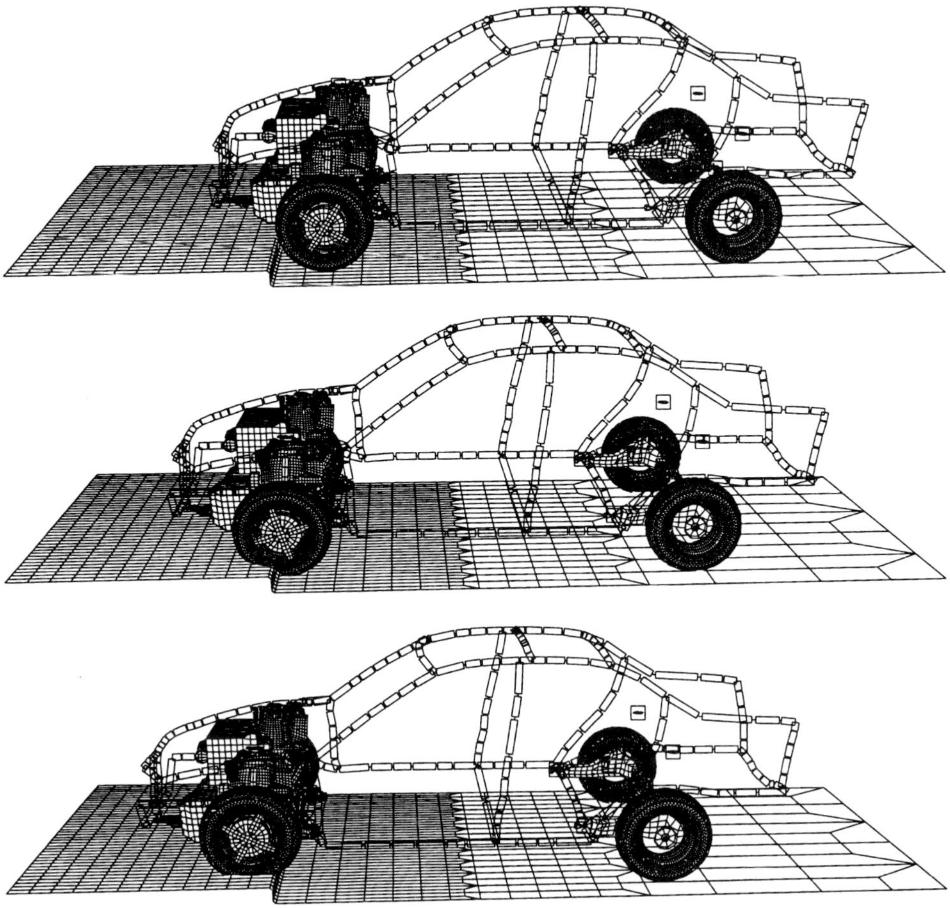

Figure 3 – Simulation of Vehicle Driving over 100mm Kerb

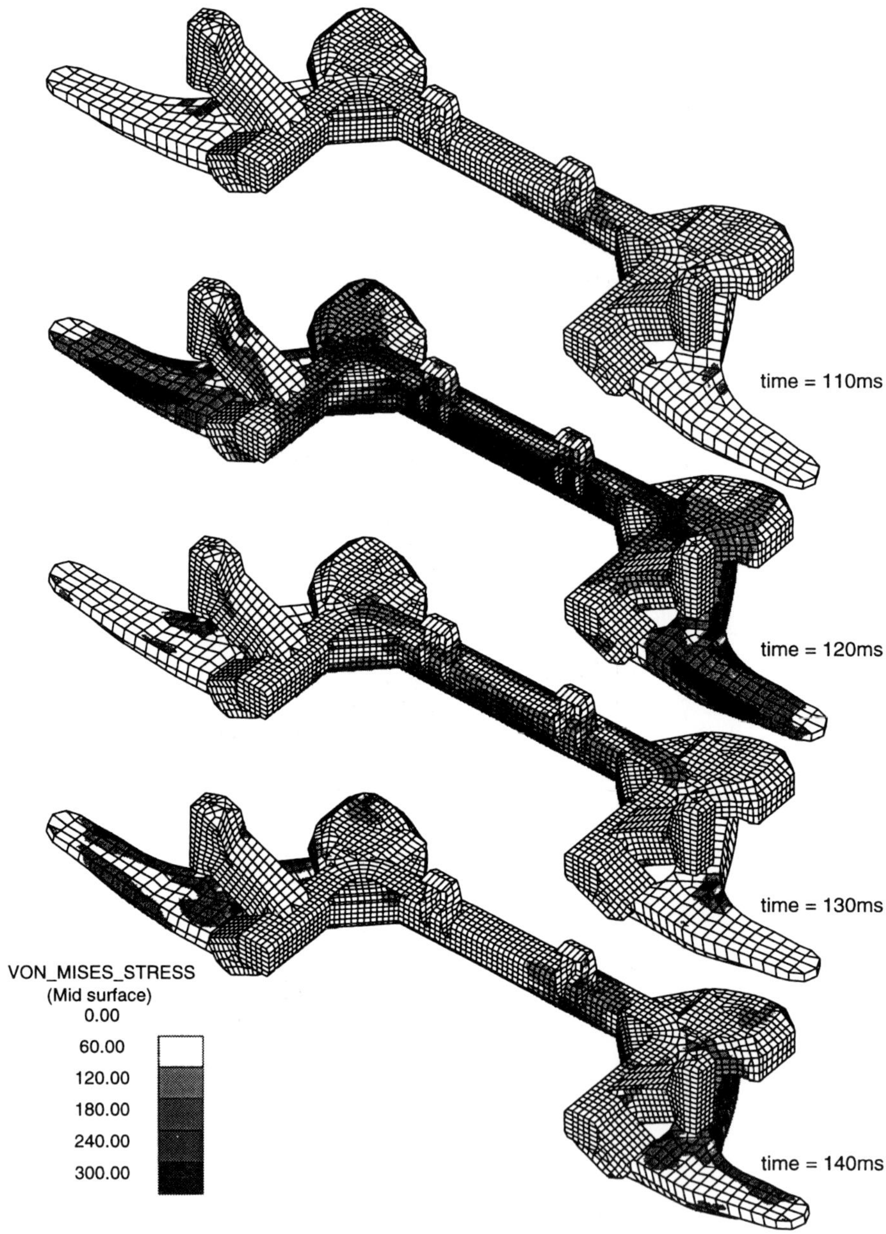

Figure 4 – Dynamic Stress Distribution in Front Suspension

C553/038 © IMechE 1998

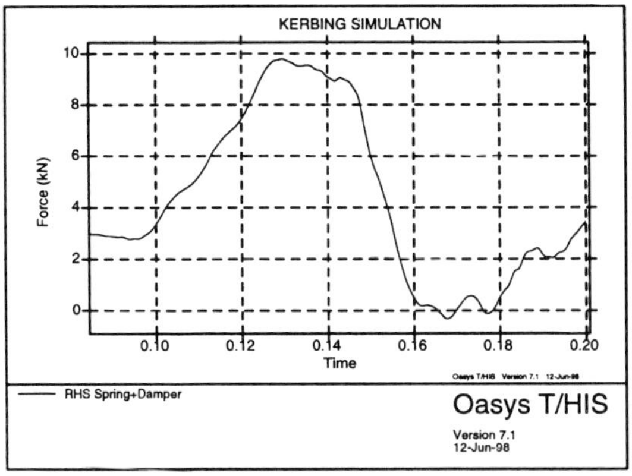

Figure 5 – Force on Shock Tower

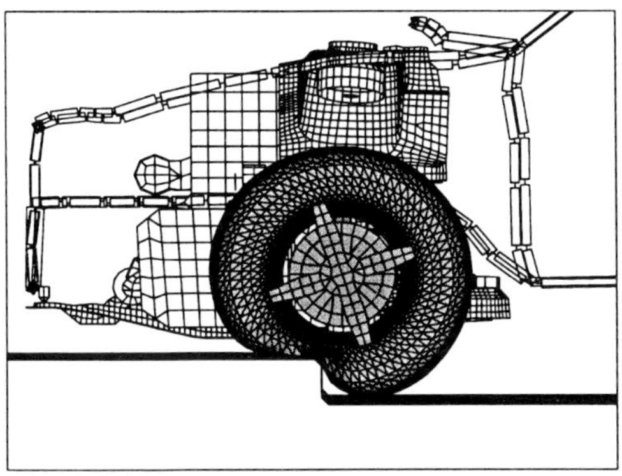

Figure 6 – Tyre Deformation

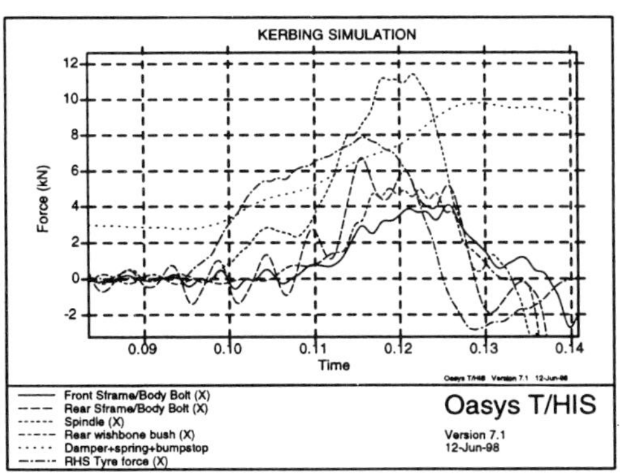

Figure 7 – Time Histories of Forces on Body and Suspension

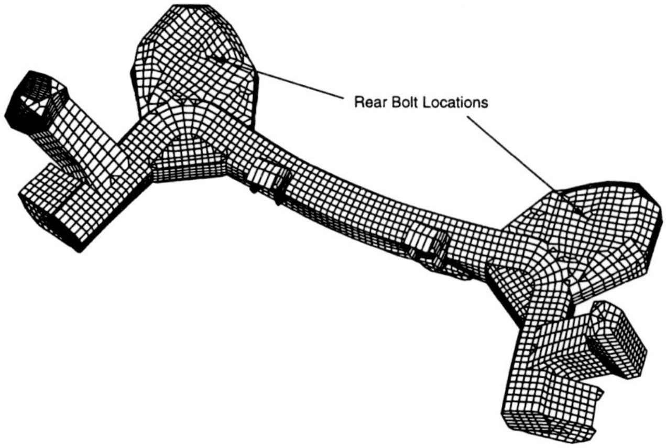

Figure 8 – Deformed Shape of Subframe (Deformations Magnified by 20)

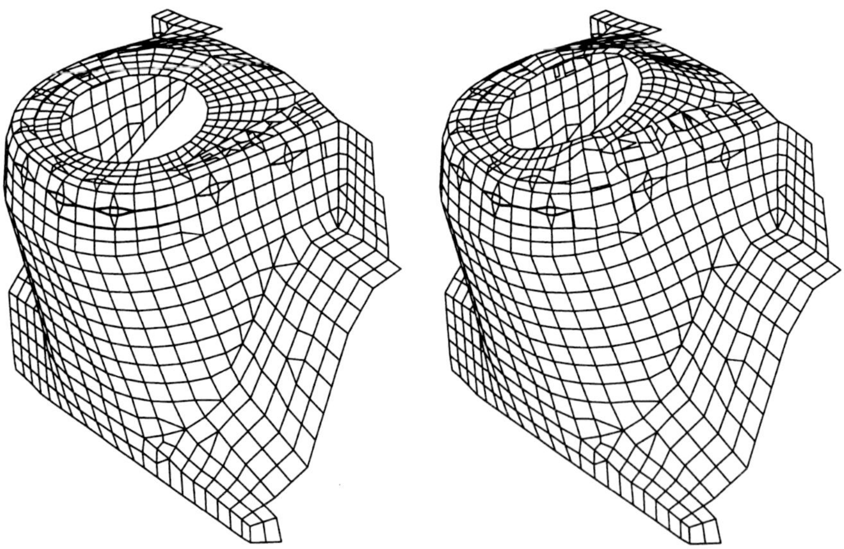

Figure 9 – Initial and Final Shape of Shock Tower (Deformations Magnified by 25). The maximum deformation at life size is 1.1mm.

C553/030/98

Optimum design of multi-body systems – the ODESIM experience

J M JIMENEZ, J G IZAGUIRRE, A SUESCUN, and **A N AVELLO**
CEIT and University of Navarre, Spain

Synopsis. This paper presents the most relevant results of ODESIM project, funded by the European Commission within ESPRIT IV programme. The project main objective has been the development of a set of software tools for the kinematic and dynamic optimum design of multi-body systems. Several types of design procedures are targeted: interactive design, designer-driven optimisation and fully automatic optimisation. In addition, ODESIM attempts to bring the world of multi-body design closer to CAD environments through a tight integration between CAD programs and multi-body solvers. This integration allows CAD designers to create and optimise multi-body designs taking advantage of the parametric approach currently used by CAD systems. ODESIM is implemented using High Performance Computing and Networking (HPCN) techniques to perform parallel computations in the most CPU-intensive optimisation problems. Results on the solution of an industrial practical problem are included.

1. INTRODUCTION

The aim of ODESIM project has been the development a complete sequence of advanced software tools for kinematic and dynamic optimal design using high performance computing and networking techniques. The project acronym stands for *Optimum Design of Multi-Body Systems*. The simulation models focus on complex mechanical systems with moving parts, integrating kinematics and dynamics of multi-body systems (MBS). Those tools can be used in production industrial environments in the mechanical engineering field, as shown by the different application examples included in this paper.

Current MBS simulation programs can simulate complex MBS with rigid and flexible bodies subject to different kinds of external forces, control laws, non-linear effects, etc. Most of those programs assume that the dimensions of the MBS are perfectly known and that the user is only interested in predicting its behaviour over time. However, this is not the case during the design phase, in which the designer starts from a draft design and has to refine its dimensions in successive iterations.

With current MBS analysis programs, the designer must perform kinematic and/or dynamic analysis at each design step, assess the performance of the system, and accordingly, take a decision on what dimensions should be modified to improve the performance. This iterative process has two main drawbacks. First, it relies heavily on previous designer's experience with similar designs and therefore can only be done by experts; and second, it is a very time consuming procedure since the designer has to repeat the assessment of performance, the decision making and the dimensional modification at every iteration step.

In the approach followed in ODESIM project, the MBS design problem is addressed from the very beginning of the design process, which starts with the CAD definition of parts. In the CAD program, the designer creates a parametric model of each element or part of the mechanical system. From this geometric model, the program creates a fully parametric multi-body model that will later be optimised according to the objective function and design constraints specified by the designer. ODESIM project has been developed using as baseline two commercial packages: a CAD modeller (Euclid Quantum, developed by Matra Datavision) and a MBS simulation package (COMPAMM, developed by CEIT).

2. OPTIMUM DESIGN OF MULTI-BODY SYSTEMS

The analysis of MBS quality and functionality is often done separately from the detailed mechanical design. Frequently the analysis is only kinematics and does not include dynamic characteristics. If an organisation has the resources, MBS analysis is traditionally performed by highly specialised analysts, while design is performed by a different team. The communication lag between the analysts and designers invariably lengthens the product design cycle time.

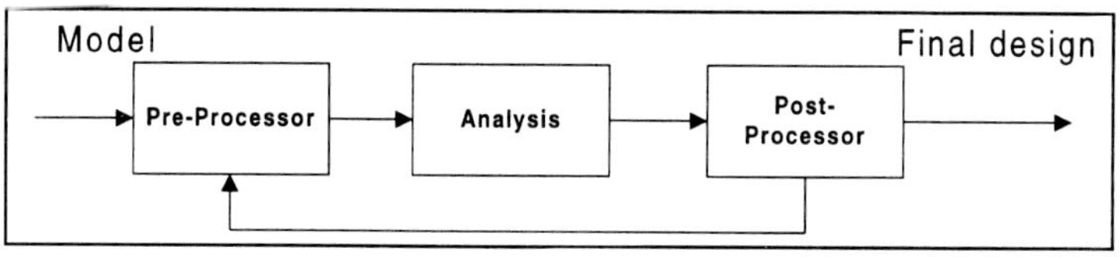

Figure 1.

Traditionally, MBS analysis codes such as ADAMS or DADS, have been structured along the lines of FEA (Finite Element Analysis) programs. Those programs typically consist of a separate pre-processor, an analysis engine and a post-processor. It is conventional for these three modules to execute sequentially each one accepting as input the data set left behind by the previous module. As a consequence, to make a change in the MBS design, the whole sequence of pre-processing, resubmitting the analysis job and post-processing the results must be repeated (see Figure 1). In fact, when wishing to design a new system that must comply with certain specifications and only analysis tools are available, the designer has to proceed in an iterative *trial-and-error* manner by means of successive analyses. This *trial-and-error* process is very slow and depends strongly on the designer's experience. This is part of the reason why MBS analysis tools have largely been used just to check final designs.

C553/030 © IMechE 1998

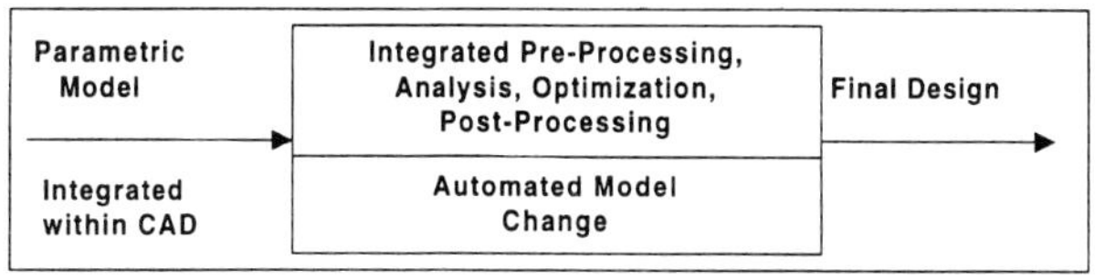

Figure 2.

A key idea in the development of CAE codes for MBS analysis is to integrate the analysis capability in the software tools that the design engineers currently use. This task requires the development of easy-to-learn graphical-oriented packages, fully integrated within a CAD system. The graphic user interface has to be defined in such a way that no special knowledge on MBS analysis is required. In addition, those software tools have to provide the ability to easily perform "*what if?*" studies, i.e., functions to change a design semi-automatically or automatically (see Figure 2), analyse it and store the comparative results for display.

Optimal design methods help to solve this problem. Those methods can lead directly, with a small or without intervention of the analyst, to a design that complies with the given specifications or that, at least, is the best available one from a certain point of view.

The optimal design of MBS is undoubtedly an important problem, although it has not yet attained the level required for a general commercial implementation. Mathematical problems arising in MBS kinematics or dynamics are of small or medium size: normally, just a few tens of degrees of freedom and a few hundreds of dependent co-ordinates and constraint equations. In spite of being of small size, MBS problems present considerable specific difficulties. Perhaps the most important one is their high degree of non-linearity, due to their large displacements and strongly changing geometry. This implies that MBS are particularly unsuitable for design methodologies relying on human intuition. This makes the computer simulation since the first design stages specially important and necessary. While design based on prototypes is an expensive, cumbersome and slow process, when it is based on interactive optimised computer models with realistic graphics it is very inexpensive and fast, allowing the solution in hours or days, of problems that would otherwise need weeks or months.

Almost all of the optimisation methods are based on the knowledge of the gradients of the goal function that measures the performance of the MBS. In dynamic optimisation of MBS, the goal function is computed as a result of a dynamic simulation. Very often, the differential equations of motion are stiff and highly non-linear, and both their formulation as well as their numerical integration large computer resources. These reasons, and the fact that the gradient calculation can be computed in parallel, make MBS optimisation especially suitable to be implemented in a HPCN environment.

3. ODESIM PROJECT OBJECTIVES

The set of software tools developed within the ODESIM project attempts to face the problem of optimum design of MBS and represents an effort to fulfil the existing gap between design engineers and MBS analysis and design tools. The main objectives in ODESIM project can be summarised in the following points:

- Development of a MBS pre-processor integrated in a parametric CAD system to build the simulation model in a easy way

- Fully integration of a general-purpose MBS simulation code with the CAD system, including graphic tools for interactive visualisation of the computed results.
- Implementation of an optimisation module for optimum kinematic and dynamic design of MBS.
- Development of a parallel version of the solver to run the optimisation problems on multiprocessor workstations and/or clusters of computers, in order to reduce the calculation time.

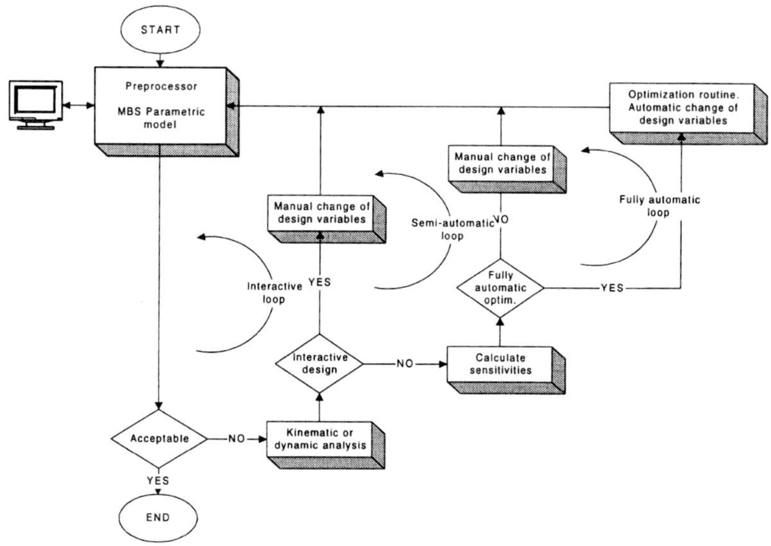

Figure 3.

A key objective is to achieve interactive performance with simultaneous parametric input definition, analysis and optimisation, and visualisation of the behaviour of the MBS at any stage of the design process. This methodology overcomes the old-fashioned but still widely used engineering software, based on independent stages for pre-processing, analysis and post-processing. The use of open standards (UNIX, C++, X-Windows, Motif, etc.) and a modular structure based on object oriented programming concepts guarantee the openness, portability and scalability of the developed software.

4. PRACTICAL IMPLEMENTATION

The study of the user needs lead to consider three different levels, in the use of the software, which are called *interactive loop* (or interactive simulation), *semi-automatic loop* (or manual optimisation) and *fully automatic loop* (or automatic optimisation). The way in which all those software integration levels or working procedures are integrated project is shown graphically in figure 3.

The logical flow chart of the interactive design mode can be seen in Figure 3. This user-driven design mode is especially well suited for kinematic design although it may be also

C553/030 © IMechE 1998

useful for dynamic design. Within the CAD environment, the user can create the bodies (or parts), defines the kinematic topology joining the bodies through kinematic joints and controls the behaviour of the mechanical system interactively through a graphic user interface. From the assessment of the motion, the user decides what design variables should be modified and how. The user changes those variables in the CAD environment and the kinematic model is automatically updated, allowing to quickly analyse the performance of the new design. This working procedure is particularly useful at the very beginning of the design process.

The semi-automatic optimisation loop is also depicted in Figure 3. At this level, kinematic and dynamic analyses are run using the parametric model created in the CAD environment. In addition, the evaluation of the objective function and design constraints is done. Optionally, the gradients of those functions can be evaluated in order to provide them to a mathematical optimisation function that computes a set of recommended values for the design parameters. The user is able to interactively modify the values of the design variables according to those recommended values, keeping always the control over the optimisation processes.

In the fully automatic optimisation loop (see Figure 3), the software will find automatically a value for the design variables that will somehow improve the system behaviour according to a given objective function and subject to some design and technological constraints. Using mathematical optimisation procedures, the software is able to compute in each iteration step the values of the design variables in the appropriate direction until the minimum of the defined objective function is reached. This automatic optimisation feature will normally be used in the final stages of the design process in order to get further refinements that are difficult to achieve using the tools previously described.

5. DESCRIPTION OF MBS CONSTRAINED OPTIMIZATION PROBLEM

Many engineers erroneously believe that optimisation is the process by which one finds the best design among all possible ones. This is not, in general, true since optimisation is most commonly used to find a better design from a given one. The certainty that the optimised design coincides with the best one almost never exists but for a limited number of simple cases in which the theoretical proof of convexity is possible. Some algorithms based on neural networks, genetic searches and stochastic analysis provide more chances of getting closer to the best design, but at the cost of increasing dramatically the computational effort.

To measure the quality of a design two mathematical entities must be used: first, a set of k design variables represented by a vector $\mathbf{b}$ and, second, an objective function ψ_0 that measures the quality of the design. Some typical design variables contained in $\mathbf{b}$ are link lengths, stiffness of springs, inertias, etc. Other parameters external to the mechanism such as control gains may also be included.

The objective function $_0$ used to assess the quality of the design depends on the design parameters $\mathbf{b}$. The objective function is written in such a way that the best design is the one that minimises it. Therefore, a design $\mathbf{b}_1$ is better than other $\mathbf{b}_2$ if $\psi(\mathbf{b}_1) < \psi(\mathbf{b}_2)$. However, not all mathematically possible designs are mechanically meaningful and, even among those, some are impossible or impractical to manufacture. To take these limitations into account, a set of inequality constraints of the form

$$\psi_i(\mathbf{b}) \leq 0 \tag{1}$$

are usually introduced. In some cases, equality constraints of the form are also considered.

The optimisation problem can be expressed in its most general way as find the value of **b** that minimises the objective function (**b**) subject to the constraints

$$\psi_i(\mathbf{b}) \leq 0 \quad i = 1,\ldots,n_c \tag{2}$$

Several methods have been proposed to find the minimum of a constrained function (see (1)). The minimisation methods follow an iterative approach that is expected to converge to a local or global minimum. In most of the minimisation methods, each of the minimisation steps is taken in such a way that the objective function is reduced. To proceed in the direction of the minimum the computation of the gradient of the objective function and the gradient of each of the constraints are needed.

The explicit computation of those gradients by the user is not strictly necessary, since can always be computed by numerical differentiation. Numerical differentiation is attractive in many optimisation problems where analysis tools are readily available. This is the case of multi-body dynamics. The alternative to numerical differentiation is the explicit calculation of sensitivities by one of the two methods described in the literature: *direct differentiation* and *adjoint variable method* (see 1).

5.1 Equations of motion

Since the n co-ordinates $\mathbf{q}$ that define the position of a multi-body system are often dependent, they are related by a set of m non-linear algebraic equations of the form

$$\mathbf{\Phi}(\mathbf{b},\mathbf{q},t) = \mathbf{0} \tag{3}$$

The Lagrangian form of the equations of motion can be written as

$$\begin{bmatrix} \mathbf{M} & \mathbf{\Phi}_{\mathbf{q}}^T \\ \mathbf{\Phi}_{\mathbf{q}} & \mathbf{0} \end{bmatrix} \begin{Bmatrix} \ddot{\mathbf{q}} \\ \boldsymbol{\lambda} \end{Bmatrix} = \begin{Bmatrix} \mathbf{Q} \\ \mathbf{c} \end{Bmatrix} \tag{4}$$

where $\mathbf{\Phi}_{\mathbf{q}}$ is the m n jacobian matrix of the constraint equations, $\mathbf{Q}$ is the n 1 vector of generalized external and centrifugal forces and **c** is a vector of dimension m that contains time derivatives of the constraint equations and, eventually, stabilisation terms.

Those equations of motion have a strong non-linear dependence on the design variables **b**. For example, the mass matrix depends on the geometry of each body which, in turn, may depend on the design variables (body length, twist, thickness, etc.). Similarly, the force vector **Q** may depend on the design variables (spring and dampers constants, control gains, etc.). Even the initial and final simulation times t_1 and t_2 may depend on the design parameters. In fact, any magnitude that appears in the equations of motion can be taken as a design parameter.

5.2 Objective function and design constraints

In the optimisation of multi-body systems, the objective function may depend explicitly on the design variables **b**, but more often it depends on the n generalised co-ordinates $\mathbf{q}$ that, in turn, depend of the k design variables **b**. Therefore, it can be written

$$\mathbf{q} = \mathbf{q}(\mathbf{b},t) \tag{5}$$

$$\dot{\mathbf{q}} = \dot{\mathbf{q}}(\mathbf{b},t) \tag{6}$$

$$\ddot{\mathbf{q}} = \ddot{\mathbf{q}}(\mathbf{b},t) \tag{7}$$

C553/030 © IMechE 1998

As mentioned before, in its most general way the objective function ψ_0 and the constraints ψ_i depend on **b** explicitly and also implicitly through $\mathbf{q}$, $\dot{\mathbf{q}}$ and $\ddot{\mathbf{q}}$, and can be written as

$$\psi(\mathbf{b}) = \phi(\mathbf{q}, \dot{\mathbf{q}}, \ddot{\mathbf{q}}, \lambda, \mathbf{b}, t) \tag{8}$$

where the subindex is dropped for the sake of simplicity. A very general way of writing the function $\succ$ that comprises virtually all the practical objective functions and constraints that arise in multi-body optimisation is the sum of an integral part plus an instantaneous value

$$\psi(\mathbf{b}) = \int_{t_1}^{t_2} \mathrm{f}(\mathbf{q}, \dot{\mathbf{q}}, \ddot{\mathbf{q}}, \lambda, \mathbf{b}, t)\, dt + \mathrm{g}(\mathbf{b}, t, \mathbf{q}, \dot{\mathbf{q}}, \ddot{\mathbf{q}})\Big|_{t_e} \tag{9}$$

where t_1 and t_2 are the initial and final times for the dynamic analysis and t_e is the instant at which the event described by function g takes place. The integral part is usually related to the mean value of a certain function, for example, to the mean value of an acceleration, which is related to comfort, or to the power consumption of the motors over a certain period. Instantaneous values can be imposed, for example, in trajectory planning problems when the end-effector must reach a given position at the end of the task.

Instantaneous values are closely related to maximum values of a function since they can be seen as instantaneous values that occur at a given instant $t = t_{max}$. They can also be written as

$$\psi(\mathbf{b}) = \mathrm{g}(\mathbf{b}, t, \mathbf{q}, \dot{\mathbf{q}}, \ddot{\mathbf{q}})\Big|_{t_{max}} \tag{10}$$

Maximum values are useful, for example, to specify that a maximum reaction force must not exceed a given limit.

The optimisation process starts with an initial guess. To proceed in the direction of a minimum, the calculation of the gradients of the goal function and the constraints is needed. The computation of gradients by direct differentiation and through the adjoint variable method is presented in (1). Within ODESIM project, this computation is performed using the finite difference approach since it is simple to implement in parallel mode.

6. APPLICATION EXAMPLE

A topic of particular relevance in the development of ODESIM project has been the evaluation of the advantages this software provides in comparison with currently used design methodologies. The industrial users participating in the project have performed this activity, using some selected test cases from the automotive, aerospace and transportation industries. In order to save space, this paper only includes the results obtained in an automotive test case.

6.1 Multi-link rear suspension design

The test case presented in this section was proposed and solved by the Centro Ricerche Fiat (CRF). This is a multi-link rear suspension that has to be optimised in order to reach the desired target curves. In the design process of such a suspension, the unknowns are the body point positions and the body bushing stiffness. The traditional design procedure for this type of suspension is performed in two steps.

The first step consists of a kinematic synthesis problem. Considering as design variables the body point positions it is sought a solution that must satisfy the specified toe-in/toe-out and camber characteristic curves, which are defined as a function of the vertical wheel displacement. In the second step, an elasto-kinematic synthesis problem has to be solved. Taking as unknowns the body bushing stiffness, a synthesis problem is solved in order to fulfil the indicated conditions of toe-in/toe-out and wheel base curves, which are given as a function of the lateral and longitudinal forces, respectively.

After the second step, the toe-in/toe-out and camber curves obtained in the first step may have been modified. Then, the designer has to solve again the first step, but this time considering the body bushing stiffness obtained in the elasto-kinematic synthesis. These steps are repeated several times until the four specified curves are perfectly satisfied. This design process requires a highly specialised analyst who needs different tools to solve the kinematic and elasto-kinematic analysis. The "convergence rate" of this iterative process depends heavily on the designer experience.

In this test example, CRF wanted firstly to evaluate the help that ODESIM software can provide to the designers without changing the usual design process for a car suspension. The approach was to use the automatic optimisation capacity to substitute the *trial-and-error* procedure typically used at each step of the design process. As a second objective, CRF modified the design process taking advantage of the optimisation tools to reduce the number of steps to be performed during the design of a new car suspension. The idea was to combine the goal functions and design variables considered in the two steps and solve the problem in a single step. In this way, the two steps performed in the traditional suspension design process are collapse in a single step.

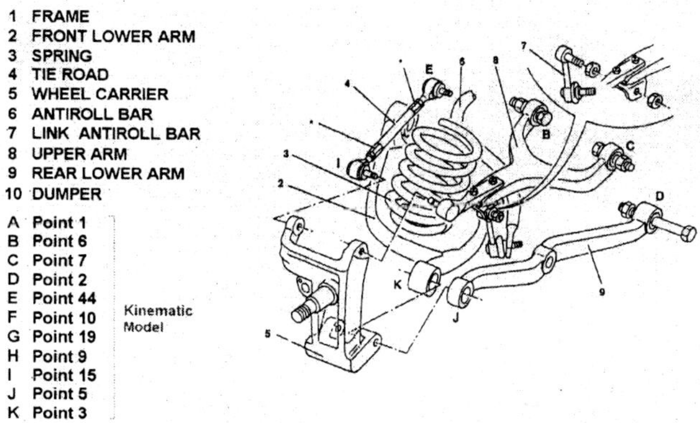

Figure 4.

Usually the multi-link suspension layout permits the control of each suspension characteristic curves because each of them is influenced only by few independent design parameters. The selected case is a four links multi-articulate suspension in which, due to roominess constraint, some design parameters influence both camber and toe simultaneously. Spring and damper are not coaxial and their forces are reacted by the lower arms. The suspension is linked to the body by a frame. The link points between arms and frame are bushings with linear stiffness (see Figure 4).

As mentioned before, ODESIM software was used to help the designer following the traditional design procedure. For each step of the design process, the goal function and the set of design variables are defined and the corresponding optimisation problem is solved.

In the design procedure that unifies both steps, the goal function is defined as a linear combination of the kinematic toe-in/toe-out and camber curves and the elasto-kinematic toe-in/toe-out and wheel base curves. The body point co-ordinates as well as the body bushing stiffness define the set of design variables, which leads to a total number of 27 design variables. Figure 5.a shows the evolution of the objective function. In this figure, they can be noticed some small peaks that appear during the optimization process. Those peaks correspond to increments in the objective function that suddenly appear when a constraint is reached. Figure 5.b depicts the kinematic toe-in/toe-out angle versus the wheel centre vertical displacement both for the 'un-optimised' and the 'optimised' suspensions. Although the response of the 'un-optimised' suspension is far from the target curve, it can be seen that the behaviour of the 'optimised' suspension nearly matches the target curve at the end of the optimisation process.

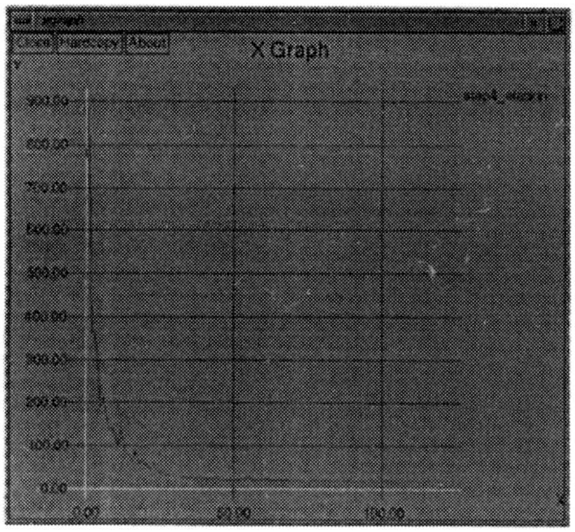

Figure 5.a

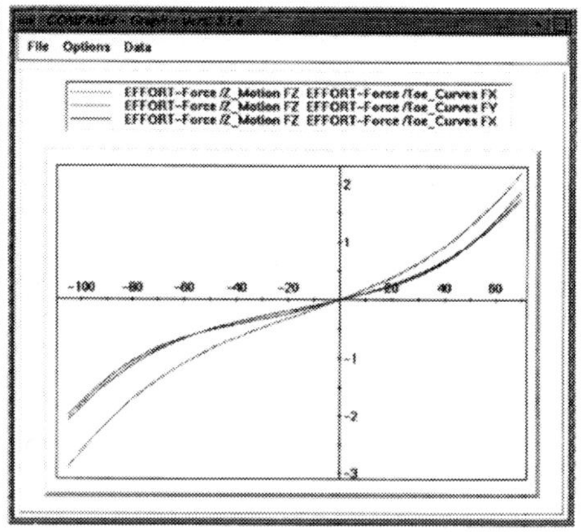

Figure 5.b

The CPU time needed to solve this optimisation problem using a single processor of a SGI ONYX machine with 4 MIPS R10000 processors is 43 hours. The minimum value of the objective function is achieved after more than 130 iterations, although it can be seen from figure 5.a. that the variation of the value of the goal function is negligible after 50 iterations.

As can be seen in the results obtained for the different test cases, ODESIM software can solve the optimisation problem so as to obtain values for the design parameters that satisfy the specified target curves. In both approaches (the traditional and the unified) the results obtained are nearly the same. In addition, this test case demonstrates that the duration of each step of the traditional design procedure can be shortened. Each step needs two days to be completed using traditional analysis and simulation tools, while using ODESIM the duration of these design steps is reduced to less than two hours for the step one (kinematic synthesis) and less than 8 hours for the elasto-kinematic optimisation. Therefore the application of ODESIM software to the traditional design process leads to significant reductions of time.

Although this contribution is very relevant, it must be pointed out that following the traditional approach the designer has still to iterate around the joint position definition and the bushings stiffness calculation because the bushing elasticity modifies the suspension bump-rebound characteristic curves. The unified approach mentioned before avoids this problem. The big advantage on the use of ODESIM software is that it allows collapsing the two traditional design steps in a single optimisation process that is run automatically. The intervention of the analyst is only sparely required to monitor the evolution of the optimisation process, while the traditional approach requires a bigger man dedication. Table 1 shows the design and designer times needed to perform the design of a new suspension using the traditional procedure with traditional tools or use the results of ODESIM project.

Table 1.

	Traditional procedure Design time = designer time	**ODESIM** Design time	**ODESIM** Designer time
Joint position definition	2 days	< 4 hours	3 hours
Bushing stiffness	2 days	< 8 hours	2 hours
Complete procedure (using ODESIM in each step)	6 days	3 days	1 day
Complete single step optimization		< 2 days	5 hours

7. CONCLUSIONS

As a general remark made by the industrial partners in ODESIM project, the resulting software was found as a very useful tool for the modelling, simulation and optimisation of the different industrial test cases. The most relevant result of the project is that the use of this software can reduce the time for the design process by a factor bigger than 200%. By the integration in a parametric CAD modeller, the building of the geometric model and the set up of the kinematic model can be improved by a factor of 100%. The most significant improvement in the design process is a result of the automatic optimisation capacity. This optimisation, which has to be done manually (a design study followed by several simulation steps) using traditional simulation methods, can be done in an automatic simulation. In this phase of the design process, the use of the developed software may result in a time reduction bigger than 500%. It is important to emphasise that this capability is completely automatic and requires nearly no intervention from the analyst. Consequently, the designer can optimise the use of his own time, which in the end is the most expensive part of the design process.

8. ACKNOWLEDGMENTS

The authors wish to thank the following people and their organisations for taking an active part and for their encouragement and advise during the development of this work: R.-L. D'Arcy (TRI-CONSULT), E. Di Pasquale (SIMTECH), J. Vilanova (CASA), L. Giraud, O. Burdakov and B. Hamma (CERFACS), D. Sternini (CRF), M. Narayanin and E. Volle (MATRA DATAVISION), J. Ulrich (SIEMENS) and R. Campo (CEE).

9. REFERENCES

1. HAUG, E.J. (Ed.), "Computer Aided Analysis and Optimization of Mechanical Systems", NATO ASI Series, Springer-Verlag, (1984).

C553/010/98

The derivation of roll reluctance rating – a measure of motorcycle agility suitable for investigation using concept level MBS models

D HARTY BEng
Prodrive Limited, Banbury, UK

It has long been recognised that the roll & yaw inertia, Centre of Gravity (CG) height and fundamental geometry are the dominant variables affecting motorcycle agility. While these effects and interactions can be captured using MBS models, useful metrics for their combined effects have been absent. The evolution of one such metric is described and discussed, with some evidence of its potential value early in the design process.

NOTATION

a	= Horizontal Distance from CG to Front Spindle
b	= Horizontal Distance from CG to Rear Spindle
k	= Radius of Gyration
m	= Total Vehicle Mass
r	= Distance From CG to Instantaneous Centre of Roll Motion
u	= Forward Speed
v	= Lateral Speed
$\mathbf{r}$	= Vector From CG to Instantaneous Centre of Roll Motion
$\mathbf{v}$	= Lateral Velocity Vector
I_{rr}	= Moment of Inertia about an Axis Passing Through Instantaneous Centre of Roll Motion and Parallel with Vehicle Longitudinal Axis
I_{xx}	= Moment of Inertia about Vehicle Longitudinal Axis
ϕ_i	= Initial Vehicle Roll Angle Prior to Transient Manoeuvre
ϕ_f	= Final Vehicle Roll Angle After Transient Manoeuvre
ϕ_u	= Vehicle Roll Angle Upper Integration Limit (see text)
ω	= Roll Rate
T	= Steer Torque

© IMechE 1998 C553/010

1 INTRODUCTION

Motorcycle behaviour has been the subject of much interest over some considerable period of time. The interested reader is referred to Sharp(1) for a useful overview of the development of analytical models. Motorcycles present a particularly interesting problem for analysis for several reasons:

- The basic vehicle is articulated.
- Side forces are generated predominantly by tyre camber, with some tyre slip angle.
- The dynamic behaviour of the rider has a large modifying effect on the underlying vehicle behaviour.

When considering the perturbation response of the rider/motorcycle system or the closed loop control of that system, MBS is a useful technique(2) to understand the behaviour of the system and its constituent parts. Eigenvalues and controller parameters can be systematically investigated using powerful Design of Experiments (DoE) methods, some of which are already built into leading MBS packages.

These detailed metrics can be helpful in calibrating a selected configuration before detailed design of components, or in troubleshooting a particular design once established. For given design data, techniques such as eigenvalue sensitivity can be used iteratively to explore the "design space" and discover optima if appropriate response variables are known and calculable (or measurable).

However if no appropriate response variables exist then there is little to go on at the concept stage. It is widely accepted that the placement and orientation of major masses, such as the engine and rider, has a considerable effect on the perceived character and objective characteristics of the motorcycle, yet useful single number metrics other than the basic inertia properties are not defined.

For four-wheeled vehicles, concept decisions on vehicle configuration are centred on the inertia properties of the vehicle, but the problem is considerably simplified by the absence of all three reasons noted above which add interest to the motorcycle problem. One common metric for four-wheelers is the ratio of k^2/ab (sometimes referred to as the dynamic index), where k is the radius of gyration in yaw or pitch and a and b are the distances from CG to front and rear wheels, respectively. This is a useful indicator of how much leverage the wheels can exert over the vehicle inertia and is anecdotally regarded as a good indicator of how much "fun" a vehicle is to drive(3).

This paper suggests a similarly general metric for motorcycles, aimed at making similar judgements and evaluating the impact of concept level decisions such as wheelbase, engine size and placement and rider position.

2 MOTORCYCLE TRANSIENT BEHAVIOUR

The transient behaviour of a motorcycle has some similarities to the transient behaviour of a four-wheeler but also some important differences. For a two-wheeler, the process of transit

C553/010 © IMechE 1998

from a straight ahead condition to a steady state cornering condition could be thought of as sequential. It is suggested that for a two-wheeler that process may be simplified as follows:

(i) Rider initiates turn by steering counter to the intended direction of turn.

(ii) After a delay associated with the relaxation length of the front tyre, side force builds at that tyre due to slip angle.

(iii) Yaw motion of the motorcycle develops counter to the intended direction.

(iv) After a delay associated with the relaxation length of the rear tyre, side force builds at that tyre due to slip angle.

(v) Side forces at ground level effect a roll moment about the CG, developing a roll rate in the intended direction.

(vi) Increasing camber angle on the tyres applies a side force in opposition to the force generated by the initial slip angle, rapidly exceeding that force and applying a net lateral force in the intended direction of turn.

(vii) Rider relaxes steer input.

(viii) Side forces at ground level effect a roll moment about the CG, eliminating the roll rate in the intended direction to zero at the equilibrium roll angle.

The final (steady state) relationship between roll angle and lateral acceleration is trivial. The sequence of events is somewhat fictitious in that the separate events described blend into one another and interact rather than being discrete, and the roll angle is managed by the rider in a rather more active manner than that suggested. Nevertheless, it is a useful sequence for discussion purposes.

Steer torque is the primary roll acceleration demand used by the rider. The ease and accuracy with which the rider is able to achieve the desired roll angle, and alter it within the turn, are clearly important. The impression of ease and accuracy is sometimes informally referred to as "flickability". Were the motorcycle/rider combination viewed as a roll angle control system, criteria used to evaluate its performance would include speed of response to a given input. The appropriate metric would be either the time constant for an overdamped response or a combination of rise time and overshoot for an underdamped response.

For any position control system, the inertia properties of the elements of the system are an important aspect in determining speed of response. For the motorcycle, there is a minor complication in that the motion in response to a control input is a combination of lateral and roll motion. There is thus a time-varying importance of the lateral inertia (the mass) and the roll inertia. This is why merely minimising roll inertia or mass alone is insufficient to optimise speed of response, and why for a given mass the positioning of items such as the rider and engine can have a critical effect on speed of response. This speed of response can be directly translated to a lap time advantage or improved feel for a road machine without necessarily compromising stability.

3 INSTANTANEOUS CENTRE OF ROLL MOTION

The motion of a body with a compound translational and rotational velocity, expressed at the CG as v and ω respectively, can be described as having an instantaneous centre of motion (Figure 1) at a distance r from the CG, where

$$r = v / \omega$$

The orientation of the vector **r** is perpendicular to the vector **v**. Thus if the CG location is known along with the velocity states, the location of the instantaneous centre of roll motion can be computed at all times when the roll velocity is not zero.

In practice the quantity calculated becomes meaningless as roll velocity approaches zero, but a small amount of ingenuity can stop the results of interest being swamped by those as the motorcycle nears the steady state.

4 MOMENT OF INERTIA ABOUT INSTANTANEOUS CENTRE

Once the location of the instantaneous centre is known, it is simple to calculate the moments of inertia about that point for the bodies in the system. In the case of the motorcycle system, since primarily roll motion is of interest, only inertias about the vehicle longitudinal axis will be considered:

$$I_{rr} = I_{xx} + mr^2$$

Although something of a simplification of the general case, this trivial expression can provide some illuminating results. If the property I_{rr} is recognised as a time varying property of the machine and the machine is subject to a constant roll torque input, then it can be seen that a minimum mean value for I_{rr} will result in a maximum value of roll acceleration.

5 SOME PRACTICAL DIFFICULTIES & THEIR SOLUTIONS

As the machine settles into the steady state, the motion tends towards purely lateral. As this happens, the instantaneous centre of rotation tends to an infinite distance away from the CG and the quantity I_{rr} becomes meaningless. It is therefore more helpful to look at the problem from the other end, and use the analysis to determine the force necessary to induce a given roll angle in the machine. It is also helpful to consider the behaviour in roll as if it were a second order response - that is to say of the form

$$Y(t) = A(1 - e^{-bt}\sin(C\omega t + \phi))$$

It is common practice with such forms of response to consider the rise time, t, at which the response, Y, is first equal to the final value, A. Since, for the motorcycle, the response may be under- or overdamped in nature then it is possible for the rise time to be infinite. Therefore the rise time for this purpose will be defined such that the target value for $Y(t)$ is equal to $A.(1 - e^{-1})$, or $0.632A$.

Although the motorcycle may or may not behave in such an ideal way, this useful upper integration limit, ϕ_u, can nevertheless be defined in terms of initial and final roll angle:

$$\phi_u = \phi_i + 0.632(\phi_f - \phi_i)$$

C553/010 © IMechE 1998

This neatly sidesteps the problem of I_{rr} becoming meaningless in the steady state. Thus, the expression

$$\int_{\phi_i}^{\phi_u} |T| \, .dt = \int_{\phi_i}^{\phi_u} |I_{rr}| \, .\dot{\omega} \, . \, dt$$

defines a response variable - a torque-duration product - in terms of a design parameter, I_{rr}, and a response variable predicted by an MBS simulation during a defined manoeuvre. It is suggested that a lower value for the torque-duration product indicates a motorcycle which will feel more manoeuvrable. It is recognised there may be some confusion over the concept of using a rider input as a response; it should be noted that the vehicle behaviour is defined and the rider input is adjusted as necessary to produce that behaviour in the vehicle.

6 SIMULATION RESULTS

MBS models of motorcycles, assembled for previous projects, were used to investigate the usefulness of the proposed measure. They are shown in figures 2 and 3. The motorcycles selected were chosen to display a clear difference in the quantity of interest, so as to give some impression of the level of resolution which might be possible with the proposed method. The models as used for the previous dynamic behaviour studies incorporated structural flexibility (using the component mode synthesis method), a user-modified tyre model, suspension characteristics and a simplified rider model. A PID controller using a combination of roll angle and path errors to determine control inputs means that the models can 'ride' themselves indefinitely in a straight line.

Initial work with the model retrieving the quantities of interest showed that the structural flexibility calculations were not germane to the prediction of the torque-duration product, and so their calculation was removed. This fits well with the original intention of using the method for 'first-sight' package and geometry modelling. Tyre behaviour including relaxation length and the rider modelling were retained, in particular the rider waist articulation. The rider waist as a force-generating element was modelled using a spring & damper element, tuned to give appropriate behaviour. For the comparative work in this and other studies, the author has found that such an approach is acceptable and computationally efficient; clearly for more detailed predictive work an approach such as that used by McGuan(2) is preferred. The final models consisted of a rigid steered mass with both sprung and unsprung components, a rigid non-steered mass with both sprung and unsprung components and a rider-above-the-waist mass. The system as a whole possesses only ten degrees of freedom - a substantial simplification over the fully flexible model with 64 degrees of freedom.

The manoeuvre selected for this exercise was a simple step roll-demand from straight running. A closed-loop feedback controller with both angle and path error terms to generate steering inputs attempting to follow the defined path was used, with paths being established to give steady state roll angles from five to 50 degrees for the speeds selected, from 10 to 40 ms^{-1}. The path was simple - a straight portion to allow the initial simulation transients to settle out, followed by a constant radius turn. This makes for a step change in idealised roll and produces sharp transients; it is analogous to the step-steer test used for four-wheelers. A typical rider steer torque trace is shown in figure 4. The typical initial high torque ramp followed by the decaying portion can be clearly seen. The models represented two different styles of roadgoing motorcycle - sports, and lightweight. The results are compared in figures 5 and 6.

When considering the results in detail, several trends are apparent. The first is that for the controller strategy adopted, the duration of the transient steer torque input is largely independent of speed for a given target roll angle. The second is that while the torque rise rate is governed by the controller strategy, for a given strategy and target roll angle, the peak level is also largely independent of speed. Finally, the torque-duration product is clearly speed and configuration dependent - the measure does indeed distinguish one motorcycle from another.

Torque-duration product is very linear with speed and moderately so with roll angle demand over the 0-30° region, and the entire dataset can be approximated by the form:

$$\int_{\phi_i}^{\phi_u} |T| \, .dt = A.u.\phi + B.u$$

For the lightweight motorcycle, A and B are 0.0148 and 0.0472, respectively when ϕ is measured in degrees, u is measured in ms^{-1} and $T.t$ is measured in Nms. For the sports motorcycle, those parameters are 0.0393 and 0.1923 respectively. The parameters A and B are thus invariants which describe the level and duration of effort which needs expending to manoeuvre the motorcycle. B may be regarded as a "mean" or overall level with A being a sensitivity to demanded roll angle. The measure B broadly reflects the difference in roll inertia between the machines, while A reflects the additional effort required for higher roll angles.

This remains some way short of the objective of a single index to quantify the behaviour of the vehicle but is nevertheless a useful step. In order to transform the torque-duration product into a more meaningful and easily understood quantity, the following treatment of the calculated product is suggested:

$$\int_{\phi_i}^{\phi_u} |T| \, .dt \, . \frac{1}{u} = \text{Roll Reluctance Rating} = A.\phi + B$$

Consideration of the units of A and B suggests they are in kg m, a quantity which has no similarities to other quantities of interest - work, momentum, and so on. Since the development of the roll includes all the fundamental geometric and inertial elements of the machine, it is proposed that a useful measure has been defined around which the effects of CG height, wheelbase, rider position and steering geometry can be explored to find an optimum.

7 CENTRE OF ROLL

Calculating the vertical location of centre of roll results in a trace like that shown in figure 7. It can be seen that for the initial transient portion of the manoeuvre there is indeed a point about which the vehicle rotates. As the roll velocity reduces the predictable numerical problems appear and it is clear that centre of roll alone is not a useful concept.

C553/010 © IMechE 1998

8 CONCLUSIONS & FURTHER WORK

The proposed Roll Reluctance Rating (RRR) does indeed distinguish one configuration of motorcycle from another. The magnitude of the numerical difference suggests it may be sensitive enough to discern differences between slightly different setups of similar motorcycles. As a numerical measure, it opens the possibility of numerical optimisation methods and Design of Experiments methods being used to "map the design space" and ensure a match between inertia properties and geometry. Since it is a measure defined using numerical methods on simplified system models, it seems reasonable to expect a useful level of robustness and resolution.

There is clearly more work to be done correlating subjective impressions of agility with calculated RRR in order to establish whether or not it maps to a subjective difference in vehicle feel, and thus whether or not it is useful. The definition of RRR may need further refinement following such an exercise.

Two-wheeler handling is about more than just agility; stability in particular may be thought of as broadly the inverse of agility. Measures exist for assessing stability in both closed and open-loop systems, and their combination with agility measures may allow a known compromise to be struck at the very earliest concept layout stage. In combination with more sophisticated Multi-Body System models, there is substantial scope for the virtual evaluation of motorcycle dynamics before physical prototyping occurs.

REFERENCES

1. Sharp, R.S., "The Lateral Dynamics of Motorcycles and Bicycles", Vehicle System Dynamics, Vol 14, 1985.
2. McGuan, S.P. "Active Human Surrogate Control of a Motorcycle: Stabilizing and De-Stabilizing", SAE 930226.
3. Williams, H. "Tools and Techniques", IGDS Lecture, Jaguar Cars/Birmingham University, 1997.

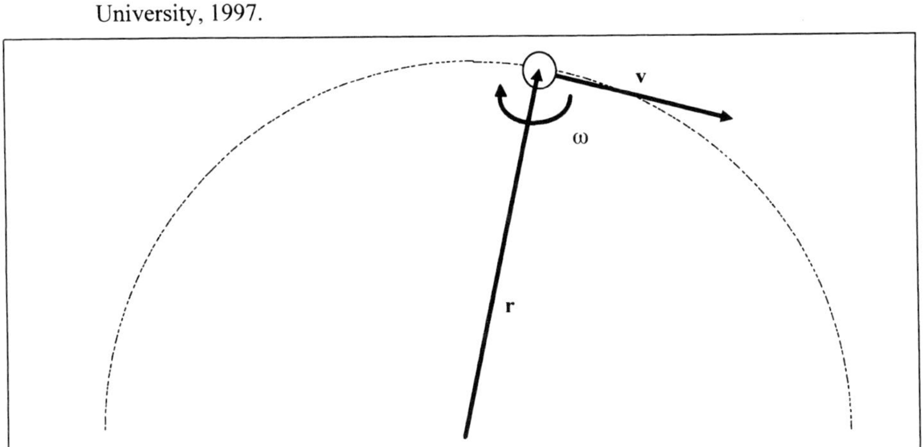

Figure 1 : A Body Travelling in an Arc

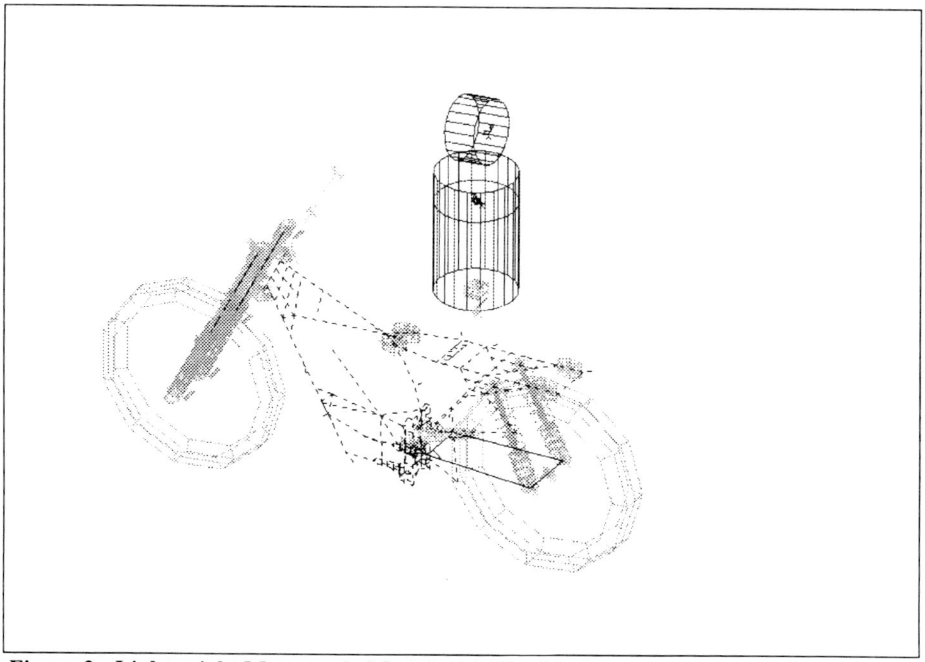

Figure 2 : Lightweight Motorcycle Model with Flexible Structure & Simplified Rider

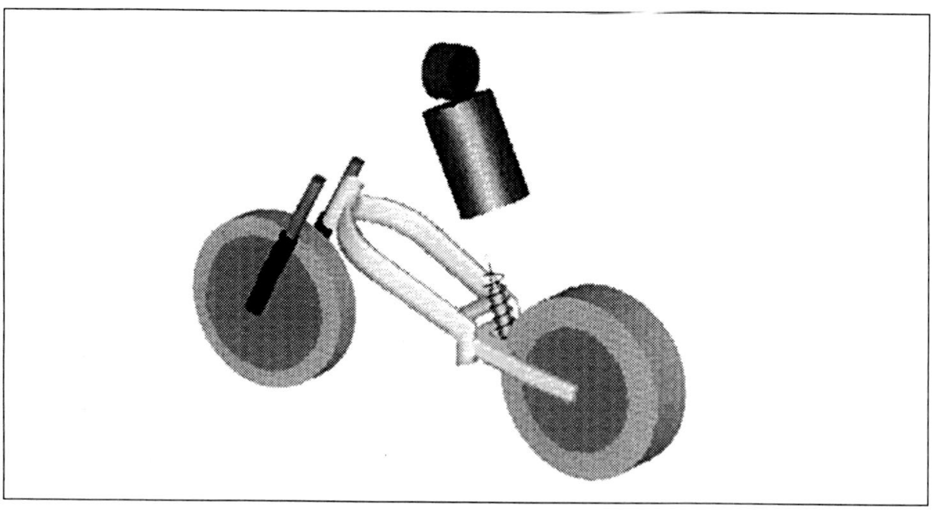

Figure 3 : Sports Motorcycle Model with Flexible Structure & Simplified Rider

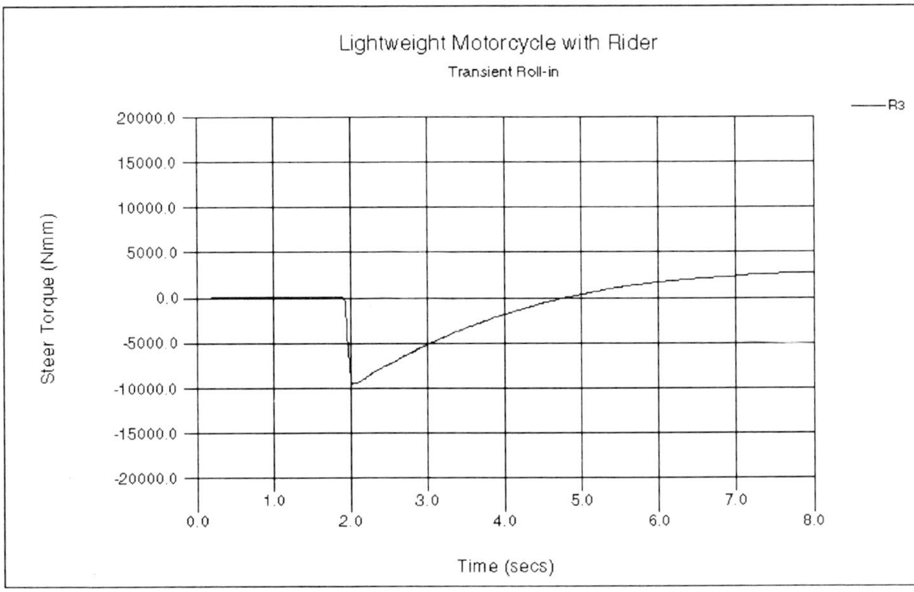

Figure 4 : Typical Steer Torque Trace

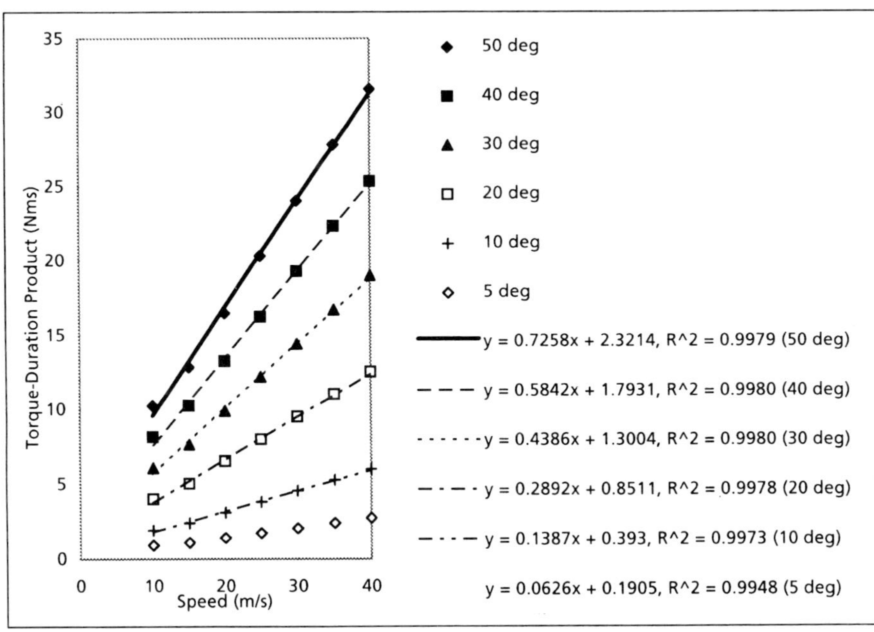

Figure 5 : Lightweight Motorcycle Torque-Duration Product Data

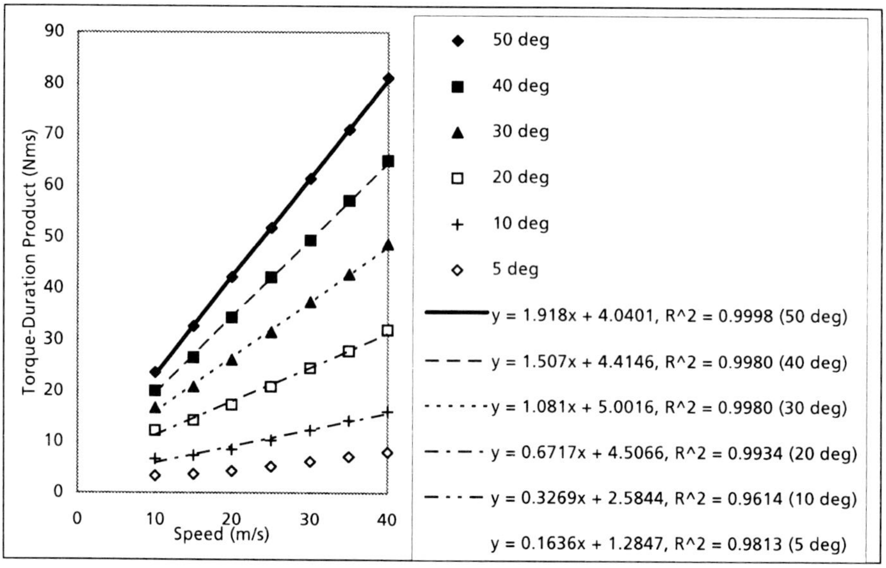

Figure 6 : Sports Motorcycle Torque-Duration Product Data

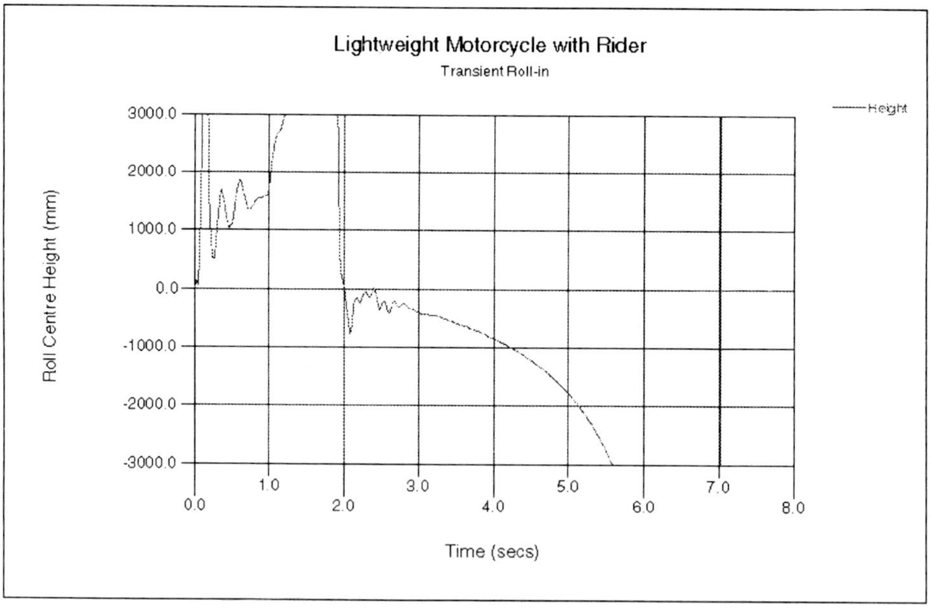

Figure 7 : Instantaneous Centre of Roll Calculation; Rider Input Starts at 2 Seconds

C553/010 © IMechE 1998

C553/017/98

A design tool for the analysis of rack and pinion steering systems

A MODJTAHEDZADEH BS, MSc, PhD, MASME, MSAE and **A AHMADI** BSc, MSc
Iran Centre for Research and Development, Tehran, Iran
M R HAIRI-YAZDI BSc, MSc, PhD, AMIMechE
Department of Mechanical Engineering, University of Tehran, Iran

A model of an entire steering system employed in a compact car has been developed which enables the chassis designer to analyze forces in various components of the mechanism. The system includes sets of steering column, a two Cardan joint system, rack and pinion, tie rods, and inner and outer ball joints. First, the kinematics of the overall system is presented and the modeling of the tie rods is discussed in detail. Second, the force balance of the overall system formed by the interaction of models of different sets is presented. The inputs to these models are steering angle, steering force and torque, and suspension displacement.

1 INTRODUCTION

The steering system design is a vital issue in chassis design. Stability, performance, and driving comfort during handling of the vehicle depend on the quality of the steering system as well as the geometry of the chassis, the kingpin axis and the suspension. The steering system outranks many safety devices employed in a vehicle and in fact this was the motivation to start this research work.

With the gaining popularity of compact front wheel drive cars, the usage of rack and pinion steering system is being increased by the automotive industries. They have not only been employed in small cars but also have replaced the conventional steering systems in some medium and large passenger cars. The main advantages of the rack and pinion systems are their compactness, low weight, lower manufacturing cost, simple construction, minimal elasticity, and easy operation due to good degree of efficiency.

The ultimate objective of this research was to develop an overall model of the rack and pinion steering system which enables the analysis of the forces and moments in the system. Obviously, linear and angular displacements of the parts in motion relative to one another were needed to determine the direction of the loads developed in the system at any steering wheel angle and suspension position. Therefore, the development of a kinematic model of the

© IMechE 1998 C553/017

steering system became the immediate objective of this research. Additionally, the kinematic analysis of the system by itself is a necessary step in the design procedure; the universal joints (u-joints), tie rod and kingpin angles must be selected in a manner to satisfy desired functions of the system.

The modeling of spatial kinematics of the tie rods is presented in detail. A force equilibrium model is developed and a numerical example is solved and the results are discussed.

The steering system, as shown in Fig. 1, consists of several parts, which are listed below:

- Steering column
- Two Cardan joint system
- Rack & pinion
- Tie rods

The connection and interaction of these parts is explained as follows. The generated angular displacement of the steering wheel by the driver causes an equal displacement of the steering shaft attached to it. This displacement is then transferred to the pinion shaft through the two Cardan joint system. The rotational movement of the pinion shaft is transformed into the linear movement of the rack that is connected to the tie rods at its ends. The other ends of the rods are connected to the steering arms of the wheels which cause turning of the wheels about the steering axes also known as the kingpin axes.

Although the model is developed for a particular type of rack and pinion steering system, similar modeling development may be applied for other types, incorporating some geometric changes and by varying the parameters accordingly.

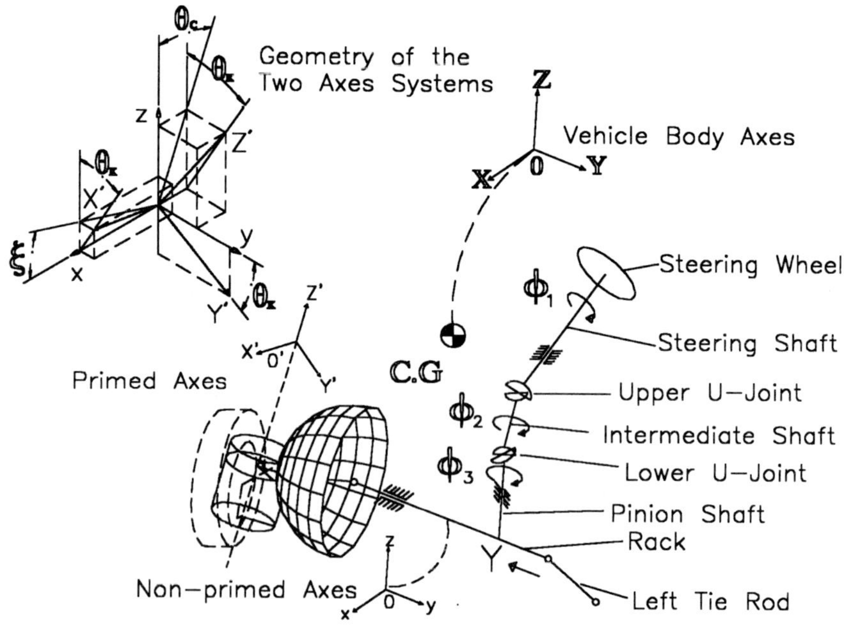

Fig. 1 Schematic of the rack and pinion steering system

 C553/017 © IMechE 1998

2 KINEMATICS

The kinematic equations for each part are separately developed and are then related to those of the adjacent ones to form the overall model of the system. Although some free play and flexibility in u-joints, rack and pinion, and ball joints exist, they are assumed negligible in this development.

The angular displacement of the steering wheel is transmitted through the steering column without any change. The axes of the pinion shaft and the steering shaft, due to the packaging limitations of the front wheel drive vehicles, usually do not fall in the same plane. To connect the two shafts together, a two Cardan joint system (which includes an intermediate shaft) is introduced between the shafts. The zigzag path of the three shafts in such designs, normally produces large joint angles, and it is well known that the output angle ϕ_2 of each u-joint is a nonlinear function of both the input angle ϕ_1 and the angle θ between the shafts as:

$$\tan \phi_2 = \tan \phi_1 \cos \theta \qquad [1]$$

However, the angles involved in the combination of the two joints may be adjusted such that the overall nonlinearity of the two joint system becomes minimal. Selection of the phase angle between yokes of the intermediate shaft in the neighborhood of the angle, formed by the planes of the input-intermediate and intermediate-output shafts, is a remedy for the reduction of the overall nonlinearity of the system (1).

Additionally, the following kinematic equation will be required to determine the orientations of the coordinate system axes for the two joint crosses (1).

$$\cos\beta_1 \cos\beta_2 = \cos \theta \qquad [2]$$

where β_1 is the angle between the axis of the input shaft and the joint cross normal., and β_2 is the angle between the cross normal and the output shaft of each joint.

The relationship for the transformation of the angular displacement of the pinion shaft into the linear motion of the rack is a simple equation involving only a constant coefficient, that is the pitch radius.

The linear movement of the rack will cause a complicated spatial motion of the tie rods, that are connected to the steering arms and the rack at their ends by four spherical ball joints. To solve for the kinematics of the rods, the following observations are made. Starting from the inner ends of the tie rods, one could visualize that if the other ends were free and unconstrained, all possible locations of the outer ends of the rods would form imaginary moving spherical surfaces with their centers attached at the ends of the rack. Hence, the back and forth motion is along the axis of the rack. Now, looking at the outer ends of the tie rods and considering the fact that in general the steering arms could both turn about and move parallel to the steering axes, two cylindrical surfaces with their axes coinciding with the steering axes would be pictured for the loci of the two outer ball joints. The rotational and linear freedoms are due to the steering and suspension play respectively. However, since the steering system forms a closed loop chain, the outer ball joints are constrained and not all the points on these surfaces are the true loci.

The two imaginary surfaces are shown in Fig. 1. The loci of the outer ball joints are a subset of the closed space curve formed by the intersection of these surfaces. The following mathematical development, in accordance to the above explanation, is presented as a solution for the loci of the outer ball joints. Equation of the sphere is written with respect to the coordinate system attached to the center of the rack at the zero steering position which is stationary with respect to the chassis, and whose axes are parallel to the vehicle body axes, as shown in Fig. 1.

$$(X_o - X_i)^2 + (Y_o - Y_i)^2 + (Z_o - Z_i)^2 = L_{tr}^2 \qquad [3]$$

where the subscripts *o* and *i* represent the coordinates of outer and inner ball joints respectively, and L_{tr} is the tie rod length. Since the amplitude of vibration of the rack relative to the chassis is negligible, X_i and Z_i may be assumed constants. Before presenting the equations of the cylinder it is helpful to point out the tilt angles associated with its axis and the selected coordinate system in which the equations are written.

The caster and the kingpin inclination angles, which are the steering axis angles with respect to the vertical in the side and front views respectively, are considered in the modeling. As it will be explained in the next paragraph the coordinate systems associated with the spheres and the cylinders must be converted into one another. Therefore, appropriate coordinate systems shall be selected for the cylinders to minimize the conversion effort. The origin of the selected coordinate system for each one of the two cylinders is placed at the point of intersection of the cylinder's axis and the horizontal plane passing through the rack. Considering this system's axes to be initially parallel to the vehicle body axes, they are then rotated as explained in the appendix, such that the z-axis of the resulting coordinate system coincides with the cylinder's axis (the primed coordinate system shown in fig. 1). Thus the equations of the cylinder may now be presented:

$$X_o'^2 + Y_o'^2 = L_{sa}^2 \qquad [4]$$

$$Z_o' = f(t) \qquad [5]$$

where L_{sa} is the steering arm length and f(t) is the function defining the motion of the tie rod along Z_o' axis due to the suspension displacement.

To be able to solve equations [3], [4] and [5] simultaneously, it is required to convert the two coordinate systems into a common one. If the coordinates of the cylinder are converted into the coordinates of the sphere (i.e. converting the non-primed coordinate system into the primed one), the following equations will be obtained:

$$G_{11}X_o^2 + G_{12}Y_o^2 + G_{13}X_o + G_{14}Y_o + G_{15}X_oY_o = G_{16} \qquad [6]$$

$$G_{21}X_o^2 + G_{22}Y_o^2 + G_{23}X_o + G_{24}Y_o + G_{25}X_oY_o = G_{26} \qquad [7]$$

The coefficients of the above equations (G_{ij}) are lengthy combinations of the arrays of the three by three conversion matrix and the coordinates of the displacement vector, defining the origin of the primed coordinate system relative to the origin of the non-primed coordinate system. These coefficients and the elements of the conversion matrix are presented in the appendix.

The system of two equations, two unknowns (i.e. eqs. [6] and [7]) are nonlinear, and therefore application of a suitable numerical method will be the solution of these equations. The non-single-valued system of equations involves two solutions. Physically, they correspond to the intersection of the plane of height Z_o' with the closed space curve resulting in two common points. However, only one of the two solutions is valid. In the numerical

C553/017 © IMechE 1998

solution, the initial conditions of the steering system will determine the valid solution for the initial step. The detection of the valid solution for the next step is made by the selection of the nearest solution to the pervious one. A similar procedure is applicable for the rest of the steps.

With the solution of the location of the outer ball joints, the angle of the front tyres will be easily determined. The angle of rotation of each wheel is equal to the angle of rotation of the corresponding steering arm about its steering axis.

Although the positions and the angles of the two front steering axes vary by small amounts due to the suspension play and the flexibility of the control arm joints connecting the knuckles to the chassis, they may be assumed fixed relative to the chassis. However, if these small variations are known exactly, they can easily be accounted for in the modeling by allowing the transformation matrics and the position vectors to be time dependent.

3 FORCE EQUILIBRIUM

With the determination of the relative positions of the parts with respect to one another, it is now possible to develop a force equilibrium model which enables the calculation of forces and moments involved in various parts of the steering system. In addition to the assumptions made in the kinematic section, it is also assumed that the system is highly efficient (frictionless), the forces involved are concentrated, and the mass and acceleration of all moving parts are negligible. The last assumption implies that the critical forces and stresses in the system are the contribution of the static loading rather than the dynamic loading. The dynamic forces (e.g. during severe cornering of vehicle) produced by the tyres are directly applied to the chassis through the control arms and only moderate moments produced by the tyre side forces acting through the pneumatic arms are applied to the steering arms. On the other hand, heavier steering torque is required during parking (2) and even more sever torque is produced if the wheels are turned against an obstacle such as a curb. The inertial forces of the parts due to their low mass and acceleration are also negligible. However, it is possible to incorporate the dynamic forces into the equations if desired.

The freebody diagrams for each part are separately developed and are then connected together by considering equal action-reaction forces between them. Fig. 2 shows the freebody diagram for the entire system. Various points of application of forces and moments on the system are numbered as shown in the figure. The steering shaft, the pinion shaft and the rack include bearing and support reactions. The ball bearings are modeled by three-dimensional reactions whereas the needle bearings and the supports are modeled by two-dimensional reactions. The reactions in the two joint system are quite involved. Each joint includes a cross member and two yokes. The reactions at the ends of the cross members form the majority of the seventy-four unknowns for this part. The modeling of the two joint system was mainly adopted from (1).

There are a total of one hundred and seventeen simultaneous linear equations involved in the mathematics of the force balance model. The number of equations associated with the steering shaft, the two u-joint system, the pinion shaft, and the rack and tie rods are eleven, seventy four, fifteen and seventeen respectively. Fourteen equations and fifteen unknowns are directly obtained from the free-body diagrams of the rack and tie rods. Three additional equations introducing two new unknowns are required to make the system of equations solvable. Two of these equations are obtained from moment relationships about the kingpin axes and the third one is the assumption for the selected distribution of moments between the right and left tyres. The development of a complete model of the tyre moments depends on several factors such as tyre normal forces and their coefficients of friction, the

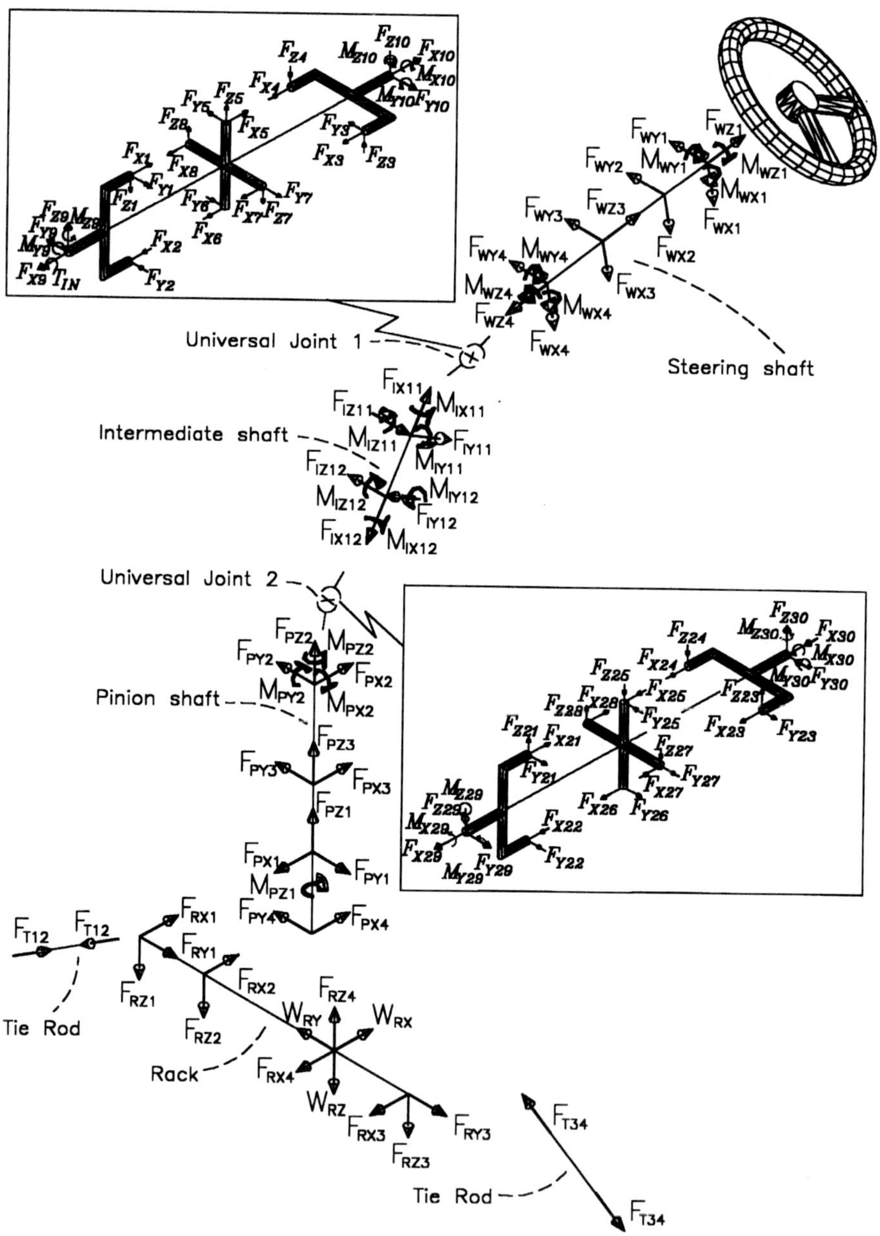

Fig. 2 Freebody diagram of the rack and pinion steering system

C553/017 © IMechE 1998

scrub radius of tyres and disturbance moments applied to the wheels. To avoid this complicated modeling work, which is beyond the scope of this research, the assumption of equal distribution of moment on the right and left tyres was made.

The coordinate systems used for the u-joints and the tie rods are fixed to these turning parts. Conversion of these coordinate systems into the adjacent non-rotating ones involves elements of rotation matrices that are functions of the changing angles between the parts. In general, any two adjacent parts form an angle with each other and their coordinates must be transformed into one another. In almost all cases, the rotations are simply about one axis only. The only exception is the tie rods, which require rotations about two axes.

4 NUMERICAL EXAMPLE

The parameters of a compact production vehicle are selected as listed in Table 1. Five out of the six input loads to the steering wheel are assumed zero and only the steering wheel torque M_{wz1} is given a 10 N.m value. The resistive road wheel torques are considered to counteract those produced by the steering wheel input, and they are assumed to be distributed evenly between the right and left wheels. Although the computer program is capable of simulating

Table 1 Parameters of a compact car used in the numerical example
(∗ Correspond to normal positions of steering wheel and suspension)

Parameter Description	Value
Projected angle of steering shaft w.r.t. the horizontal in the side view	29°49′
Projected angle of steering shaft w.r.t. the vertical in the front view	2°
Pinion shaft angles w.r.t. the vertical in the side and front views	17°, 15°
Intermediate shaft angles w.r.t. the vertical in the side and front views	45°, 45°
Yoke phase angle on the intermediate shaft	83.5°
Length of the rods	294 mm
Pitch diameter of pinion	10.6 mm
Lock to lock rotation of steering wheel	1350°
Projected angle of the tie rods w.r.t. rack axis on the plan view ∗	3.3°
Projected angle of the tie rods w.r.t. rack axis in the front view ∗	0.8°
Kingpin inclination angle	14°
Caster angle	1°35′
Wheel base	2345 mm
Track	1385 mm
Segment lengths on the steering shaft (points 1-2, 2-3, 3-4 in Fig. 2)	42, 495, 23 mm
Yoke torque diameter	35 mm
Length of intermediate shaft	60 mm
Distance between center of cross to the middle of yoke serration	36 mm
Pressure angle of pinion	20°
Helix angle of pinion	30°
Segment lengths on the pinion shaft (points 2-3, 3-1, 1-4 in Fig. 2)	77, 30, 21 mm
Helix angle of rack	15°
Segment lengths on the rack (points 1-2, 2-4, 4-3 in Fig. 2)	178, 400, 222 mm

the suspension play, but no such vibrations are considered in this example. Therefore, the simulations correspond to the steering of the stationary vehicle (e.g. in the parked position).

The program initially calculates the two u-joint angles as well as the angle formed by the planes of the shafts. For the given data they are 37.19^0, 32.67^0 and 83.65^0 respectively; these constant values will be used for the calculation of the kinematics of u-joints at each step of the computer run. Although the program yields a high volume of information rapidly, only a selected number of graphs are presented here to demonstrate the capability of the computer modeling while some important characteristics of the system are also pointed out and discussed.

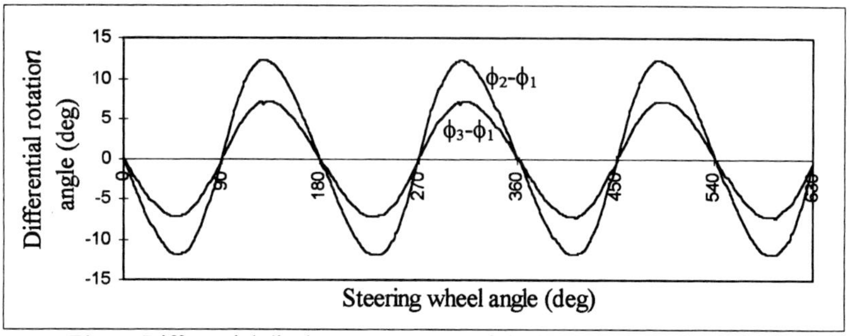

Fig. 3 Differential displacement angles of the intermediate and output shafts relative to the input shaft for the two u-joint system

Fig. 3 shows the differential angles between the input (steering) shaft and the intermediate and the output (pinion) shafts of the joint system. The cyclic variation of the output displacement of each joint with a period of 180^0 is observed from this figure. The variation amplitude of the two joint system is less than that of the single joint system. Referring to the data printed in Table 1, the phase angle of the yokes of intermediate shaft is selected almost equal to the angle of the planes formed by the shafts of the joint system. The examination of various phase angles proved that the selected angle is the optimum value.

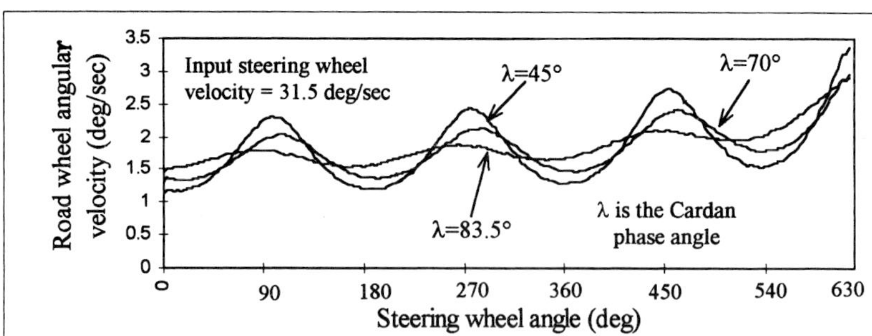

Fig. 4 Average of right and left road wheel angular velocities for several Cardan phase angles

To demonstrate the effects of Cardan phasing on the system, plots of average road wheel angular velocities verses steering wheel angle for a constant steering wheel velocity and for

C553/017 © IMechE 1998

three different phasing angles are presented in fig. 4. It can be seen that the more phase angle deviates from the optimum value (i.e. 83.5 deg.) the larger the amplitude of variations becomes, and also the more non-harmonic these variations appear. The slight inclinations of the curves are due to the drop in the steering ratio as shown in fig. 8.

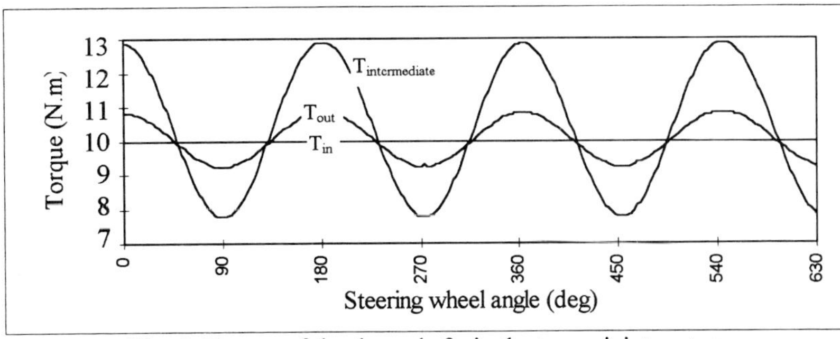

Fig. 5 Torque of the three shafts in the two u-joint system

Torque outputs of the shafts of the two joint system is shown in Fig. 5. Discussion of the torques is parallel to that of the joint displacements, i.e. a period of 180^0 is seen for the cyclic variations, and a reduction in the amplitude is obtained for the two joints as compared to the single joint. If the shaft velocities were to be plotted, it could be seen that the torque and velocity variations are exactly 180^0 out of phase. Further if the products of the two variables were to be plotted, it could be seen that powers associated with the input and output shafts are equal. In fact this is a method for checking the accuracy of the results which was performed in this research.

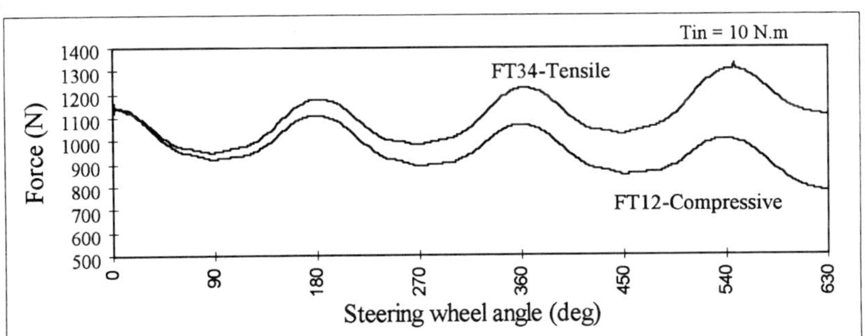

Fig. 6 Axial forces of the left and right tie-rods

Due to the spherical joint attachments at the ends of the rods, they only withstand axial loads as shown in Fig. 6; one rod is in tension whilst the other one is in compression. As the steering wheel angle increases the direction of the rods change as well and the magnitude of force in each rod inclines or declines slightly. Additionally, the force curves include a cyclic component that are caused by the torque variation introduced by the u-joints and have the same phase and frequency characteristics as the output shaft of the joint system.

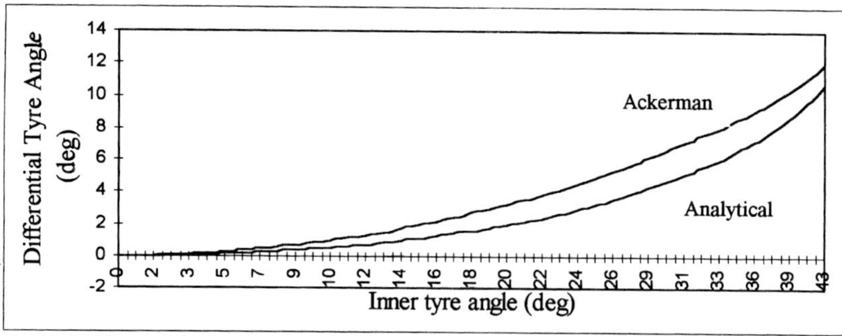

Fig. 7 Differential steer angle as compared to that of the theoretical Ackerman steer

Differential steer angle of the front wheels (i.e. the difference between the front tyre angles) is shown in Fig. 7. It is well known that for the elimination of tyre scrubs during turning of a vehicle at low speeds the inner tyre must have a larger angle than the outer one according to the Ackerman steering geometry (3).

$$\cot \delta_0 - \cot \delta_i = B/L \qquad [8]$$

where δ_o and δ_i are the outer and inner tyre angles respectively and B and L are track and wheelbase lengths respectively. However, less differential tyre angles as compared to that of the Ackerman steering is generally desirable to increase the lateral force on the front axle (4) while cornering at higher speeds.

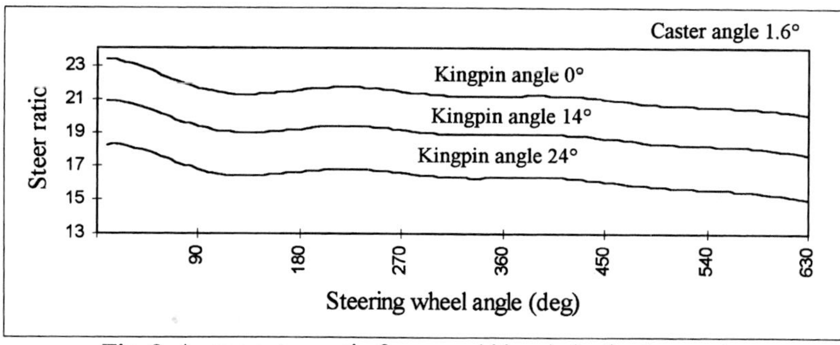

Fig. 8 Average steer ratio for several kingpin inclination angles

The average steer ratio, which is the ratio of the steering wheel angle to the average angle of the front tyres, is shown in Fig. 8 for several kingpin geometries. In general, the ratio decreases with the increase of steering wheel angle and as a result the required steering wheel torque increases. This is a characteristic of the rack and pinion steering systems which is considered a disadvantage and may be reduced by the addition of a nonlinear hydraulic pump or by the application of the new variable ratio rack bar technology (5).

C553/017 © IMechE 1998

5 CONCLUSION

An overall model of a complete rack and pinion steering system has been developed. The model is capable of analyzing the kinematics as well as the developed forces in the entire system. The computer program enables the chassis designer to vary a number of parameters and observe the effects on the system. The program yields a high volume of information in a relatively short period of time.

In the next phase of the research, the obtained force and moment information will be employed for the stress analysis of the members by finite element methods. Further, it is envisaged to compare the results of the model with those obtained from an instrumented vehicle. Although the results of the computer program have been studied and evaluated in depth, but comparison with the true results will be the ultimate validation method. Further, the impact behavior of steering column shall be studied for safety issues (6).

REFERENCES

1. **Cowles, J. H., Chiu, Jr.** and **Chiu, Y. P.** Computer-aided analysis of a steering column using Cardan universal joints. SAE paper 841204, 1985.
2. **Bastow, D.** Car suspension and handling. Pentech Press, London, 1990.
3. **Wong, J.Y.** Theory of ground vehicles. 2nd edition, Wiley, New York, 1993.
4. **Reimpell, J.** and **Stoll, H.** The automotive chassis. Arnold, 1996.
5. **Bishop, A. E.** and **Baxter, J.** New rack and pinion steering design and manufacturing technology. SAE paper 841199, 1985.
6. **Park, Young-Sun et. al.** Optimum design of a steering column to minimize the injury of a passenger, Int. J. of Vehicle Design, Vol. 17, No. 4, 1996.

APPENDIX

The formula for the transformation of the primed coordinate system into the non-primed coordinate system is:

$$p = A^{t}(\theta_c, \theta_K)\, p' + P$$

where, P is the vector defining the origin of the primed axes in the non-primed axes and θ_C and θ_K are the caster and kingpin inclination angles respectively. Based on the geometry shown in fig. 1, the projections of the primed axes onto the planes of the non-primed coordinate system form the three θ_c , θ_K and ξ angles. The arrays of the transformation matrix, which are functions of these angles, are derived as given below.

$$A^{t}_{11} = \frac{1}{\sqrt{1+\tan^2\xi+\tan^2\xi.\tan^2\theta_K}} \quad , A^{t}_{12} = 0 \quad , \quad A^{t}_{13} = \frac{-\tan\theta_C}{\sqrt{1+\tan^2\theta_C+\tan^2\theta_K}}$$

$$A^{t}_{21} = \frac{\tan\theta_K}{\sqrt{1+\cot^2\xi+\tan^2\theta_K}} \quad , A^{t}_{22} = \cos\theta_K \quad , \quad A^{t}_{23} = \frac{\tan\theta_K}{\sqrt{1+\tan^2\theta_C+\tan^2\theta_K}}$$

$$A'_{31} = \frac{1}{\sqrt{1+\cot^2\xi+\tan^2\theta_K}} \quad , \quad A'_{32} = -\sin\theta_K \quad , \quad A'_{33} = \frac{1}{\sqrt{1+\tan^2\theta_C+\tan^2\theta_K}}$$

θ_c and θ_K are known angles and ξ may be calculated from the equation derived from the geometry of fig. 1 as:

$$\tan\xi = \tan\theta_C.\cos^2\theta_K$$

The coefficients of equations [6] and [7], describing the points of intersection of the spherical and cylindrical imaginary surfaces are:

$$G_{11} = H_1 + H_3 \times H_{11}^2 + H_8 \times H_{11} \quad , \quad G_{12} = H_2 + H_3 \times H_{12}^2 + H_9 \times H_{12}$$

$$G_{13} = 2 \times H_3 \times H_{10} \times H_{11} + H_4 + H_6 \times H_{11} + H_8 \times H_{10}$$

$$G_{14} = 2 \times H_3 \times H_{10} \times H_{12} + H_5 + H_6 \times H_{12} + H_9 \times H_{10}$$

$$G_{15} = 2 \times H_3 \times H_{11} \times H_{12} + H_7 + H_8 \times H_{12} + H_9 \times H_{11}$$

$$G_{16} = L_{sa}^2 - X_{10}^2 - Y_{10}^2 - H_3 \times H_{10}^2 - H_6 \times H_{10}$$

$$G_{21} = 1 + H_{11}^2 \quad , \quad G_{22} = 1 + H_{12}^2 \quad , \quad G_{23} = -2 \times X_0 + 2 \times (H_{10} - Z_0) \times H_{11}$$

$$G_{24} = -2 \times Y_0 + 2 \times (H_{10} - Z_0) \times H_{12} \quad , \quad G_{25} = 2 \times H_{11} \times H_{12}$$

$$G_{26} = L_{tr}^2 - X_0^2 - Y_0^2 - (H_{10} - Z_0)^2,$$

where H_{ij} are given below:

$$H_1 = A_{11}^2 + A_{21}^2 \quad , \quad H_2 = A_{12}^2 + A_{22}^2 \quad , \quad H_3 = A_{13}^2 + A_{23}^2$$

$$H_4 = 2 \times (X_{10} \times A_{11} + Y_{10} \times A_{21}) \quad , \quad H_5 = 2 \times (X_{10} \times A_{12} + Y_{10} \times A_{22})$$

$$H_6 = 2 \times (X_{10} \times A_{13} + Y_{10} \times A_{23}) \quad , \quad H_7 = 2 \times (A_{11} \times A_{12} + A_{21} \times A_{22})$$

$$H_8 = 2 \times (A_{11} \times A_{13} + A_{21} \times A_{23}) \quad , \quad H_9 = 2 \times (A_{12} \times A_{13} + A_{22} \times A_{23})$$

$$H_{10} = (ZZ - Z_{10}) \big/ A_{33} \quad , \quad H_{11} = {-A_{31}} \big/ {A_{33}} \quad , \quad H_{12} = {-A_{32}} \big/ {A_{33}}$$

it shall be reminded that A is orthogonal. Thus $A^t = A^{-1}$

C553/032/98

Fuzzy logic control of a novel suspension system

H HEUMANN BEng, **B B HALL** BSc, PhD, CEng, MIMechE, and **W WESTON** BSc, PhD, CEng, MIMechE, FIEE
Department of Mechanical Engineering and Manufacturing Systems, University of Huddersfield, UK

ABSTRACT

A novel Self-Energising Self-Levelling Semi-Active (SESLSA) suspension strut has been developed together with a digital controller. In previous work this controller employed non-linear flow and linear optimal control elements. In order to make the controller adapt to varying operational conditions with a reduced number of sensors, an investigation has been carried out into the feasibility of replacing the linear optimal controller with a fuzzy logic element.

The aim of this paper is to establish the feasibility of employing fuzzy logic control in this context and to compare the performance of a fuzzy logic controller with other forms of suspension control. Preliminary findings are favourable as indicated in the results presented. Proposals are made for integrating the fuzzy logic controller with the non-linear flow controller element in the SESLSA system.

© IMechE 1998 C553/032

NOTATION

A	system coefficient matrix
ACC	vertical body acceleration
B	control coefficient matrix
b_s	suspension damping coefficient
C_d	discharge coefficient
DTL	dynamic tyre load
E	expectation operator
f	frequency
FLC	fuzzy logic controller
IFF	inverse flow function
J	performance index
K	gain vector
k_c	coil overdrive ratio
k_s	suspension spring stiffness
k_t	tyre stiffness
LVDT	linear variable displacement transducer
m_b	body mass
m_w	wheel mass
P	non-negative definite solution of the algebraic Riccati equation
$P_{2\text{-}3}$	chamber pressures
$q_{1\text{-}2}$	weighting matrix coefficients (output)
q_d	demand flow
R	weighting matrix (control)
S	vertical spectral density of road surface
SD	standard deviation
SD ACC	standard deviation of vertical body acceleration
SD DTL	standard deviation of dynamic tyre load
SD SWS	standard deviation of suspension working space
SWS	suspension working space
T	time interval
u	control force
U	vehicle forward speed
V	non-negative definite symmetric matrix
v	vertical velocity
W	spool valve area gradient
x	state vector
x_{vd}	demand valve opening
x_{vdsat}	saturated valve opening demand
x_{vmax}	maximum valve displacement
x_0	ground displacement (road input)
x_1	wheel displacement
x_2	body displacement
δ	Dirac delta function
κ	road roughness coefficient
υ	wave number
ω	slope exponent
ξ	white noise disturbance input

C553/032 © IMechE 1998

1 INTRODUCTION

The main purpose of a vehicle suspension is to provide a comfortable ride for the occupants while maintaining good road holding capability. Ride is considered as one of the most important characteristics defining the quality of a vehicle and is related to vertical body acceleration. Road holding is influenced by the deformation of the tyre due to ground excitations also referred to as dynamic tyre load. The design requirements are to achieve low levels of vertical body acceleration, whilst minimising the dynamic tyre load. These requirements have to be fulfilled effectively under a range of operating conditions including braking and acceleration, cornering at speed and traversing rough roads and lead to the classic passive suspension design compromise:

- To improve ride the passive elements must be soft
- To improve road holding the passive elements must be hard

Using passive elements, it is only possible to design one optimal suspension system for one combination of vehicle forward speed and road roughness. Therefore, when operating conditions change the suspension set-up will not remain optimal. Recent advances in automotive engineering made electronically controllable active systems technically feasible. Using a hydraulic actuator that exerts the suspension force between wheel and body according to a control law, it was shown that a performance superior to a passive system can be achieved [1]. The main disadvantage is the need for expensive hardware such as an external power source, hydraulic actuators and servo valves.

The Self-Energising Self-Levelling Semi-Active (SESLSA) system developed by Hall and Tang [2, 3, 4] avoids the need for an external power source and the hydraulic actuators are replaced by continuously controllable dampers. This approach offers a commercially more viable solution. The work carried out previously has also resulted in the completion of a quarter vehicle test rig, implementation of rig instrumentation and data capture software, the design and manufacture of a prototype damper and the development of a damper controller. Full details of this work are contained in the following publications [5, 6 ,7].

The existing control strategy is based on linear optimal control with fixed gain feedback and is not suitable for a non-linear suspension system because it cannot adapt to different conditions such as a change in vehicle forward speed or road roughness. Adaptation schemes and controller modifications to take account of non-linearities include gain scheduling and self-tuning regulators [8] but these inevitably increase the complexity of the whole system.

Fuzzy logic enjoys great popularity in the field of control engineering. It is therefore hardly surprising that major car manufactures, including BMW, FORD and VW, have investigated in active and semi-active vehicle suspensions controlled by fuzzy logic [9, 10, 11, 12]. If it is the aim to deploy a non-linear controller that reduces the complexity of the SESLSA system, fuzzy logic must be taken into consideration. Various control techniques have been investigated and it has been concluded that the one offering the highest potential is based on fuzzy logic [13]. This is mainly due to the inherently non-linear nature of fuzzy logic control, which requires no additional modifications. Furthermore hardware requirements can be reduced by using fewer transducers.

The remainder of this paper reviews the concept of the SESLSA system and the existing control strategy. Furthermore it presents model descriptions of vehicles, tyres and road surfaces. Control strategies, such as fuzzy logic and LQR are explained in more detail and simulation results are presented. Finally implications for SESLSA control are outlined and conclusions are drawn.

2 REVIEW OF SESLSA CONCEPT AND EXISTING CONTROLLER

The hardware of the current prototype SESLSA damper is shown schematically in Figure 1. The strut (2) itself consists of three interconnected chambers separated by valves. Two pressurised air springs (1, 5) are located at chamber 1 and chamber 3. The semi-active performance is achieved by the damper valve (6), which is a direct acting solenoid valve. Continuous suspensions force control is achieved by controlling the oil flow through the solenoid valve. The levelling operation is performed by the changeover valve (7) and the check valve (8). The bump blow-off valve and the rebound blow-off valve (3, 4) act as safety devices in case the pressure in the strut is excessively high.

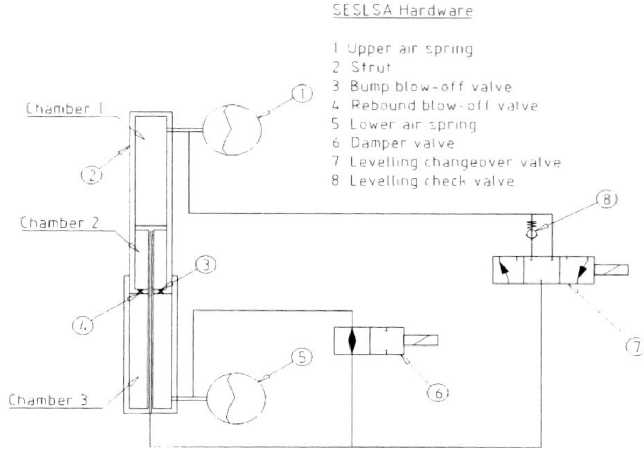

Figure 1: Main components of the SESLSA system

Figure 2 shows the layout of the existing multi-level controller, which consists of a linear optimal demand force controller (top-level controller) and a non-linear strut force controller. The non-linear strut force controller consists of a flow rate controller and a valve position controller.

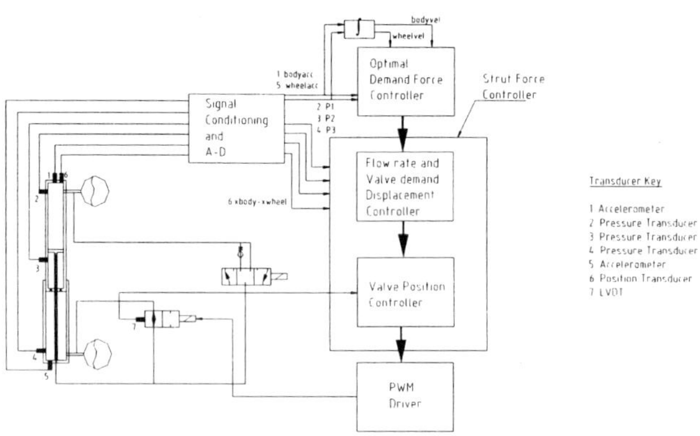

Figure 2: Multi-level controller layout

C553/032 © IMechE 1998

The problem is approached by defining a linear controller model, which is described by the state equations of the LQR controlled active system:

$$\dot{x} = Ax + Bq_d + \xi \qquad q_d = -Kx \tag{1}$$

K is the optimal gain vector, found by solving the discrete Riccati equation [14]. The valve demand is generated as the saturated product of flow demand and an inverse flow function (IFF),

$$x_{vd} = q_d(kT)IFF \qquad x_{vdsat} = \begin{cases} x_{v\max}k_c, & x_{vd} > x_{v\max}k_c \\ 0, & x_{vd} < 0 \end{cases} \tag{2}$$

where the upper saturation level is determined by the maximum coil driver current ($x_{vmax}k_c$).

The inverse flow function (IFF) is given by:

$$IFF = \frac{1}{\left(\operatorname{sgn}((P_3 - P_2))C_d W \sqrt{\frac{2}{\rho}}\sqrt{|(P_3 - P_2)|}\right)} \tag{3}$$

The current controller is based on fixed feedback gain LQR techniques, which have been modified in order to take non-linearities into account. These modifications resulted in a rather complex controller applied to a quarter vehicle suspension. Further refinements will include the need for adaptation and full vehicle modelling. Alternative techniques have to be investigated and it is the aim to design a less complex controller with adaptation capabilities.

3 SIMULATION OF SYSTEM PERFORMANCE

Reduced development times dictated by a more and more competitive market, forces the automotive industry to launch their new cars as quickly as possible [15]. This requires the suspension engineer to design a refined suspension with a minimum of hardware testing (rig and vehicle testing) prior to product launch. Consequently, considerable emphasis is placed on simulation work using powerful software packages such as MATLAB [16], SIMULINK [17] and ADAMS [18].

3.1 Vehicle models

Three basic models of varying complexity are in use for vehicle suspension design. The most realistic is the full vehicle model, which also is the most complex one incorporating all four wheel stations. The 7 DOF full vehicle model comprises 3 DOF for the vehicle body (pitch, bounce and roll) and a further vertical DOF for each wheel station. Crolla and Abdel-Hady [19] have shown that the effect of including the cross-correlation information between left and right tracks is of little benefit in the resulting control law, and hence the full vehicle model can be reduced to a 4 DOF half vehicle model. Despite the loss of 3 DOF this model is widely used to analyse vehicle suspensions.

Another common representation of the vertical dynamics of a car is the 2 DOF quarter vehicle model, which represents a proportion of the car body (sprung mass) and the wheel assembly (unsprung mass). Hall and Gill [20] state that this model can be justified for bounce motions for vehicle with inertia couplings close to unity (typified by a conventional automobile). The quarter vehicle model is a significant simplification of the suspension problem and cannot provide information about pitch and roll, but it also has advantages over more complex models in terms of;

- being described by fewer design parameters,
- having fewer performance parameters,
- having only a single input, leading to ease of application of optimal control theory to derive control laws,
- ease of understanding of the relationships between design and performance [21].

The following features are not part of the quarter vehicle analysis;

- coupling between the four wheel stations,
- vehicle manoeuvring forces.

If the assumptions made above are acceptable, the quarter vehicle model contains the basic features of the real problem and allows the conflict between ride and grip to be explored. Quarter vehicle representations of ideal passive and active suspensions are shown in Figure 3.

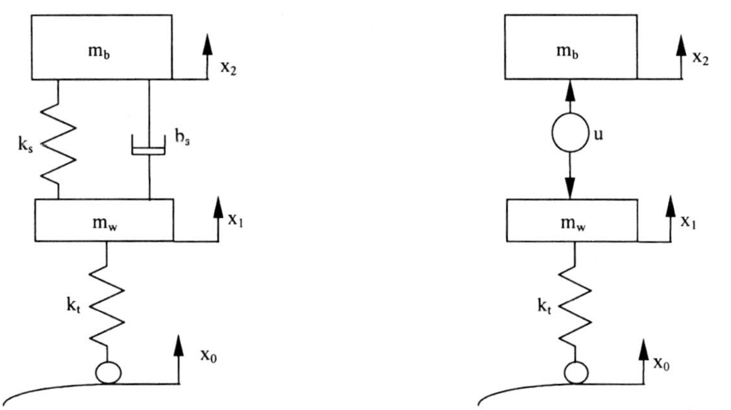

Ideal passive quarter vehicle model　　Ideal active quarter vehicle model

Figure 3: Quarter vehicle models

The wheel mass, m_w, represents tyres, wheels, axles, brakes and the parts of the suspension that move vertically with the wheels. The linear spring with stiffness, k_t, represents the tyre. In practice the body mass, m_b, is approximately 25 % of the actual body mass, for passenger cars. The spring rate, k_s, and damping coefficient, b_s, are related to suspension spring and shock absorber, respectively. The control force, u, is produced by the actuator in the active system.

3.2 Tyre models

A vehicle tyre is essentially a visco-elastic toroid which mainly serves the following four functions:

- It supports the vertical load (i.e. vehicle mass and payload).
- It cushions against road shocks.
- It develops lateral forces for cornering.
- It develops longitudinal forces for accelerating and braking.

Modern refinement and optimisation have made the vehicle tyre a very complex, highly non-linear dynamic system that is difficult to quantify [22]. Many approaches to accurately describe the behaviour of a real tyre have been made [23]. One of the most popular is known as the "magic formula" or "Pacejka formula" [24, 25]. This method involves the fitting of analytical functions to experimental side force, longitudinal force and alignment moment data.

For a simplified analysis it is advisable to find a reasonable compromise between model complexity and simulation realism [26, 27]. The interesting aspects of tyre behaviour for the quarter vehicle model are stiffness and damping. Since damping in the tyre is relatively small compared to damping in the suspension it is a reasonable simplification to neglect tyre damping. Tyre stiffness is dependent on the type of tyre and its inflation pressure. The tyre stiffness is at its maximum when the vehicle is stationary (static stiffness) and varies with rotational speed (dynamic stiffness), but linear approximations may be made [28].

For the quarter vehicle model, the point follower model is frequently used to calculate vertical forces. The tyre itself consists of a linear spring with constant spring rate, k_t, and damping is neglected (Figure 3).

3.3 Model excitation

Random road models

A realistic representation of road surfaces is necessary for conclusive simulations. Road surfaces can be measured directly and recorded [29], but most of them have been described in the frequency domain by Gaussian random processes [30]. The excitation of a linear model can be carried out in the frequency domain [31]. A method of generating a time series of displacements by using the inverse FFT is presented in [32] and can be used to excite non-linear models.

Although, a bi-linear (log-scale) model is available [30], Robson [33] suggest that the simpler linear model,

$$S(\nu) = \kappa \nu^{-\omega} \qquad \textbf{(4)}$$

gives satisfactory results.

The roughness coefficient κ varies, depending on the quality of the road surface. The following table shows the chosen values for road roughness κ and slope exponent ω [34].

Table 1: Road roughness coefficients and slope exponents

	κ in 10^{-8} cycles/m	ω
minor road	135	2.5
major road	44	2.5
motorway	7	2.5

For a constant vehicle forward speed, U, the spectrum can be written in terms of frequency, f, rather than wavenumber, ν:

$$S(f) = \kappa U^{(w-1)} f^{-w} \quad (5)$$

In this study the time series of road profiles were generated from a given displacement spectrum shape using the inverse FFT, saved as ASCII files and loaded into the MATLAB workspace. The road profile of a minor road (Figure 4) was used to excite simulation models. A vehicle forward speed U=20 m/s was chosen for all simulations.

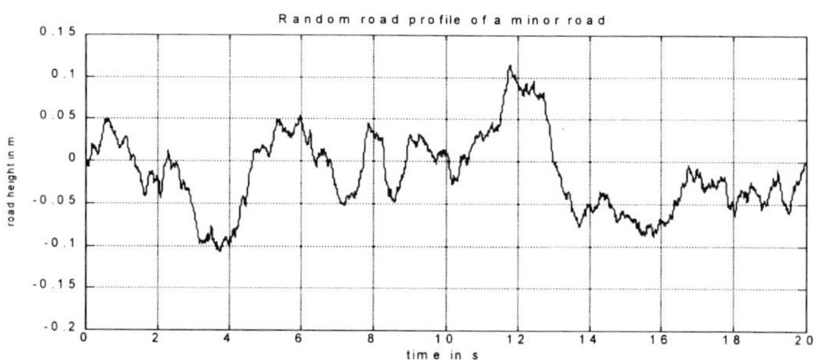

Figure 4: Time series of the random road profile of a minor road

Step inputs

The effects of large obstacles cannot be described by random processes, and must be treated separately. To simulate transient events such as potholes a simple step input of 50 mm served as the road input model.

3.4 LQR control

A linear system with full state feedback can be described by:

$$\begin{aligned} \dot{x} &= Ax + Bu + \xi \\ z &= Cx \end{aligned} \quad (6)$$

with n states x, p control inputs u, q controlled variables z and a white noise disturbance input ξ satisfying

$$E[\xi(t)\xi^{T}(\tau)] = V\delta(t-\tau) \quad (7)$$

 C553/032 © IMechE 1998

The resulting linear quadratic regulator problem consists of determining the control force u, which minimises the performance index J below:

$$J = \lim_{t \to \infty} E[z^T(t)Qz(t) + u^T(t)Ru(t)] \qquad \textbf{(8)}$$

The optimal gain vector K can be found by solving the algebraic Riccati equation:

$$A^T P + PA + C^T QC - PBR^{-1}B^T B = 0 \qquad \textbf{(9)}$$

The optimal control law is given by:

$$u = -Kx \qquad \text{and} \qquad K = R^{-1}B^T P \qquad \therefore\ u = -R^{-1}B^T P\ x \qquad \textbf{(10)}$$

Linear optimal control techniques are especially useful for problems where the dynamic model of the plant is linear or can be linearised. State space description of the dynamic systems was most suitable for the MATLAB software used during this project, which is based on matrix computation. For complex, non-linear suspension systems the basic theory is not valid and needs to be extended.

A number of expensive transducers such as accelerometers and LVDTs are required for the full state feedback system. It must also be emphasised that a full state feedback controller needs to measure the road elevation, which is major problem in practice. This can be avoided using limited state feedback and an observer, which can estimate state variables that are difficult or expensive to measure. However, this leads to a performance reduction of such a design. In order to control real suspensions it must be the target to design a non-linear controller and simultaneously reduce the number of transducers.

3.5 Fuzzy logic control

Fuzzy logic is a method of describing analogue processes on a digital computer. Fuzzy logic is a general calculation system that its inventor, Lotfi Zadeh, claims is a superset of traditional logic (Boolean logic), just as Newton's physics is a special case of Einstein's physics [35]. Fuzzy logic and fuzzy expert control systems are one of the most active research and development areas of artificial intelligence (AI) at the present time [36, 37].

A fuzzy system can be subdivided in the following five stages:

1.) Fuzzification of the inputs
2.) Application of the fuzzy operator
3.) Application of implication method
4.) Aggregation of all outputs
5.) Defuzzification

The first step of a fuzzy logic system is the generation of membership values (degrees of fulfilment) between 0 and 1 for a fuzzy variable using membership functions. The fuzzy operators AND, OR and NOT are applied to connect the input membership values. The implication method in step 3 is the process of shaping the fuzzy set in the consequent according to the results of the antecedent. For the type of fuzzy systems used for this research

(Mamdani) the output of the aggregation is one fuzzy set for each output variable. Defuzzification is the process of converting the final aggregated fuzzy set into a crisp value that can be sent to the outside world (Figure 5).

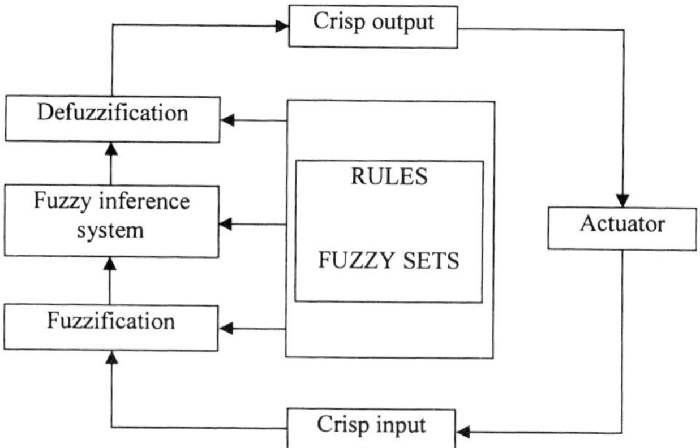

Figure 5: A typical fuzzy system

Fuzzy logic controllers are defined in terms of the desired relationship between inputs and outputs rather than to determine the parameters that describe the dynamics of the system. The advantages of fuzzy logic control (FLC) compared to conventional control approaches are substantial. FLC is a non-linear control approach and non-linear functions of arbitrary complexity can be modelled. FLC can also be blended with conventional control techniques and can simplify their implementation. Expert knowledge can be incorporated in fuzzy logic. Fuzzy logic is based on natural language, and consequently this control method proves to be very flexible and tolerant to imprecise data [38].

Many of the FLC properties described above are invaluable for the control of real suspensions. It is expected that fuzzy logic can deal with the non-linearities of advanced suspensions and achieve good performance compared to ideal, linear benchmark models. The inherently non-linear nature of FLC will ideally avoid the need for additional adaptation schemes. However, if it is necessary or desirable FLC can be extended using ANFIS (Adaptive-Neuro Fuzzy Inference Systems), which modifies the shapes of the membership functions using an artificial neural network.

3.6 Software

Most of the simulation work was carried out using MATLAB, SIMULINK and the relevant Toolboxes from the Mathwork Inc. This section describes the software implementation of the models.

Passive quarter vehicle model

A passive model was simulated in order to find the combination of parameters (k_s and b_s) that offered the best performance. A SIMULINK block diagram is shown in Figure 6.

C553/032 © IMechE 1998

Parameters used: k_s = 5 ... 50 kN/m in steps of 5 kN/m; b_s = 500 ... 5000 Ns/m in steps of 500 Ns/m; m_b = 250 kg; m_w = 50 kg; k_t = 200 kN/m.

The input was a random road profile generated in VDAS [39]. The time series was stored as an ASCII files and loaded into the MATLAB workspace as a matrix (roadmin), from where it could be called by SIMULINK. The plant is a state space description of a linear quarter vehicle model. The outputs of the simulation were sent to a multiplexer (Mux) and stored as a matrix in the MATLAB workspace (minor). Standard deviations of ACC, DTL and SWS were calculated and hence it was possible to determine the best set of suspension parameters.

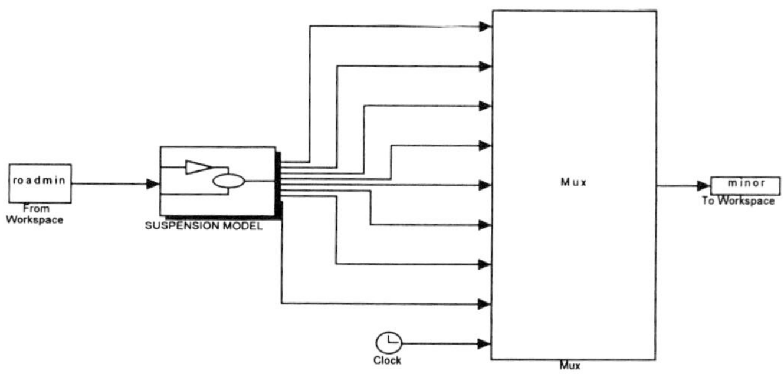

Figure 6: Passive suspension model using SIMULINK

Optimal control of an active quarter vehicle model

The purpose of this simulation was to analyse an active full state feedback model using linear optimal control. The following model was set up using MATLAB, SIMULINK and additionally the CONTROL SYSTEM TOOLBOX [40] was employed to implement LQR techniques. The basic layout is similar to Figure 7, except for the fuzzy logic controller is replaced by a constant feedback matrix K with pre-calculated controller gains. The state vector is sent to the multiplexer (Mux) and multiplied by the matrix K before it is fed back to the suspension system. The feedback gains are found by minimising a quadratic performance index and solving the algebraic Riccati equation.

Fuzzy logic control of a quarter vehicle model

A fuzzy logic controller (FLC) was developed to control the active suspension system. The following model, Figure 7, was set up using MATLAB, SIMULINK and the FUZZY LOGIC TOOLBOX [41]. The basic idea was to measure several system outputs (e.g. body velocity, suspension velocity etc.) and feed them into the FLC. The controller output is the control force, u, which is fed back into the suspension system. An important part of this heuristic approach is to find a combination of system outputs that achieves good performance and keeps transducer costs low.

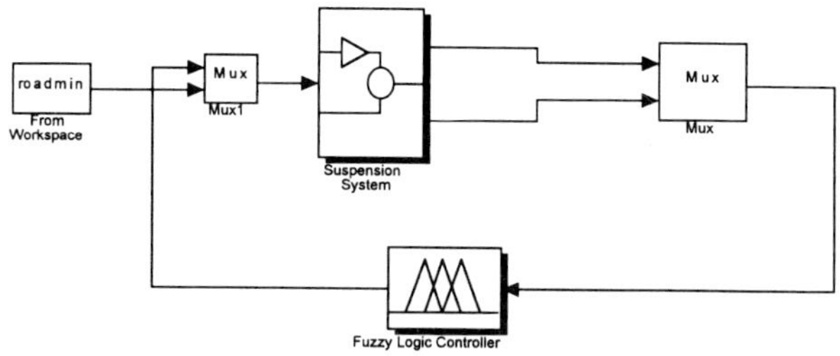

Figure 7: Suspension block diagram featuring a FLC

4 RESULTS

The main purpose of the previous simulation work was to identify the "ideal" passive and active systems (benchmark models) and to compare these with the fuzzy logic controlled system in order to prove the feasibility of the FLC approach. The results of random road excitation and step inputs are given below.

4.1 Random road results

The random profile of a minor road was used to excite the models and standard deviation results (SD) of the relevant system outputs were calculated.

Passive system

Parameters used: $k_s = 5$ kN/m; $b_s = 1500$ Ns/m; $m_b = 250$ kg; $m_w = 50$ kg; $k_t = 200$ kN/m

Results: SD ACC = 1.49 m/s^2; SD DTL = 777 N; SD SWS = 14.3 mm

It has to be emphasised that this simulation only served as a benchmark model. The very low suspension stiffness ks = 5 kN/m resulted in a good performance in terms of SD ACC. For a real suspension a spring stiffness of approximately 20 kN/m is more appropriate. However, using a stiffer suspension resulted in an increase of sprung mass acceleration. Furthermore Hall [42, 43] found that non-linear damping effects result in a significant reduction in ride comfort compared to linear models. Therefore performance comparisons between controlled and linear passive system must be made with great care.

The passive suspension was simulated with the same parameters as in the previous simulation, except for the suspension stiffness, k_s was increased to 20 kN/m.

Results: SD ACC = 1.81 m/s$^{2;}$ SD DTL = 765 N; SD SWS = 13.0 mm

Using the stiffer suspension spring resulted in an increase of 17.7% in sprung mass acceleration SD ACC compared to the "ideal" passive system.

C553/032 © IMechE 1998

Active system

Parameters: m_b = 250 kg; m_w = 50 kg; k_t = 200 kN/m

Results: SD ACC = 1.21 m/s^2; SD DTL = 893 N; SD SWS = 14.3 mm

The best set of controller gains was found by varying the output weighting parameters q_1 and q_2 and keeping the control weighting R constant. For every combination a set of controller gains was calculated, implemented in the active model and simulated.

Weighting parameters selected: $q_1 = 4.05*10^9$; $q_2 = 1*10^8$; R = 1
Controller gains calculated: K = [-18513 10000 -976 2283]
The body acceleration was reduced by 19% compared to the "ideal" passive suspension. Compared to the more realistic model with a stiffer suspension spring (k_s = 20 kN/m), a reduction in SD ACC of 33% could be achieved.

Fuzzy logic system

Parameters: m_b = 250 kg; m_b = 50 kg; k_t = 200 kN/m

Results: SD ACC = 0.92 m/s^2; SD DTL = 1127 N; SD SWS = 14.8 mm

The chosen membership functions (MFs) were piecewise linear and define how each point in the input space is mapped to a membership value between 0 and 1. The trapezoidal MFs were formed using straight lines. These functions have a flat top and are truncated triangles. These straight-line MFs have the advantage of simplicity. The necessary link between membership functions and fuzzy operators (in this case the AND operator was chosen) is provided by the use of IF-THEN rules. Nine rules were necessary to map the two inputs (three MFs each) to the output. The rules were chosen using a heuristic method based on engineering knowledge about the dynamics of the suspension system. The MFs were optimised by truncating the initial triangular MFs and by determining the optimal level of overlap by trial and error. The output MFs (control signal) are shown in Figure 8.

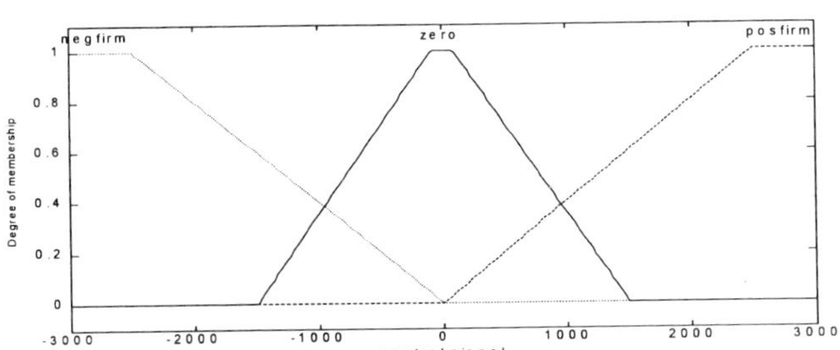

Figure 8: Membership functions for control signal (in N)

© IMechE 1998 C553/032

Rules:

1. If (BV is positive) and (SV is positive) then (CS is neg_firm)
2. If (BV is positive) and (SV is zero) then (CS is neg_firm)
3. If (BV is positive) and (SV is negative) then (CS is neg_firm)
4. If (BV is zero) and (SV is positive) then (CS is pos_firm)
5. If (BV is zero) and (SV is zero) then (CS is zero)
6. If (BV is zero) and (SV is negative) then (CS is neg_firm)
7. If (BV is negative) and (SV is positive) then (CS is pos_firm)
8. If (BV is negative) and (SV is zero) then (CS is pos_firm)
9. If (BV is negative) and (SV is negative) then (CS is pos_firm)

Rule 1 says that if the body velocity (BV) is positive and the suspension velocity (SV) is positive then the control signal (CS) is negative firm (neg_firm).

Using the fuzzy logic controller, it was possible to reduce body acceleration by 49% compared to the real passive suspension. Relative to the active system a reduction of 24% was achieved.

4.2 Step response results

An alternative method to evaluate the performance of a particular suspension was to compare the sprung mass step responses. Decreases in displacement, velocity and acceleration levels indicate an increase in ride comfort. Similarly, step responses of DTL and SWS were compared in order to assess system performance.

On the following page a step of 50 mm served as the road input. The body acceleration step responses of the passive, active and fuzzy system are shown to allow the reader to visually compare the results (Figure 9). Performance criteria, such as percent overshoot, settling time, peak ACC, peak DTL and peak SWS were used to quantify the step input results (Table 2).

The peak ACC of the active system was significantly lower than the passive one. The fuzzy system reduced body acceleration by 46% and 30% compared to the passive and active model, respectively.

Table 2: Step response results.

	PASSIVE SYSTEM	ACTIVE SYSTEM	FUZZY SYSTEM
Percent overshoot (x_2)	24 %	6 %	0 %
Settling Time (x_2)	≈ 2s	≈ 1s	≈ 2s
Peak ACC	≈ 13 m/s^2	≈ 10 m/s^2	≈ 7 m/s^2
Peak DTL	≈ 10500 N	≈ 10000 N	≈ 10000 N
Peak SWS	≈ 63 mm	≈ 73 mm	≈ 78 mm

C553/032 © IMechE 1998

Body acceleration step responses

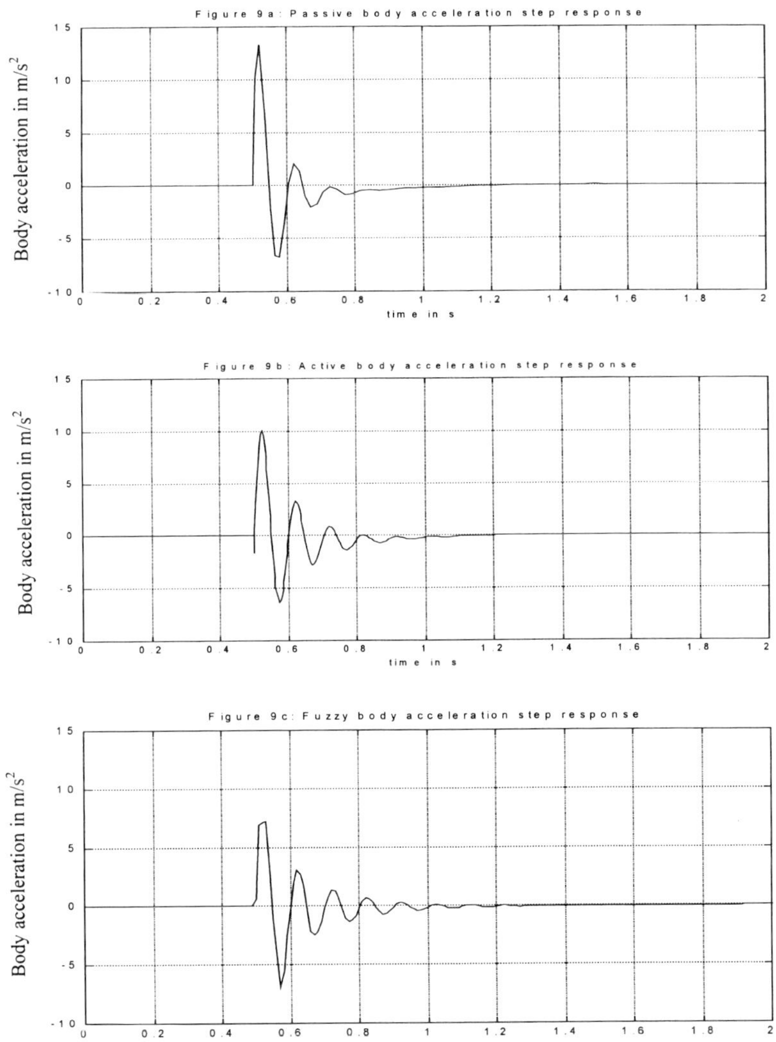

Figure 9: Step response results

4.3 Discussion of results

The simulation results indicate how superior performance can be achieved using controlled suspensions. The active model, which was based on a fixed feedback gain linear controller (LQR), reduced SD ACC by 33%. The design constraint of the random road simulations was SD SWS, which was kept constant at approximately 15 mm.

The step responses confirmed the superiority of controlled active suspensions. All relevant system outputs (body displacement, body acceleration, dynamic tyre load) showed improvement over the passive model. The reason why the active system utilised more suspension working space stems from the fact that the weighting parameters were chosen to reduce body acceleration and not to minimise suspension working space.

The fuzzy logic controlled model reduced SD ACC by 49%. The rather high SD DTL was mainly a result of the method employed to design the controller. The emphasis was to reduce SD ACC, which caused an increase in SD DTL. Moreover, since the membership functions were tuned to improve the system's step response a further deterioration of SD DTL was the result. Comparing the step response results, the fuzzy logic controller showed significant performance improvements (Table 2). An increase in peak SWS originated from the same reasons as mentioned for the active system.

5 IMPLICATIONS FOR SESLSA CONTROL

This research is mainly concerned with the development of non-linear fuzzy logic controllers in order to replace the existing LQR-based controller. The fuzzy logic controller used for the simulation work carried out so far had basic features, such as equally spaced trapezoidal membership functions. Further work will include a heuristic approach to determine the optimal number and shape of the membership functions. Furthermore, the relative position of the membership functions in their universes of discourse will be investigated. The most time-consuming part of fuzzy logic controller development is the fine-tuning, especially since no methodology exists to find the 'optimal' controller set-up. Improving controller performance using heuristic methods is equivalent to the refinement of mathematical model descriptions of conventional approaches. The enhancement and fine-tuning of the existing fuzzy logic controller and its integration in the SESLSA concept will form the next part of this work. On line and off line neuro-fuzzy techniques will be investigated in order to make the controller adaptive. Future research projects will include the incorporation of the new controller in the real-time C code environment and practical testing on the quarter vehicle suspension rig.

It is the aim to develop a fuzzy logic controller that performs similarly to the ideal active LQR controller using linear quarter vehicle models. Afterwards various non-linear features will be included in the suspension models and the performance of the fuzzy logic controller will be evaluated. Since the existing top-level controller is based on linear methods the advantages of the new fuzzy controller applied to a non-linear quarter vehicle model are evident.

For a full vehicle model a number of additional aspects have to be taken into consideration. For the quarter vehicle model a fuzzy logic control strategy can be based on just a few measurements. Since a full vehicle model has 7 DOF (four wheel-stations plus pitch, bounce and roll of the sprung mass) the derivation of an effective control strategy becomes more complex and more measurements (transducers) are required. The coupling between the four wheel-stations has to be considered. Overall, an extension of the fuzzy logic controller to a full vehicle model should cause fewer problems than modifying controllers based on linear techniques. The fuzzy logic controller applied to the quarter vehicle model can be extended and modified to control more complex models (i.e. half- and full vehicle models). This can be carried out by adding more information to the input space and re-tune the controller. Recommendations regarding the application of the controller to full vehicle models will be given.

C553/032 © IMechE 1998

6 CONCLUSIONS

Although the simulation results were very encouraging, further improvement could be achieved using more sophisticated controllers. When operating conditions change (vehicle forward speed, payload, road roughness) the fixed feedback gain LQR controller would no longer be optimal. A controller that can adapt the gains as operating conditions change is required. Furthermore, the controller must be capable of dealing with non-linear phenomena, such as semi-active operation, bump stops, damper non-linearities etc.

One option is to use an adaptive LQR controller (e.g. gain scheduling, model reference adaptive control etc.) capable of choosing a set of optimal feedback gains when operating conditions change. Since LQR controllers are based on linear techniques rather complicated, modifications are necessary in order to take account of system non-linearities.

It was concluded that an inherently non-linear method such as fuzzy logic control offers the highest potential to deal with highly non-linear suspension systems. Furthermore fuzzy logic controllers are less sensitive to modelling errors and uncertainties because the model of the suspension is only used as an evaluation tool and does not form an integral part of the controller design.

REFERENCES

[1] Thompson A.G.: "An active suspension with optimal linear state feedback"; Vehicle System Dynamics, Vol. 5, pp 187-203, 1976

[2] Tang J.S.; Hall B.B.: "A damper unit"; U.K. patent application No. 9016202.3, Jul. 1990

[3] Hall B.B.; Tang J.S.: "A valve control strategy and its application in semi-active suspensions"; IMechE, Oct 1990

[4] Tang J.S.; Hall B.B.: "A combined self-energising self-levelling semi-active damper unit for use in vehicle suspensions"; EAEC paper 91064, 3rd EAEC Conference on Vehicle Dynamics and Powertrain Engineering, Strasbourg, Jun. 1991

[5] Underhill T.R.: "The implementation of a novel damper"; Transfer Document, Huddersfield University, Jul. 1994

[6] Hall B.B.; Underhill T.R.: "Active suspension systems - some aspects of digital controller design"; Paper No. 94ME045, 27th International Symposium on Automotive Technology and Automation, Aachen, Germany, Nov. 1994

[7] Underhill T.R.; Hall B.B.: "Digital control of a semi-active suspension featuring flow control and a state observer"; Autotech '95, Birmingham, Nov. 1995

[8] Fan Yu: "Adaptive and self-tuning control of vehicle suspension"; PhD Thesis, The University of Leeds, Apr. 1996

[9] Yester J.L.; McFall R.H.: "Fuzzy logic controller for active suspension"; SAE Paper 92C030, 1992

[10] Cai B.; Konik D.: "Intelligent vehicle active suspension control using fuzzy logic"; IFAC 12th Triennial World Congress, Sidney, Australia , 1993

[11] Heinrich A.; Kreft J.; Doerrscheidt F; Boll M.: "Fuzzy-Regelung eines semiaktiven Pkw-Fahrwerks"; Automatisierungstechnische Praxis, Vol. 36; 1994

[12] Nizar AL-Holou; Shout A.: "The development of fuzzy logic based controller for semi-active suspension system"; IEEE, 1995

[13] Heumann H.: "Adaptive control of a novel suspension system"; Transfer Document, Huddersfield University, Feb. 1998

[14] Arnold W.F. III; Laub A.J.: "Generalized eigenproblem algorithms and software for algebraic Riccati Equations"; Proc. IEEE, pp 1746-1754, 1984

[15] Braess H.H.: "Selbstorganisation und Selbstoptimierung – Chancen einer fraktalen Entwicklung der Mess- und Versuchstechnik im Automobilbau? (Selforganisation and Selfoptimisation – Opportunities for a 'fractal' evolution especially in the field of measurement and test engineering in the automotive industry?)"; VDI Berichte Nr. 1189, 1995

[16] MATLAB: "User's guide"; The Math Works Inc., Feb. 1993

[17] MATLAB: "SIMULINK [Dynamic System Simulation Software] -user's guide"; The Math Works Inc., Feb. 1993

[18] ADAMS: "Automatic Dynamic Analysis of Mechanical Systems"; Mechanical Dynamics Inc, Ann Arbor, USA

[19] Crolla D.A.; Abdel-Hady M.B.A.: "Active suspension control; performance comparisons using control laws applied to a full vehicle model"; Vehicle System Dynamics, Vol. 20, pp 107-120, 1991

[20] Hall B.B.; Gill K.F.: "Performance evaluations of motor vehicle active suspension systems"; Proceedings of the Institution of Mechanical Engineers, Vol. 201, 1987

[21] Sharp R.A.; Hassan S.A.: "The fundamentals of passive automotive suspension design"; Society of Environmental Engineers, Conference on Dynamics in Automotive Engineering, pp 219-228, 1984

[22] Clark S.: "Mechanics of pneumatic tyres"; U.S. Department of Transportation, Washington D.C., p 931, 1981

[23] Pacejka H.; Sharp R.: "Shear force development by pneumatic tyres in steady state conditions: A review of modelling aspects"; Vehicle System Dynamics, Vol. 20, No. 3-4, pp 121-176, 1991

[24] Pacejka H.; Sharp R.: "Shear force development by pneumatic tyres in steady state conditions: A review of modelling aspects"; Vehicle System Dynamics, Vol. 20, No. 3-4, pp 121-176, 1991

[25] Bakker E.; Nyborg L.; Pacejka H.: "Tyre modelling for use in vehicle dynamics studies": SAE Paper 870421, 1987

[26] Dugoff H.; Fancher P.S.; Segel L.: "Tyre performance characteristics affecting vehicle response to steering and braking control inputs"; The University of Michigan Highway Safety Research Institute, p 105, Aug. 1969

[27] Sharp R.S.; El-Nashar M.A.: "A generally applicable digital computer based mathematical model for the generation of shear forces by pneumatic tyres"; Vehicle System Dynamics, Vol. 15, No. 4, pp 187-209, 1986

 C553/032 © IMechE 1998

[28] Wong J.Y.: "Theory of ground vehicles"; Wiley-Interscience, 1978

[29] La Barre R.; Forbes R.; Andrew S.: "The measurement and analysis of road surface roughness"; MIRA Report No. 1970/5, 1970

[30] Dodds C.J.; Robson J.D.: "The description of road surface roughness"; Journal of Sound and Vibration, Vol.31, No.2, pp 175-183, 1973

[31] Robson J.D.; Dodds C.J.: "Stochastic road inputs and vehicle response"; Vehicle System Dynamics, Vol. 5, pp 1-13, 1975/76

[32] Cebon D.; Newland D.E.: "The artificial generation of road surface topography by the inverse FFT method"; IAVSD - IUATM 8th Symposium on the Dynamics of Vehicles on Roads and Rails, Cambridge Mass., USA, pp 160-165, 1983

[33] Robson J.D.: "Road surface description and vehicle response"; International Journal of Vehicle Design, Vol. 1, No. 1, pp 25-35, 1979

[34] Cebon D.; Besinger F.; Cole D.: "Control strategies for semi-active lorry suspensions"; Proceeding IMechE, Vol. 210, 1996

[35] Johnson C.: "That fuzzy feeling"; Datamation, pp 39-43, Jul. 1989

[36] Lee C.: "Fuzzy logic in control systems: Fuzzy logic controller, part I"; IEEE Transactions on Systems, Man, and Cybernetics, Vol. 20, No. 2, pp 404-418, Mar. 1990

[37] Lee C.: "Fuzzy logic in control systems: Fuzzy logic controller, part II"; IEEE Transactions on Systems, Man, and Cybernetics, Vol. 20, No. 2, pp 419-435, Mar. 1990

[38] Earl Cox: "Fuzzy fundamentals"; IEEE Spectrum, pp 58-61, Oct. 1992

[39] Horton D.: "VDAS - Vehicle Dynamics Analysis Software, version 3.2"; Department of Mechanical Engineering, Leeds University

[40] MATLAB: "CONTROL SYSTEM TOOLBOX - user's guide"; The Math Works Inc., Jul 1992

[41] Gulley N.; Jang J.; "FUZZY LOGIC TOOLBOX - user's guide", The Math Works Inc., 1995

[42] Hall B.B.: "The effect of vehicle suspension characteristics on ride performance"; PhD Thesis, The University of Leeds, 1985

[43] Hall B.B.; Gill K.F.: "Performance of a telescopic dual-tube automotive damper and the implications for vehicle ride prediction"; Proceedings of the Institution of Mechanical Engineers, Vol. 200, 1986

C553/027/98

On the dynamics and control of flexible multi-body non-linear space systems

A G KELKAR PhD, MASME, MAIAA
Mechanical and Nuclear Engineering Department, Kansas State University, USA

Abstract

This paper presents an overview of our selected work in the area of dynamics and control of a class of nonlinear multibody flexible space structures. The equations of motion presented consider only rotational motion of the spacecraft and assume a cluster configuration of flexible bodies. That is, a spacecraft is assumed to have a large central body with multiple chains of appendage bodies. The equations of motion are obtained using Euler-Langrange formulation. The control problem addressed is that of obtaining globally stable closed-loop rotational maneuvers of multibody flexible spacecraft. Although the control law development is primarily focused on the multibody space systems the results are directly applicable to a wide range of terrestrial systems, as well. The control laws presented include both linear and nonlinear controllers and are shown to be robust to unmodeled dynamics and parametric uncertainties.

1 Dynamic Model

A class of systems considered consists of a branched configuration of flexible bodies where each branch by itself could be a serial chain of flexible structures. Dynamic model of such systems can be obtained using Newtonian mechanics, calculus of variations, or Lagrangian approach. In this paper, however, an Euler-Lagrange formulation is presented from [1]. Being an energy-based approach, this formulation is less cumbersome compared to the Newtonian approach which deals with vector quantities. It is assumed that the bodies deform only due to the elastic motion and the deformations are in the linear range. A detail derivation of the rotational dynamic model can be found in [1]. Only selected steps of the derivation are given here due to the space limitations. The coordinate systems used for derivation of kinematics of the spacecraft are given in Fig. 1. As a first step, the

© IMechE 1998 C553/027

expressions for kinematic quantities are derived for a representative partical mass dm in ij-th link of the system. Then the expressions for kinetic and potential energy can be obtained for the entire system and Euler-Lagrange equations of motion can be derived.

The velocity of particle mass dm due to rotational motion of the spacecraft can be written in a compact form as

$$\overline{v}_{dm}^{ij} = \overline{v}_{dm}^{rot} = N\dot{p} \tag{1}$$

where $\overline{v}_{dm}^{rot}$ indicates the contributions to the velocity vector of particle mass dm from the rotational-plus-flexible motion of the spacecraft. In equation (1), N is $3 \times n$ matrix, n is the total number of rotational-plus-flexible degrees of freedom, and $\dot{p}$ is $n \times 1$ vector of generalized velocities corresponding to these degrees of freedom. Matrix N has the form: $N = [N_{rigid} \quad N_{flex}]$, where, N_{rigid} is $3 \times 3(n_c m + 1)$ matrix related to the rigid degrees of freedom and N_{flex} is $3 \times s(n_c m + 1)$ matrix related to the flexible degrees of freedom. For the sake of notational simplicity, and without loss of generality, it is assumed that there are n_c number of chains of appendage bodies, each of which has m bodies, and that each body has s flexible degrees of freedom. Then vector $\dot{p}$ has the form

$$\dot{p} = [\overline{\omega}_c, \overline{\omega}_{11}, \cdots \overline{\omega}_{1m}, \cdots \overline{\omega}_{ij}, \cdots \overline{\omega}_{n_c m}, \dot{q}_{c1}, \cdots \dot{q}_{cs}, \dot{q}_{111}, \cdots \dot{q}_{ijk}, \cdots \dot{q}_{n_c ms}]^T \tag{2}$$

The matrices N_{rigid} and N_{flex} contain terms which are nonlinear functions of configuration variables and correspond to the rigid and flexible degrees of freedom, respectively. Having obtained the expression for velocity, the kinetic energy of the entire spacecraft can be given by

$$T = \int_{\Omega} dT_{dm} = \frac{1}{2}\int_{\Omega} \overline{v}_{dm}^{rot\,T}\,\overline{v}_{dm}^{rot}\mu d\Omega \tag{3}$$

where, μ is the mass density, dT_{dm} is the kinetic energy of mass dm, and Ω denotes the spatial domain of integration. Substituting equation (1) into equation (3), yields

$$T = \frac{1}{2}\int_{\Omega}(N\dot{p})^T N\dot{p}\mu d\Omega = \frac{1}{2}\int_{\Omega}\dot{p}^T(N^T N)\dot{p}\mu d\Omega = \frac{1}{2}\dot{p}^T M(p)\dot{p} \tag{4}$$

where, $M(p)$ is the mass-inertia matrix of the system, given by

$$M(p) = \int_{\Omega}(N^T N)\mu d\Omega. \tag{5}$$

$M(p)$ is a symmetric and positive definite matrix. It is to be noted that although the kinetic energy expression derived here assumes only rotational-plus-flexible dynamics for the spacecraft, the effects of translational motion can be easily incorporated by modifying the term $\overline{v}_{dm}^{ij}$ to yield the kinetic energy expression which has both translational and rotational components.

1.1 Potential Energy

The potential energy of the system could be due to many sources, such as elastic displacement, thermal deformation, etc. The deformations due to thermal effects is not considered

C553/027 © IMechE 1998

here, however, it can be easily included in the formulation if desired. Thus, it is assumed that the potential energy has contribution only from the elastic strain energy. Also, it is assumed that the materials under consideration are isotropic in nature and that they obey Hooke's law.

For the isotropic materials obeying Hooke's law, the strain energy differential is given as

$$\delta V = \int_{\Omega} \sigma^T \delta \epsilon d\Omega \tag{6}$$

where the terms σ and $\delta\epsilon$ represent the stress and differential strain, respectively. Now, using the stress-strain relationship, strain-displacement relationship, and displacement equation in modal coordinates the strain energy differential can be rewritten after some simplification as

$$\delta V = \int_{\Omega} \sigma^T \delta \epsilon d\Omega = q^T \int_{\Omega} (\mathcal{D}\Phi)^T E \mathcal{D}\Phi d\Omega \delta q = q^T K \delta q \tag{7}$$

where, $\mathcal{D}$ is the differential operator used to define the strain tensor and K is the stiffness matrix of the system given by

$$K = \int_{\Omega} (\mathcal{D}\Phi)^T E \mathcal{D}\Phi d\Omega. \tag{8}$$

The potential energy of the system is, then given by

$$V = \frac{1}{2} q^T K q \tag{9}$$

Using equations (4) and (9) the Lagrangian of the system can be defined as: L=T-V. The Euler-Lagrange equations for the system can then be derived from the following equation.

$$\frac{d}{dt}\left(\frac{\partial L}{\partial \dot{p}_k}\right) - \frac{\partial L}{\partial p_k} = F_k \quad (k = 1, 2, \cdots, n) \tag{10}$$

where, F_k are generalized forces from non-conservative force field. Evaluating the derivatives and simplifying yields

$$\sum_j M_{kj}\ddot{p}_j + \sum_{i,j} C_{ijk}\dot{p}_i\dot{p}_j + D_{kj}\dot{p} + K_{kj}p_j = F_k \qquad (k = 1, 2, ..., n) \tag{11}$$

where D is the inherent structural damping matrix and $D\dot{p}$ is the vector of nonconservative forces. This term is added posteriori to reflect inherent damping in the structural members. In the matrix-vector notation, the equations (11) can be rewritten in a compact form as

$$M(p)\ddot{p} + C(p, \dot{p})\dot{p} + Dp + Kp = F \tag{12}$$

For the sake of simplicity, and without loss of generality, consider a spacecraft with only one branch of appendage bodies. It is assumed that the spacecraft configuration consists of a central flexible body and single chain of $(k-3)$ flexible links. The central body has three rigid rotational degrees of freedom, and each link is connected by one rotational degree of freedom to the neighboring link. The rotational dynamics of this system can be represented by Eq.(12). However, since all degrees of freedom in Eq. (12) are not actuated for flexible spacecraft, equations of motion take the following modified form.

$$M\left(p\right)\ddot{p}+C\left(p,\dot{p}\right)\dot{p}+D\dot{p}+Kp=B^{T}u \tag{13}$$

where F in Eq.(12) is replaced by $B^T u$. In Eq. (13), $\dot{p}=(\omega^T,\dot{\theta}^T,\dot{q}^T)^T$, ω is the 3×1 inertial angular velocity vector (in body-fixed coordinates) for the central body, $\theta=(\theta_1,\theta_2,\ldots,\theta_{(k-3)})^T$ (θ_i denotes the joint angle for the ith joint expressed in body-fixed coordinates), q is the $(n-k)$th vector of flexible degrees of freedom (modal amplitudes), $M(p)=M^T(p)>0$ is the configuration-dependent mass-inertia matrix. $p=(\gamma^T,\theta^T,q^T)^T$, $\dot{\gamma}=\omega$, and $C(p,\dot{p})$ corresponds to Coriolis and centrifugal forces. $B=[I_{k\times k}\quad 0_{k\times(n-k)}]$ is the control influence matrix, and u is the k vector of applied torques. The first three components of u represent the attitude control torques applied to the central body about its X-, Y-, and Z-axes; the remaining components are the torques applied at the $(k-3)$ joints. The symmetric positive semidefinite stiffness and damping matrices K and D are given by

$$K=\operatorname{diag}\left[0_{k\times k},\ \widetilde{K}\right],\quad D=\operatorname{diag}\left[0_{k\times k},\ \widetilde{D}\right] \tag{14}$$

where $\widetilde{K}$ and $\widetilde{D}$ are $(n-k)\times(n-k)$ symmetric matrices corresponding to the flexible degrees of freedom. $\widetilde{K}$ is inherently positive definite, and $\widetilde{D}$ is assumed to be positive-definite. This assumption is realistic because all physical systems have some inherent damping.

The angular measurements for the central body are Euler angles (not the vector γ), whereas the remaining angular measurements between bodies are relative angles. An important inherent property of such systems that is crucial to the stability results to be presented is given next.

Property $\mathcal{S}$: For the system in Eq. (13), the matrix $(\frac{1}{2}\dot{M}-C)$ is skew-symmetric $\forall\ t$.

The central-body attitude (Euler angle) vector η is given by $E(\eta)\dot{\eta}=\omega$, where $E(\eta)$ is a 3×3 transformation matrix. (See [2].) The sensor outputs consist of three central-body Euler angles, the $(k-3)$ joint angles, and the angular rates (i.e., the sensors are collocated with the torque actuators). The sensor outputs (assuming perfect sensors) are then given by

$$y_p=B\widehat{p}\qquad y_r=B\dot{p} \tag{15}$$

where $\widehat{p}=(\eta^T,\theta^T,q^T)^T$ and η is the Euler angle vector for the central body. The measured angular position and rate vectors are $y_p=(\eta^T,\theta^T)^T$ and $y_r=(\omega^T,\dot{\theta}^T)^T$, respectively. The body rate measurements ω are assumed to be available from rate gyros. The system in (13), with state vector $\overline{x}=(p^T,\dot{p}^T)^T$ is affine in control and is internally passive (with respect to output y_r), as given by the next theorem stated without proof [1].

Theorem 1 *The system in (13) (with output y_r) is internally passive, i.e., there exists a storage function $E_s(\overline{x})\geq 0$ such that*

$$\int_0^T y_r^T(t)u(t)\geq E_s[\overline{x}(T)]-E_s[\overline{x}(0)]\ \ \forall\, T\geq\ 0,\ \forall\ u\in\mathcal{L}_{2e}^k. \tag{16}$$

C553/027 © IMechE 1998

(where $\mathcal{L}_{2e}^{k}$ denotes extended Lebesgue space.)

It can be verified that $\frac{d}{dt}E_s[\overline{x}(t)] = -d(\overline{x}) + y_r^T u,$ where $d(\overline{x}) = \dot{p}^T D\dot{p}$. Note that $E_s(\overline{x})$ is only positive semidefinite because K is positive semidefinite.

1.2 Quaternion as Measure of Attitude

An attitude of a free-floating body can be minimally represented by a non-unique three-dimensional Euler angle vector. However, this representation has a singularity problem for certain orientations. An effective way of overcoming the singularity problem is to use the quaternion formulation. (See [3, 4, 5].)

The unit quaternion (Euler parameter vector) α is defined as follows:

$$\alpha = \left(\overline{\alpha}^T, \alpha_4\right)^T \quad \overline{\alpha} = [\widehat{\alpha}_1,\ \widehat{\alpha}_2,\ \widehat{\alpha}_3]^T \sin\left(\frac{\phi}{2}\right) \alpha_4 = \cos\left(\frac{\phi}{2}\right) \tag{17}$$

where $\widehat{\alpha} = (\widehat{\alpha}_1, \widehat{\alpha}_2, \widehat{\alpha}_3)^T$ is the unit vector along the eigenaxis of rotation and ϕ is the magnitude of rotation. The quaternion is subject to the norm constraint

$$\overline{\alpha}^T\overline{\alpha} + \alpha_4^2 = 1. \tag{18}$$

The quaternion can also be shown [Har.64] to obey the following kinematic differential equations:

$$\dot{\overline{\alpha}} = \frac{1}{2}\left(\omega \times \overline{\alpha} + \alpha_4\omega\right); \quad \dot{\alpha}_4 = -\frac{1}{2}\omega^T\overline{\alpha}. \tag{19}$$

The attitude control of a single-body rigid spacecraft with quaternion feedback has been thoroughly investigated in [6], [3], [4], and [5] . The quaternion representation is used here for the central-body attitude. The quaternion can be computed from Euler angle measurements given by equations (15). (See [7])

The open-loop system, given by equations (13) and (19) with $u = 0$, has multiple equilibrium solutions $(\overline{\alpha}_{ss}^T, \alpha_{4ss}, \theta_{ss}^T)^T$, where the subscript '$ss$' denotes the steady-state value; the steady-state value of q and $\dot{p}$ is zero. By defining $\beta = (\alpha_4 - 1)$ and denoting $\dot{p} = z$, equations (13) and (19), can be rewritten as

$$M\dot{z} + Cz + Dz + K\left(0_{1\times 3}, \theta^T, q^T\right)^T = B^T u \tag{20}$$

$$\begin{bmatrix} \dot{\theta} \\ \dot{q} \end{bmatrix} = [0_{(n-3)\times 3} I_{n-3}]z \tag{21}$$

$$\dot{\overline{\alpha}} = \frac{1}{2}\left[\omega \times \overline{\alpha} + (\beta + 1)\,\omega\right]; \quad \dot{\beta} = -\frac{1}{2}\omega^T\overline{\alpha}. \tag{22}$$

In equation (20), the matrices M and C are functions of p and $(p, \dot{p})$, respectively. Note that the first three elements of p associated with the orientation of the central body can be fully described by the unit quaternion. Hence, M and C are implicit functions of α, and therefore the system represented by equations (20-22) is autonomous and can be expressed in the state-space form as follows:

$$\dot{x} = f(x) + g(x)u \tag{23}$$

where $x = (\overline{\alpha}^T, \beta, \theta^T, q^T, z^T)^T$ and $g(x)$ is a $(2n+1) \times k$ matrix. Note that the dimension of x is $(2n+1)$, one more than the dimension of the system in equation (13). However, the constraint of equation (18) is now present. Verification that the constraint of equation (18) is satisfied for all $t > 0$ if it is satisfied at $t = 0$ easily follows from equation (19).

In the following development, some key stability results (mostly from [1]) are given without proofs.

2 Nonlinear Dissipative Control Laws

Consider the control law u, given by:

$$u = -G_p\tilde{p} - G_r y_r \tag{24}$$

where $\tilde{p} = (\overline{\alpha}^T, \theta^T)^T$. Matrices G_p and G_r are symmetric positive definite $(k \times k)$ matrices and G_p is given by:

$$G_p = \text{diag}\left[\frac{1}{2}\{(\tilde{\overline{\alpha}} + I_3\alpha_4)G_{p1} + \nu(1-\alpha_4)I_3\}, G_{p2}\right] \tag{25}$$

where ν is a positive scalar, $\tilde{\overline{\alpha}}$ represents a cross product operator matrix $(\overline{\alpha}\times)$, and G_{p1}, G_{p2} are (3×3) and $(k-3)\times(k-3)$ symmetric positive definite matrices, respectively. Note that the feedback $\tilde{p}$ can be computed using the sensor measurements y_p using quaternion equations (17). The following Lemma gives the closed-loop equilibrium solutions [8]. Note that this is an extension of Lemma 1 in [9] to multibody flexible spacecraft.
Lemma 1. *Suppose G_{p1} and G_{p2} are symmetric and positive definite, and $0 < \lambda_M(G_{p1}) \leq 2\nu$, where $\lambda_M(\cdot)$ denotes the largest eigenvalue. Then the closed-loop system given by (23) and (24) has exactly two equlibrium solutions:* $[\tilde{p} = q = \dot{p} = 0, \beta = 0]$ *and* $[\tilde{p} = q = \dot{p} = 0, \beta = -2]$.

It was shown [1] that these two equilibrium points correspond to the same equilibrium point in the physical space; thus, there is a unique equilibrium point for the closed-loop system. The global asymptotic stability (g.a.s.) of the physical equilibrium state (taken to be the origin of the state-space) is given by the next result [1].

Theorem 2 *Suppose G_{p1}, G_{p2}, and G_r are symmetric and positive definite, and $0 < \lambda_M(G_{p1}) \leq 2\nu$. Then the closed-loop system given by Eqs. (23), (24) and (25) is g.a.s.*

The significance of this result is that using the nonlinear control law of Eq. (24), the multibody space system undergoing large-angle rotational maneuvers can be globally re-orientated with guranteed asymptotic stability.

C553/027 © IMechE 1998

3 Systems in Attitude-Hold Configuration

Consider a special case where the central-body attitude motion is small. This can occur in many real situations. For example, in cases of space station-based or shuttle-based manipulators, the inertia of the base (central body) is much greater than that of any manipulator link or payload. In such cases, the rotational motion of the base can be assumed to be in the linear region, although the payloads (or links) attached to it can undergo large rotational and translational motions. The attitude of the central body is simply γ (the integral of the inertial angular velocity ω) and the use of quaternions is not necessary. The equations of motion (1) can now be expressed in the state-space form simply as

$$\dot{\overline{x}} = \begin{bmatrix} 0 & I \\ -M^{-1}K & -M^{-1}(C+D) \end{bmatrix} \overline{x} + \begin{bmatrix} 0 \\ B^T \end{bmatrix} u \tag{26}$$

where $\overline{x} = (p^T, \dot{p}^T)^T$, $p = (\gamma^T, \theta^T, q^T)^T$. Note that M and C are functions of $\overline{x}$ and the system is still nonlinear. Flexible terrestrial manipulator can also be represented in this manner with the addition of gravity terms.

3.1 Stability With Static Dissipative Controllers

The static dissipative control law u is given by

$$u = -\overline{G}_p y_p - G_r y_r \tag{27}$$

where $\overline{G}_p$ is symmetric positive definite $k \times k$ matrix and

$$y_p = Bp \qquad y_r = B\dot{p} \tag{28}$$

where y_p and y_r are measured angular position and rate vectors, respectively. The following theorem addresses the closed-loop stability.

Theorem 3 *Suppose $\overline{G}_{p_{k\times k}}$ and $G_{r_{k\times k}}$ are symmetric and positive definite. Then, the closed-loop system given by equations (26)-(28) is globally asymptotically stable.*

The significance of Theorems 2 and 3 is that any nonlinear multibody system belonging to these classes can be robustly stabilized with the dissipative control laws given. In the case of manipulators, this means that any terminal angular position can be achieved from any initial position with guaranteed asymptotic stability.

4 Numerical Example: Two link flexible robot

In order to demonstrate the applicability of the theoretical development given in earlier sections, an example system is considered for simulation analysis. The example system consists of a two-link flexible space robot with revolute joints. It is also assumed that the base link of the robot is fixed to a stationary body with large inertia. That means, for simplicity, it can be assumed that the base link of the robot is attached to an inertial frame via revolute joint. This system is a representative of a class of space systems which are said to be in the "*attitude-hold configuration*". A finite dimensional approximation model obtained by assuming first two modes as significant modes yields a 12th order state-space model for this nonlinear system. Figure 2 shows a schematic of the example system with the values of system's inertia parameters.

The problem addressed is that of controlling a set-point maneuver using a static dissipative control law given by Eq. (27). By the proper choice of the controller gains, consistent with the conditions of Theorem 3, asymptotic stability of the closed-loop maneuver is ensured. The set-point maneuver to be achieved was to reorient the robot from the initial configuration of $\theta_1 = 30$ deg and $\theta_2 = 15$ deg to the final configuration of $\theta_1 = 45$ deg and $\theta_2 = 25$ deg. The objective was to achieve this maneuver while keeping the link vibrations as small as possible. The fully populated gain matrices G_p and G_r that were chosen to accomplish this task are given by: $G_p = [10.04; 0.040.05]$ and $G_r = [1.70.05; 0.050.066]$. The closed-loop θ-responses are shown in Fig. 3 whereas the modal amplitude responses are shown in Figs. 5 and 6. The control effort time history is shown in Fig. 4. As can be seen form the responses, the static dissipative controller gave a stable closed-loop response with asymptotically damped oscillations.

5 Concluding Remarks

An overview of a class of globally stabilizing nonlinear dissipative control laws was presented for a class of nonlinear multibody flexible space systems. A brief outline of dynamic modeling of such systems was also presented. The stability proofs use the Lyapunov approach and exploit the inherent passivity of such systems. The stability is robust to unmodeled dynamics and parametric uncertainties. A numerical example was given to demonstrate some of the analytical results on the control of set point rotational maneuvers.

6 Acknowledgements

The author acknowledges the support of NASA Grant NCC-1-226 and NSF Grant CMS-9703092 for the work presented in this paper. The author is also thankful to Dr. Bruno Siciliano from University of Naples, Italy, for providing the parameters used in the example system of Section 4.

7 References

[1] Kelkar, A.G. and Joshi, S. M.: *Control of Nonlinear Multibody Flexible Space Structures.* London: Springer-Verlag, 1996 (Vol. 221, Lecture Notes in Control and Information Sciences).

[2] Greenwood, D. T.: *Principles of Dynamics.* New Jersey: Prentice-Hall, 1988.

C553/027 © IMechE 1998

[3] Kane, T. R.: Solution of Kinematical Differential Equations for a Rigid Body, *Journal of Applied Mechanics*, pp. 109-113, March 1973.

[4] Ickes, B. P.: A New Method for Performing Control System Attitude Computation Using Quaternions, *AIAA Journal*, Vol. 8, pp. 13-17, January 1970.

[5] Harding, C. F.: Solution to Euler's Gyrodynamics, *Journal of Applied Mechanics*, pp. 325-328, June 1964.

[6] Wen, J. T. and Kreutz-Delgado, K.: The Attitude Control Problem, *IEEE Transactions on Automatic Control*, Vol. 36, No. 10, pp. 1148-1163, 1991.

[7] Haug, E. J.: *Computer-Aided Kinematics and Dynamics of Mechanical Systems.* Allyn and Bacon Series in Engineering, 1989.

[8] Kelkar, A. G. and Joshi, S. M.: Global Stabilization of Flexible Multibody Spacecraft Using Quaternion-Based Nonlinear Control Law, *Journal of Guidance, Control, and Dynamics*, Vol. 19, No. 5, pp. 1186-1188, 1996.

[9] Joshi, S. M., Kelkar, A. G., and Wen, J. T.-Y.: Robust Attitude Stabilization of Spacecraft Using Nonlinear Quaternion Feedback, *IEEE Transactions on Automatic Control*, Vol. 40, No. 10, pp. 1800-1803, October 1995.

[10] Joshi, S. M.: *Control of Large Flexible Space Structures.* Berlin: Springer-Verlag, 1989 (Vol. 131, Lecture Notes in Control and Information Sciences).

[11] Hill, D., Ortega, R., and van der Schaft A.: *Nonlinear Controller Design Using Passivity and Small-Gain Techniques*, Notes from 1994 IEEE CDC Tutorial Workshop.

[12] Joshi, S. M. and Gupta, S.: On a Class of Marginally Stable Positive-Real Systems, *IEEE Transactions on Automatic Control*, Vol. 41, No. 1, pp.152-155, January 1996.

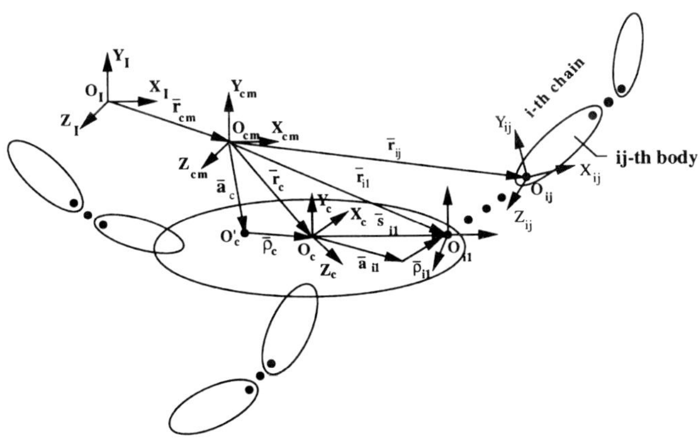

Figure 1: Coordinate Systems

© IMechE 1998 C553/027

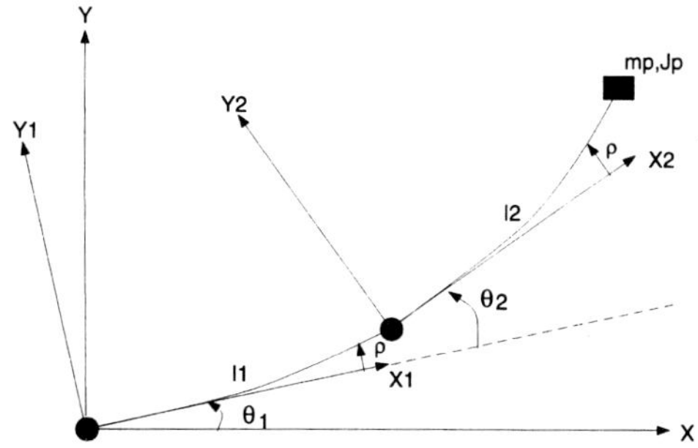

Parameters:

link length, l1 = l2 = 0.5 m,
diameter, d1 = d2 = 0.25 m
mass, m1 = m2 = 0.5 kg
payload mass, mp = 0.1 kg
mass of hub, mh2 = 1 kg,
link Inertia Jo1 = Jo2 = 0.0083 kg-m^2
hub inertia, Jh1 = Jh2 = 0.1 kg-m^2
payload inertia, Jp = 0.0005 kg-m^2
flexural rigidity, EI1 = EI2 = 10 N-m^2
freq. ω_{11}= 8.79 Hz, ω_{12}=32.04 Hz
ω_{21}= 32.73 Hz, ω_{22}=203.95 Hz

ρ= elastic displacement

Figure 2: Two-link flexible robot

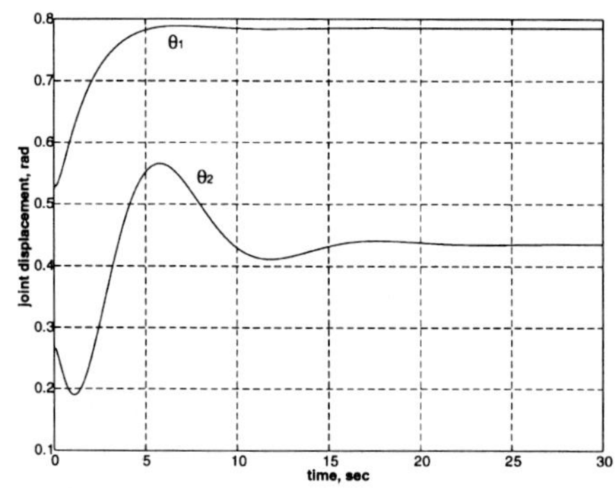

Fig. 3 Angular position response

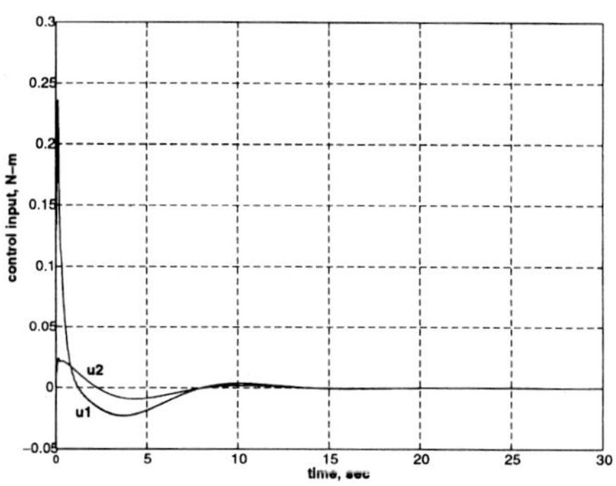

Fig.4 Control input time history

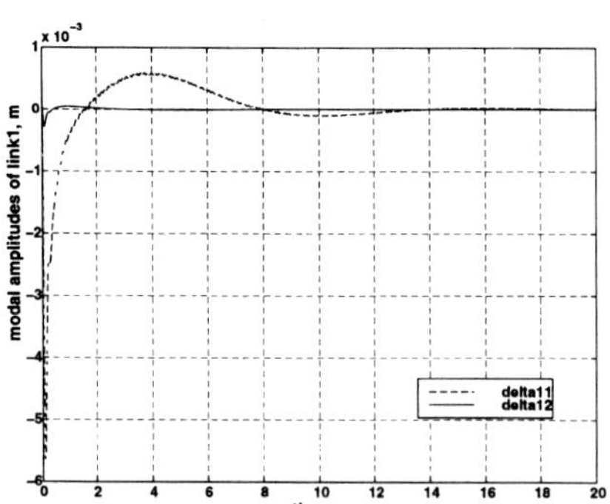

Fig. 5 Modal amplitudes for link 1

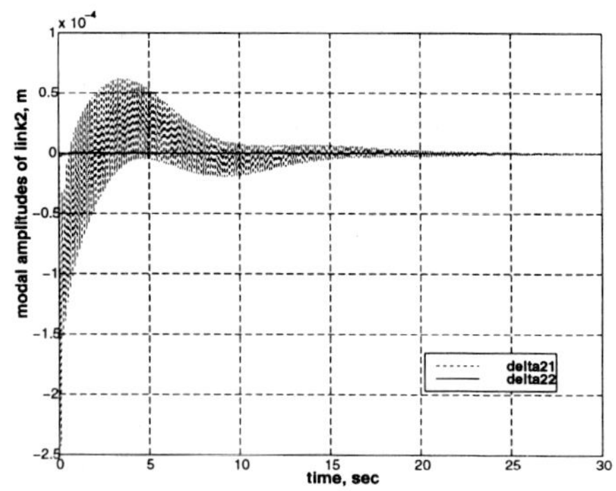

Fig.6 Modal amplitudes for link 2

C553/027 © IMechE 1998

C553/043/98

Multi-physics modelling for design optimization and manufacturing on high-performance parallel systems

M CROSS BSc, PhD, DSc, FIMA, **C BAILEY** BSc, PhD, FIMA, **K PERICLEOUS, N CROFT, K McMANUS, G TAYLOR, S BOUNDS, G MORAN,** and **D WHEELER**
Centre for Numerical Modelling and Process Analysis, University of Greenwich, London ,UK

ABSTRACT

An overview of the design and implementation of a software environment for the computational modelling of engineering processes is described, where the coupling amongst phenomena is strong. It is asserted that such a framework is necessary because of the numerical consistency and minimal data exchange required to enable such multi-physics calculations to be made effectively and in a practical time frame. The object based software framework uses 3D finite volume techniques on an unstructured mesh that facilitates free surface flows, heat transfer with phase change and non-linear solid mechanics. High parallel efficiencies are demonstrated and applications of the software framework to a range of problems with various degrees of coupling amongst the physical phenomena are highlighted. Particularly, in the context of shape casting, it is demonstrated how that simulation of the interactions is necessary to predict the defects of key interest to the target manufacturing sectors.

INTRODUCTION

It has become clear over the past few years that in the assessment of manufacturing processes and the optimisation of the design for performance phase, simulation involving the interaction of various continuum phenomena has become an enduring theme. Obvious examples include aircraft flutter and the shape casting of high integrity components. There are, of course, a wide range of problems involving the interactions amongst various phenomena and in many instances, the coupling is low or one way. In this case, the simulation can be managed through the solution of a field variable vector in one code and the passing of results to another for use in the calculation of another field variable vector in the second code. An example of this kind of behaviour is typified by magneto-hydro-dynamics (MHD) where the liquid flow field is heavily dependent on the magnetic field, but often the latter is virtually independent of the former. This kind of simulation, typically known as multi-disciplinary can be reasonably accomplished through the exchange of data between two phenomena specific codes (eg. CFD and electromagnetics) providing:

- the volume and frequency of data exchanged by the codes is not a significant component of the overall computational effort,

- the data filtering required to use the information (typically a field variable vector) does not unreasonably compromise the accuracy of the 'coupled' aspect of the calculation.

There are currently a range of research programmes aimed at providing the data exchange/information filtering environments to enable a range of phenomena-specific computational mechanics (CM) codes to be used in multi-disciplinary simulations.

However, when the coupling is two way and/or the exchange of information between phenomena specific "calculations" is large, then the data traffic is too significant a factor in the solution process, and it could be argued that such simulations can only be achieved practically when the field variable vectors for each of the phenomena solved for, are stored in a common database and map onto a single mesh. These multi-physics calculations are very challenging and involve 3D meshes with 10^5-10^6 nodes/elements, 10^3-10^5 timesteps and 20+ field variables. Such problems can only be realistically solved on high performance parallel systems - in this case, the need to reduce data exchange between tasks on the same and other processors to an absolute minimum is vital.

The key problem with designing software tools which are targeted at multi-physics processes is that a high degree of consistency is required in the discretisation methods and solution strategies for a wide range of phenomena (and consequent partial differential equations). From a computational perspective

- Navier-Stokes turbulent in/compressible flows with heat transfer involving combustion and/or phase change, have conventionally been solved by finite volume methods with iterative solvers using single- and block-structured meshes.

- Structural analysis of all kinds, including thermo-mechanical behaviour, is almost exclusively solved using finite element methods often with direct solvers (on unstructured meshes, of course).

- Electromagnetic calculations have been solved effectively using both FV and FE methods.

At the time the research programme reported here was initiated almost a decade ago, it was clear that FV methods were able to address the very non-linear flow problems (involving free surface, for example). As such, in the attempt to design and develop a single software framework for multi-physics simulation FV methods were selected, but were to be extended to fully unstructured meshes. At this stage, it was straightforward enough to conceive of the extension of flow algorithms to unstructured meshes - a number of workers in the aerospace sector had signalled the way with FV solution strategies for Euler and Navier-Stokes equations on unstructured meshes [see for example, 1,2]. However, key developments had to take place in FV based procedures for solid mechanics. After considerable efforts 3D procedures for the full range of non-linear structural and thermomechanical analysis with large deformation and contact analysis have been achieved. In recent years substantial work has also gone into the development of effective strategies to enable multi-physics codes using FV methods on unstructured meshes (FV-UM), to run effectively in parallel on high performance systems.

The results of all the above research have been integrated into the multi-physics modelling software environment, PHYSICA[3]. In the following pages we provide a brief overview of the modelling software environment and briefly describe its use in a number of industrially focused applications, illustrating various levels of phenomena coupling.

OVERVIEW OF THE MULTI-PHYSICS MODELLING SOFTWARE ENVIRONMENT

Finite volume-unstructured mesh (FV-UM) context

When seeking to utilise a FV solution strategy it is convenient to recognise that the range of continuum phenomena highlighted above can be expressed in a common naturally conservative form:

$$\frac{\partial}{\partial t}\int_v \rho A \varphi dv = \int_s \Gamma_\varphi \operatorname{grad}\varphi \underline{n} ds + \int_v Q_v dv - \int_s Q_s \underline{n} ds \qquad (1)$$

where for each physical variable the specific form of equation (1) is summarised in Table 1 for Navier-Stokes flow, heat transfer, electromagnetics and solid mechanics.

In the work described here a cell-centred approx-imation has been used for the fluid flow, heat transfer and electromagnetics solvers whereas for the solid mechanics solvers the discretisation has been based upon a cell vertex. Certainly, in the pressure correction based flow solutions a Rhie-Chow type correction is applied to prevent the pressure checker-board effect on the mesh[4]. The methods originally developed for incompressible flows[5] have been extended to include compressibility, combustion[6], free surfaces[7] and a range of two equation turbulence models[8]. The flow combined with the heat transfer has included chemical reactions and melting/ solidification phase change simultaneously with free surfaces[9] The electromagnetics equations for AC and DC fields have been solved with both one-way and two-way coupling to both fluid flow and heat transfer with phase change[10]. The solid mechanics was originally developed for thermoelasticity[11,12] and then extended to elasto-visco-plastic, creep and elasto-plastic[13] with contact[14] and recently for large deformation also[15].

In summary, we now have a set of solution procedures for a range of phenomena based on a similar discretisation and solution strategy, which can employ the same mesh and associated data structure. This sets us in a solid position from which to conceive of a simulation software environment which enables interactions amongst the separate phenomena once the coupling relations have been included.

Key design issues for multi-physics modelling software

The software design has involved an object oriented strategy, at least, a far as this can be exploited through the use of FORTRAN77, since a key issue in multi-physics simulation is processor speed and with these kinds of calculations and data-structures this language yields the best performance on current technology processors; FORTRAN77 is currently a factor of 2-4 faster than C++. The object orientation is achieved through the concept of a mesh as constructed of a hierarchy of objects - nodes, edge, faces and volumes, see the illustration of the concept in 2D in Figure 1. Once the memory manager has been designed as an object hierarchy, then all other aspect of discretisation and solution procedures can be related to those 'spatial' objects. Given that the procedures can all employ a common mesh structure, this enables the software to be structured in a highly modular fashion and leads to four levels of abstraction:

- **Model** - for the implementation of multi-physics analyses by the coupling of 'phenomena' specific modules

- **Control** - to enable the creation of new modules an/or solution control strategies

- **Algorithm** - tools for discretisation, interpolation, source construction, solvers, etc

- **Utility** - for file input-output, memory manager, database manager, etc.

With the above abstraction framework it is relatively straightforward to implement discretisation and solution procedures for a range of distinct continuum phenomena, in a manner that particularly facilitates interactions.

The above strategy has been implemented as the PHYSICA framework and has used the FV-UM discretisation and solution procedures described above to solve for a range of continuum phenomena. The initial core version of PHYSICA has:

a) tetrahedral, wedge and hexahedral cell/element shapes,
b) full adaptivity implemented in the data structures which are consistent for refinement/coarsening,
c) a range of linear solvers (including PC-CG)

with the following 'core' models:

- single phase transient compressible Navier-Stokes flow with k-ε turbulence models

- convection -conduction heat transfer with solid-ification phase change and simple reaction kinetics

- non-linear solid mechanics (elasto-visco-plastic, creep and elasto-plastic)

and their interactions. Recently, a focused version of the environment, PHYSICA_SCM, has been produced which additionally includes:

- free surface flows

C553/043 © IMechE 1998

- contact analysis

and a prototype has been extended to include dynamic solid mechanics.

Implementation in parallel systems

The computational effort involved in multi-physics simulation can be prodigious and so it is essential that such software must be able to exploit high performance parallel hardware. As such, PHYSICA has been parallelised using the conventional SPMD approach, whereby each processor runs the same task, but operates on its own set of data. Most often, PHYSICA is implemented assuming a distributed memory model, and PVM, MPI and shmem library calls are available to ensure dynamic data exchange to enable a processor to use data it does not own. A key feature in the parallel implementation of PHYSICA is the mesh partitioning tool, JOSTLE, employed to ensure an appropriately high quality load balance of effort amongst all the processors. In fact, JOSTLE has been extended to enable dynamic load balancing as the multi-physics simulation progresses(16,17). This software has been evaluated on a number of parallel systems and produces high efficiencies. For example, on DEC Alpha 466 Mhz processors the run times vary from 10^{-3}-10^{2} sec/node/time step. As such, even for modest scale problems with 70,000 nodes and 10,000 time steps take 150 hours of elapsed single processor time. With the high parallel efficiencies realised from PHYSICA such run times can be reduced by a factor of 10-100, depending on the size of parallel system.

MULTI-PHYSICS SIMULATION WITH INCREASING DEGREES OF COUPLING

In the above sections we have described the key features which have underpinned the development and implementation of a modelling software environment to facilitate multi-physics simulation. In this section we will show some examples of problems with increasing degrees of coupling and indicate how they have been solved in the multi-physics framework.

Lead ingot casting

In the casting of lead ingots, a cascade of cast iron moulds are filled with liquid metal at 1200°C and then left to cool and solidify. A frustrating problem of some importance in this industry sector has been the cracking of the moulds during the cooling phase. Because of the size of the ingots, the residual convection of the liquid metal is an important factor in the cooling process. Hence, the process modelling has to involve(18):

- residual convective Navier-Stokes flows in the metal,
- heat transfer in the metal and the mould,
- solidification in the metal,
- development of stress in the solidified metal and the mould.

Although, there is strong two way coupling between the fluid flow, heat transfer and solidification, and whilst the heat transfer drives the stress development, the stress itself has a negligible effect upon the other continuum variables in this case. Hence, the solution can proceed without any necessity to feedback the impact of stress or deformation upon the heat transfer, etc. Figure 2 shows the geometry of a quarter of the ingot and the mould plus an illustration of the mesh used. Figure 3 shows the residual flow patterns at an early stage of the cooling. Figure 4 shows the advancing solidification pattern with time and Figure 5 illustrates the development of stress at the exact location where cracks are observed in practice.

Shape casting-gap formation
Shape casting of metals has been modelled for many years. One of the factors that make it difficult to model some shape casting processes is the changing nature of the metal-mould heat transfer coefficient. This is caused by a gap formation at the metal-mould interface as the metal solidifies and shrinks. Recently, a benchmark test was published with reproducible experimental results to measure gap formation and assess its impact on cooling. The test is illustrated in Figure 6 along with the geometrical components and mesh to model the process[(19)].

Because of the nature of this test it was clear that residual convection would be negligible. Hence, the process involved two way interactions amongst:

- heat transfer by conduction
- solidification
- elasto-visco-plastic solid mechanics.

Most existing casting simulation tools only model the first two phenomena. However, if the gap formation due to the solid mechanics behaviour is ignored then the cooling rates are over-predicted, as shown in Figure 7(a). But, if the solid mechanics analysis is included, then the deformation of the structure and hence the gap at the metal-mould interface can be predicted. This then feeds into the heat transfer coefficient calculation (and reducing its value) so that the cooling rates are then well predicted. Unsurprisingly, the gap formation development is also well predicted, see Figure 7(b).

Solder joint formation
In the context of electronic packaging the process whereby solder melts, reflows, and then solidifies to bond electronic components to PCB's is challenging. In particular, as the solder reflows its surface tension forces coupled with the associated Marangoni flow determine its shape in the context of the component leg - PCB context. One way to characterise the behaviour of solder material in different contexts is the 'wetting balance test'. In this test, the surface tension effects in the solder induce Marangoni flow which gives rise to the initial 'flat' liquid solder surface changing its shape in the neighbourhood of the lead; see Figure 8. To model this problem we need to account for[(20)]

- the impact of the surface tension,.
- the Navier-Stokes flow field,
- the heat transfer by convection and conduction,
- the movement of the free surface.

C553/043 © IMechE 1998

It turns out that fluid free surface is not changing substantially (as in mould filling, for example). However, the prediction of its development is the key measurement in the test! One convenient way to address this problem is to calculate, at every time step, the new location of the free surface, and then use the solid mechanics procedure to adapt the mesh for the flow and heat transfer calculations, in much the say way as Farhat et al do in their work on dynamic fluid-structure interaction[21].

In Figure 9 we show the temperature and flow velocity magnitude distributions for a part of the solder volume, and also a cross section of the induced Marangoni flows.

CONCLUSIONS

As the demand for engineering analysis moves inexorably towards multi-disciplinary and strongly coupled multi-physics modelling then software tools are going to be required which concurrently enable the simulations to proceed on parallel high performance systems. This functionality will be required to enable the simulation run time to fit into the context of the engineering analysis time-frame.

In this paper we have provided a brief overview of one such attempt to conceive, design and develop a software framework for the modelling of closely coupled multi-physics phenomena. A suite of consistent numerical procedures for a range of physical phenomena have been implemented in a software framework that is essentially object oriented in style. A strategy for parallelising such 'multi-featured' software is outlined, and its use in a variety of applications with various levels of coupling is described.

REFERENCES

1. A Jameson, T J Baker and N P Weatherill, Calculation of inviscid transonic flow over a complete aircraft, AIAA, Paper AIAA-86-0103 (1986).

2. O Hassan, K Morgan, E J Probert and J Peraire, Unstructured tetrahedral mesh generation for 3D viscous flows, Int Jnl Num Methods in Engg, 39, 549-567 (1993).

3. PHYSICA, University of Greenwich, London - see http:/www.physica.gre.ac.uk.

4. C M Rhie and W L Chow, A numerical study of the turbulent flow past an isolated aerofoil with training edge separation, AIAA, J, 21, 1525-1532 (1982).

5. P M Y Chow, Control volume-unstructured mesh procedure for convection-diffusion solidification processes, PhD thesis, Univ of Greenwich, London (1993).

6. N Croft, K Pericleous and M Cross, "PHYSICA: a multiphysics environment for complex flow processes", in Num Meths in Laminar and Turbulent Flow (Eds C Taylor et al), Vol 9, Pineridge Press - Swansea, 1269-1280 (1995).

7. K Pericleous et al, Free surface Navier-Stokes flows with simultaneous heat transfer and solidification/melting, Adv Comp Maths, 6, 205-308 (1996).

8. T N Croft, Unstructured Mesh -Finite Volume Algorithms for Swirling, Turbulent, Reacting Flows, PhD Thesis, Univ of Greenwich, London (1998).

9. K Pericleous, G J Moran, S Bounds, P Chow and M Cross, 3D free surface modelling in an unstructured mesh environment for metal processing applications, Applied Math Modelling (in press).

10. K Pericleous et al, Modelling the electromagnetic stirring of solidifying alloys and its influence on solid front progression, in Modelling of Casting, Welding and Advanced Solidification Processes VII (Eds M Cross and J Campbell), TMS - Warrendale, PA, 213-221 (1995).

11. Y D Fryer et al, A control volume procedure for solving elastic stress-strain equations on an unstructured mesh, Appl Math Modelling 15, 639-645 (1991).

12. C Bailey and M Cross, A finite volume procedure to solve elastic solid mechanics problems in three dimensions on an unstructured mesh, Int Jnl Num Methods Engg, 38, 1757-1776 (1995).

13. G A Taylor et al, Solution of the elastic/ viscoplastic constitutive equations: a finite volume approach, Appl Math Modelling 19, 746-770 (1995).

14. G A Taylor, C Bailey and M Cross, A 3D FV approach to thermo-mechanical modelling of the shape casting of metals, in Modelling of Casting, Welding and Advanced Solidification Processes - VII (Ed B G Thomas and C Beckermann), pub TMS, 755-767 (1998).

15. N Fallah, C Bailey and M Cross, in preparation.

16. C Walshaw and M Cross, Load balancing for parallel adaptive unstructured grids, in Num Grid Generation in Computational Field Simulations, (Eds M Cross et al) pub ISGG-MSU Press, 781-792 (1998).

17. JOSTLE, University of Greenwich, see http://gre.ac.uk/jostle.

18. C Bailey, Computational modelling of mould design in lead ingot casting, in Modelling of Casting, Welding and Advanced Solidification Processes VIII (Ed B G Thomas and C Beckermann), pub TMS, 827-836 (1998).

19. G A Taylor, C Bailey and M Cross, in preparation.

20. D Wheeler, C Bailey and M Cross, in preparation.

21. C Farhat et al, Mixed implicit/explicit time integration of coupled aeroelastic problems, Int Jnl for Num Methods in Fluids, 21, 807-835 (1995).

 C553/043 © IMechE 1998

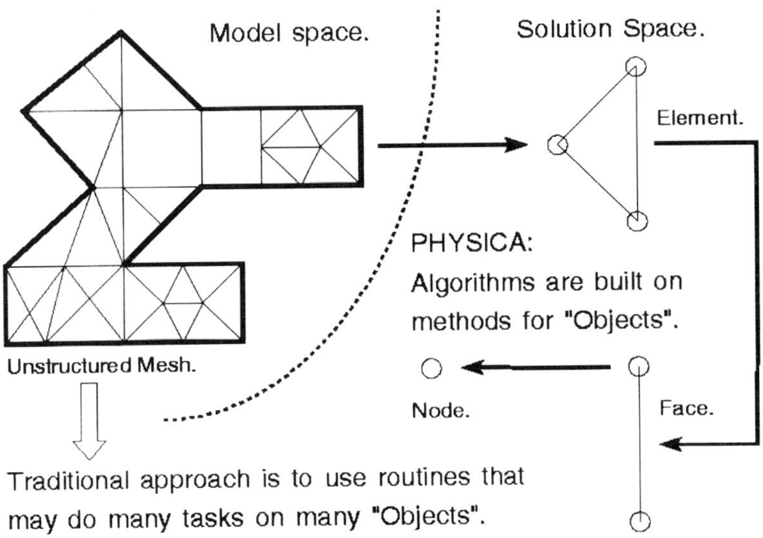

Figure 1: Object design concept in PHYSICA.

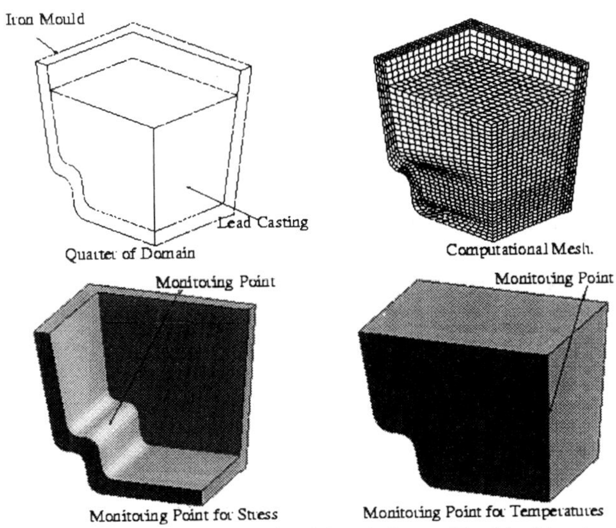

Figure 2 Computational domain for mould and lead ingot

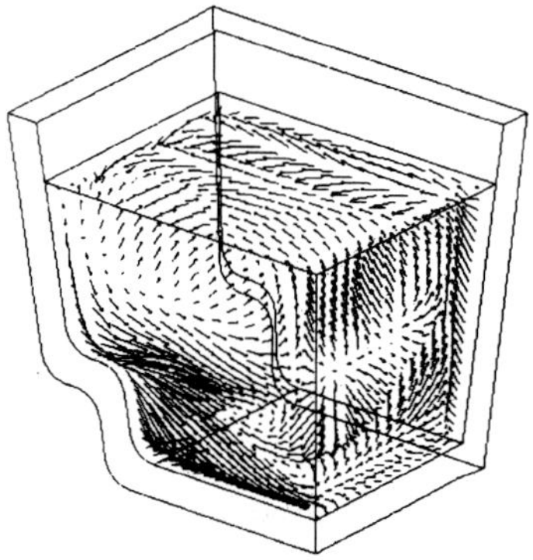

Figure 3: Thermal convection after 30 minutes of cooling.

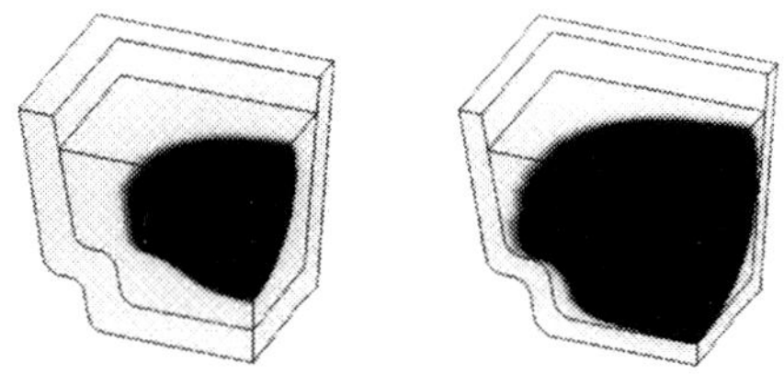

Figure 4 Solidification profiles after 100 minutes for thick and thin moulds.

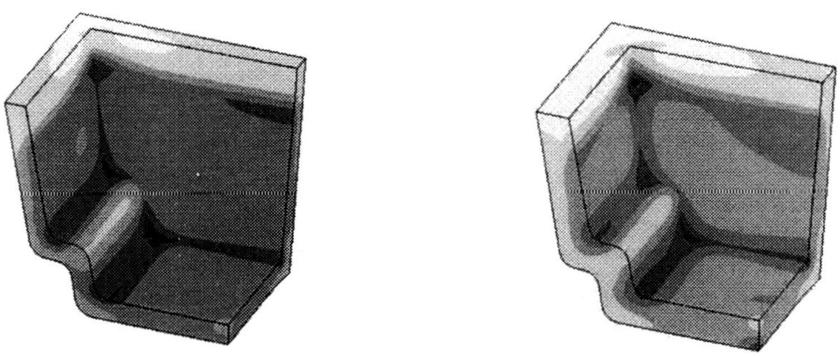

Figure 5: Stress profiles after 100 minutes of cooling for thick and thin moulds.

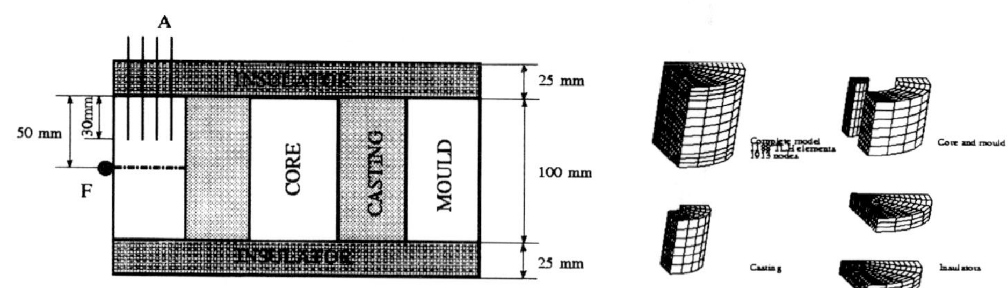

A Thermocouples

F Displacement transducers

Figure 6 Benchmark gap formation test (a) geometry and (b) FV mesh

C553/043 © IMechE 1998

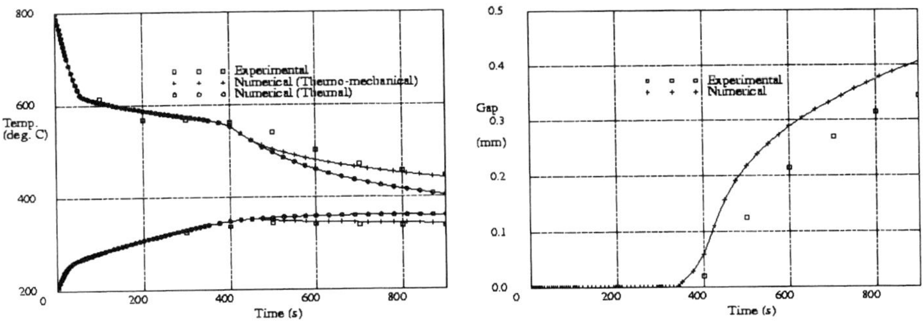

(a) Temperature profile. (b) Gap formation.

Figure 7 Results for benchmark gap formation test.

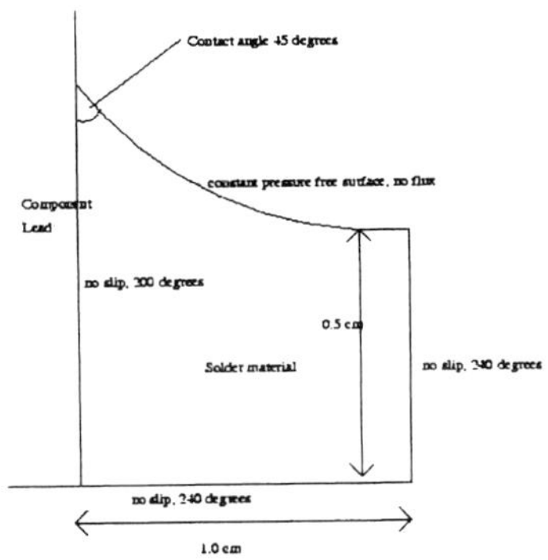

Figure 8: Wetting balance boundary conditions.

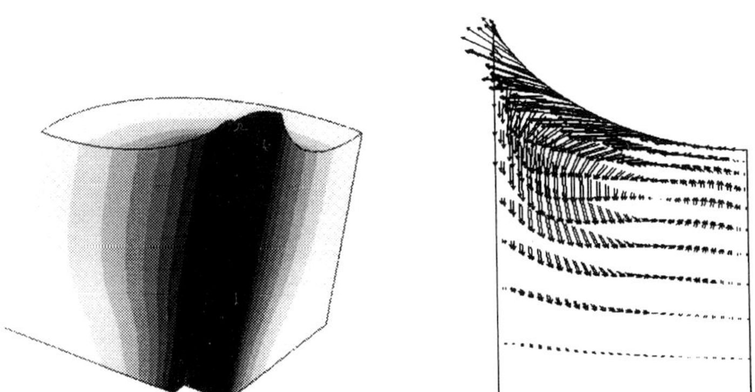

Figure 9 Solder joint simulation showing (a) temperatures across the domain and (b) a cross-section of the Marangoni velocity field

C553/033/98

The use of finite element analysis in solving highly non-linear multi-body dynamics problems

M R DORN BSc
Frazer–Nash Consultancy Limited, Dorking, UK

This paper describes how finite element analysis (FEA) has been used by Frazer-Nash Consultancy Limited (FNC) to solve a variety of multi-body dynamics problems. The paper outlines how four different components of multi-body dynamic events have been modelled using the non-linear FEA code DYNA3D and shows validation of the models, where possible. The components covered include structural response, joints and linkages, failure of joints and the interaction of water. The applications discussed range from vehicle crashworthiness, through to human injury and building demolition. Whilst the applications discussed here typically make use of only one of the four components, each of the components can be included in a single model. This makes FEA, and in particular DYNA3D, a powerful tool for examining multi-body dynamics problems.

1. INTRODUCTION

Frazer-Nash Consultancy Limited (FNC) has been using finite element analysis (FEA) for over fifteen years to solve a huge variety of complex problems. These range from investigations into the mechanisms of human injury to the analysis of bird strike events on an aircraft. Much of the dynamic analysis work has been undertaken using DYNA3D as it has a number of advanced features which make it applicable to a large range of problems. Other FEA solvers are also used by FNC where applicable (e.g. ABAQUS, NIKE3D and IDEAS).

DYNA3D is a non-linear dynamic finite element analysis (FEA) code. It is developed by the Lawrence Livermore National Laboratory (LLNL) for the simulation of large deformation transient dynamic situations. It has complex algorithms for determination of contact surfaces and a simple inexpensive element formulation which makes it ideal for this type of analysis. In addition, it has many complex material models and loading conditions, allowing the user to examine the response of most types of structures. FNC is supplied with the source code for DYNA3D and as such modifications can be made to the code where necessary. Modifications that have been made by FNC include the development of new material models, the development of rotational joints and the inclusion of new subroutines to represent buoyancy and drag forces.

This paper describes a number of projects, where advanced DYNA3D modelling has been used. Not all of the applications described are multi-body problems. However, they have been included to illustrate the wide range of events that can be modelled with DYNA3D and to illustrate the various components that can be included in a multi-body dynamics problem.

The paper is presented in four sections which outline four different types of components that can be modelled in DYNA3D. These components are as follows:

- Structural response.
- Joints and linkages.
- Failure of linkages.
- Fluid interaction loads.

The point that should be remembered when examining these examples, is that all the analyses were conducted using DYNA3D. As such any or all of the components shown in this paper can be combined in a single analysis and hence extremely complex problems can be assessed using this system.

2. STRUCTURAL RESPONSE

Structural response modelling, as the term suggests, involves the examination of the deformation and rigid body dynamics of an object subjected to external forces. The need to conduct these types of analyses was the primary reason for developing FEA codes.

Two problems are discussed here. The analysis of the response of a crashworthy train design to a quasi-static test loading and the analysis of a bird strike on an aircraft leading edge.

The use of DYNA3D in the design process for crashworthy trains has been discussed elsewhere (1). Hence, only a brief overview will be given here. A train designed for UK lines must meet the British Railways Board Structural Design Load Cases. These load cases include a requirement for carriage bodyends to meet crashworthiness design criteria. These criteria place limits on the maximum collapse load (to limit the deceleration of the passenger compartment) and the minimum energy of collapse (to ensure that crash energy is absorbed by crushing the bodyends and not adversely affect the passenger compartment). The design of the bodyends is verified by full scale static crush tests. These tests are extremely expensive and hence computer modelling is used by FNC to assess designs during the actual design process and prior to construction of any body sections. In this way computer modelling greatly reduces the cost of the design process for new bodyends.

Figure 1 shows a DYNA3D model used in the development of the European Over-Night Stock bodyends. This picture shows the structural response of the model in an overriding crush test. Figure 1 also shows a comparison between the crush force predictions made using DYNA3D and the actual test results. Good agreement can be observed.

The second example of structural response examined, using DYNA3D, is a truly multi-body problem. The problem examined is that of a bird strike on an aircraft leading edge and the analysis work was concerned with predicting the damage to the structure. As in the case of the train crashworthiness problem discussed above, testing to assess the vulnerability of aircraft to bird strike is very costly. Hence, computer modelling is used to reduce the cost of the design process.

Due to the high speed of the aircraft in flight, the impact speed in a bird strike can be anywhere between approximately 60 m/s and 700 m/s. In the particular case which is discussed here DYNA3D was used to examine the response of a leading edge to a bird strike at a speed of approximately 500 m/s. The aircraft leading edge was modelled explicitly (i.e. the skin and all internal stiffeners were represented in detail and with appropriate material characteristics). The end of the leading edge which would normally be fixed to the main aircraft structure was then fixed in space to represent the relatively rigid structure of the aircraft. The bird was represented in less detail, but the mass, density and overall dimensions were appropriate for a typical bird. The bird "material" was represented using a fluid material model since the previous experience of the client had suggested that in high speed impacts a bird's body will respond in a similar way to water. The response of the aircraft leading edge and the bird model can be seen in Figure 2. The "bird" was observed to be cut in half by the leading edge and deform considerably. The aircraft structure suffers some distortion, but overall it remains intact.

Other structural response events modelled by FNC using DYNA3D include rollover and collision events involving road petroleum tankers, the impact of helicopters with water and the response of elements of the human body to impact and blast.

3. JOINTS AND LINKAGES

Complex systems of joints can also be represented in DYNA3D. There are joints built into the code, such as ball and sliding joints. In addition, FNC has extended the code to include rotational joints with friction and user defined torque resistance.

One of the most complex joint and linkage systems modelled using the DYNA3D system is a human being. FNC has developed a technique called DYNAMAN which allows the response of humans to be examined in any dynamic environment. The use of a finite element code in this application has advantages over the more traditional lumped mass models, in that contact surfaces and body deformation can be more accurately modelled.

A human is represented by a series of body sections linked by joints. The body sections, such as a thigh or abdomen, can be complex anatomical representations or simpler rigid shapes, which greatly reduces computer run-time. The joints linking the body sections can be assigned any characteristic and thus the model can represent various levels of human attentiveness and strength, such as a fully braced or unconscious person. Any size of adult or child can be modelled. Previous applications of DYNAMAN have included models of; vehicle occupants in crash events, human response to blast, passenger response on escalators during emergency braking and the response of wearers of lifejackets during self-righting. Clients for DYNAMAN work have included The Railways Inspectorate, the Defence Evaluation and Research Agency and the Health and Safety Executive.

Figure 3 shows a typical DYNAMAN model. In this analysis FNC was asked to provide predictions of typical head impact speeds of military personnel in armoured personnel carriers when traversing a rough terrain. This information was required as an input into helmet design and testing procedures. A model of the interior of the personnel carrier was constructed and DYNAMAN was placed inside. The model was then excited by a vehicle motion that was measured on an actual vehicle. The response of DYNAMAN was then examined and factors such as the use or non-use of seat belts were evaluated. The use of computer modelling was the only real method of obtaining such information. The alternative would be to subject human volunteers to potentially injurious head impacts.

4. FAILURE OF LINKAGES AND MATERIALS

There are several methods of representing failure of joints and materials in DYNA3D. For DYNAMAN work such as that discussed above a user defined spring is generally used. The spring is assigned a non-linear force-displacement characteristic where the force drops to zero once a strength and displacement limit has been reached. This was used in the DYNAMAN vehicle motion model to represent the strength of the man's hand grip on the seat front.

If it is a material that is required to have a representation of failure there are two methods available. Firstly, a number of material models are available in DYNA3D which include failure algorithms. Failure criteria such as the Von-Mises criteria are embedded into the code and once the failure criterion is reached the failed elements are automatically deleted from the model solution.

The alternative approach is to model failure by the failure of links between elements. This is similar to the discrete element solution technique being developed in other codes. With this technique the elements representing the material are linked together by a tied interface which will fail at a specified failure criterion. Failure criteria can include normal or shear stress or a combination of the two. This approach was used in a recent study by FNC in which the client (the Defence Evaluation and Research Agency) required a prediction of the charge size required to demolish a masonry building. The results of one of the models used in the study is shown in Figure 4 compared with the actual results of the demolition test. The test was conducted subsequent to the modelling work and in all cases the test results were very similar to the modelling predictions. This work is reported in more detail elsewhere (2).

5. FLUID INTERACTION

For the analysis of the interaction of fluids and structures a new version of DYNA3D has been developed by FNC called DYNASWIMMR. Subroutines have been added to the code to allow the effects of buoyancy, drag, impact (wave slap) and added mass to be modelled as pressure loads rather than being explicitly modelled. This greatly reduces the man and computer cost of such models and increases the number of applications which can be examined. The code has proven to be an accurate and economic tool and has been validated for a number of applications including:

- Lifejacket freeboard and self righting tests.
- Drop tests used to assess lifeboat dynamics.
- Boat motion (sink and resistance) through flat water.

Much of the development work has been conducted under contract to RNLI. The code has also been used to examine the kinematics of free-fall lifeboat launch. Any structure can be examined using the system and both the overall kinematics and detailed structural response can be analysed. The system can also be used to provide structural loads in both calm and rough sea states. A picture of a DYNASWIMMR model of a lifeboat is shown in Figure 5.

The correlation between DYNASWIMMR predictions and a water tank test with a scale model are also shown.

DYNASWIMMR can also be used in an analysis of boat collision, where the inclusion of appropriate boundary conditions (i.e. the water buoyancy, drag etc.) is likely to be critical in obtaining an accurate prediction.

6. CONCLUSIONS

This paper has demonstrated the advanced capabilities of DYNA3D and how they can be applied to solve complex multi-body dynamics problems. One of the particular strengths of the version of the code used by FNC is that the source code is available and hence more advanced algorithms can be introduced to extend the code and reduce the cost of the analyses. The role of computer modelling in the design process has also been highlighted and in particular it has been shown that the correct application of modelling can dramatically reduce the costs of the design process.

7. REFERENCES

(1) Mr T J Edwards and Mr M W Anderson. Development of the Heathrow Express Crashworthy Vehicle. Proc. of Railtech 96. Birmingham. 1996. Publ. by the Institution of Mechanical Engineers.

(2) Nash M and Dorn M.R. "The Numerical Prediction of the Collapse of a Complex Brick Building Due to an Internal Explosion". Structures Under Extreme Loading - 1997, ASME Pressure Vessels and Piping Conference, Florida. USA, July 1997.

ACKNOWLEDGEMENTS

The building demolition has been carried out with the support of the Defence Evaluation Research Agency. The study involving the prediction of head injury to personnel in military vehicles has been carried out with the support of the Defence Clothing and Textile Agency. The author would also like to thank all the staff at FNC who contributed to the various work programmes.

COPYRIGHT

The Copyright in this work is vested in Frazer-Nash Consultancy Limited. However, the Institute of Mechanical Engineers are assigned the rights to copy this work as required for the purposes for which it is intended.

© With Author 1998

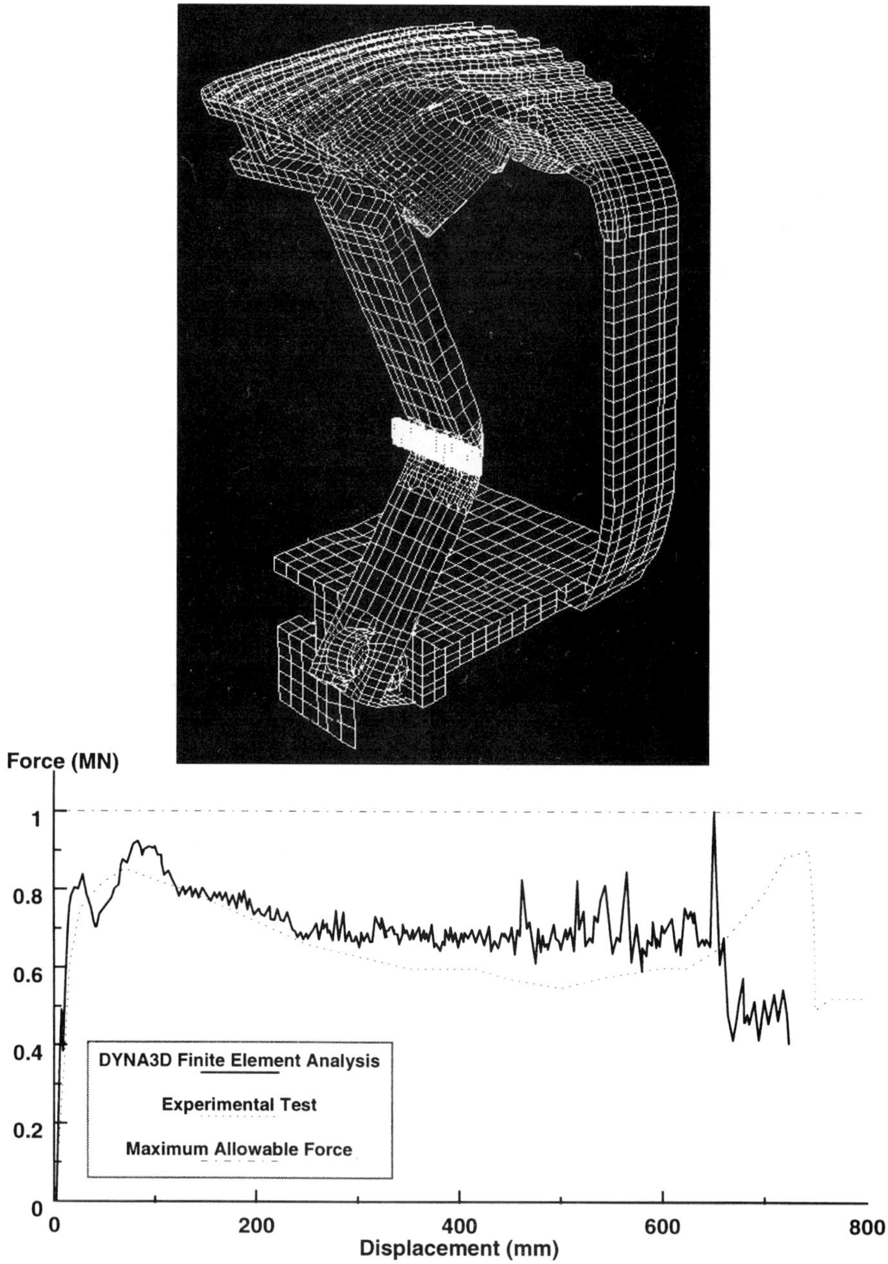

Figure 1: A model of a railway carriage bodyend undergoing an overriding crush test.

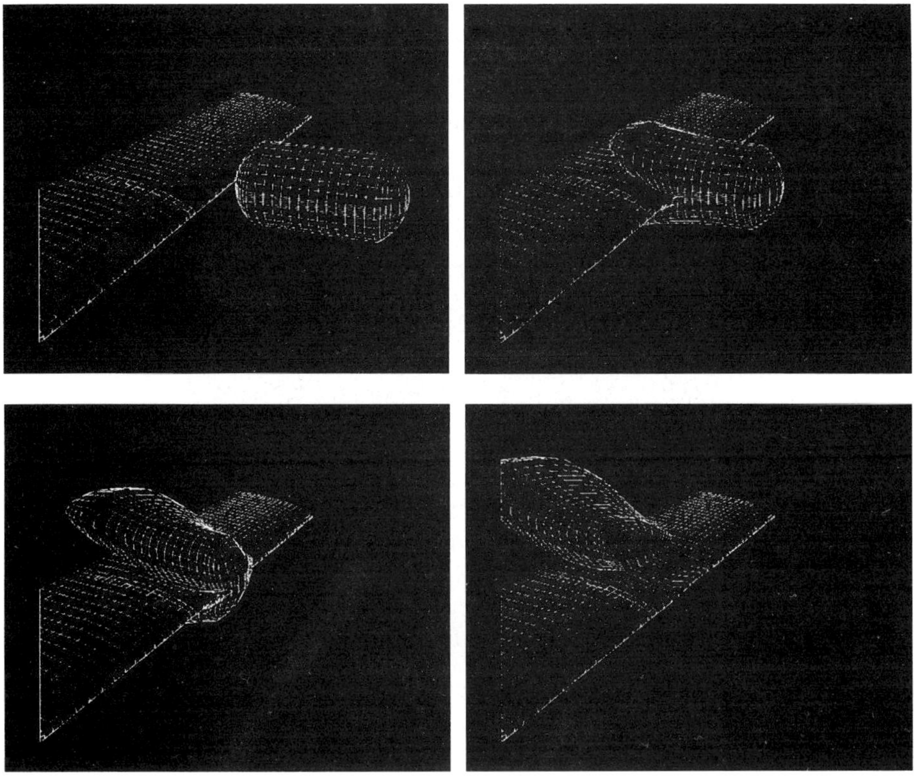

Figure 2: A model of a bird strike on an aircraft leading edge.

Figure 3: DYNAMAN used to examine the response of passengers in an armoured personnel carrier to severe off-road motion. Courtesy of DCTA.

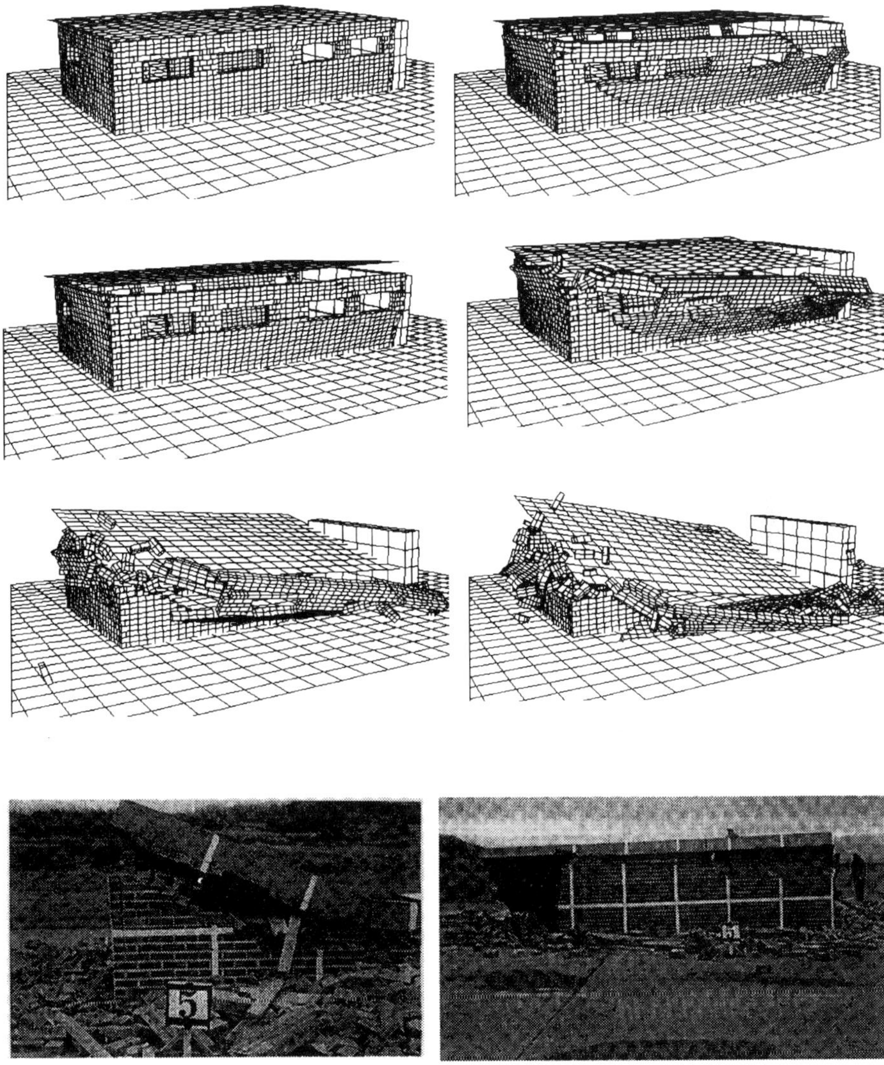

Figure 4: Comparison of model predictions and actual results of a building demolition.

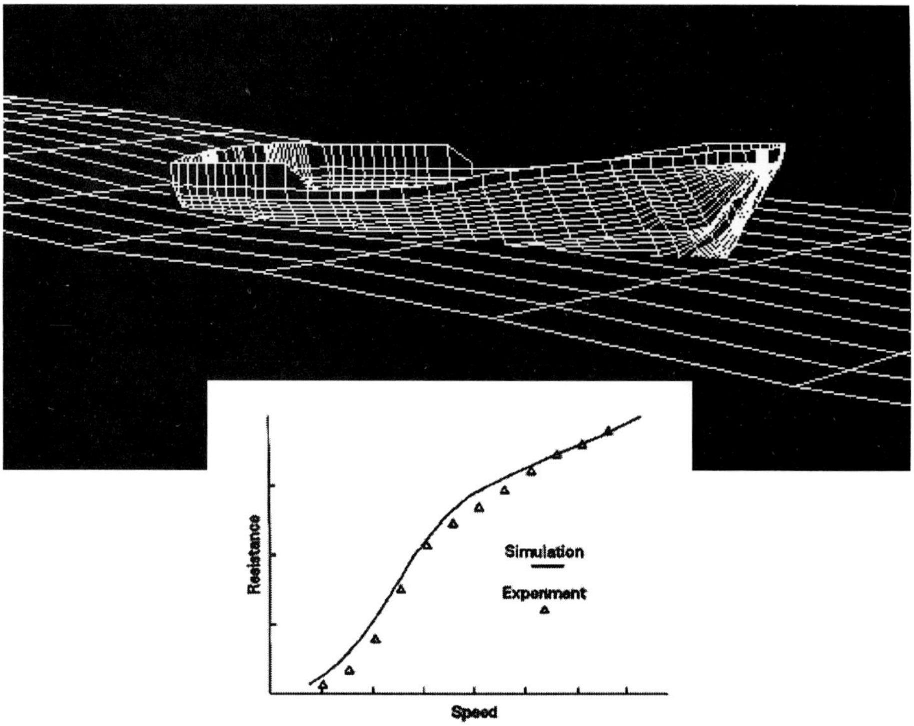

Figure 5: Example of a DYNASWIMMR model and comparison between model prediction and a water tank test.

C553/034/98

Dynamic simulation – an essential design tool

G J LOCK BSc, CEng, MIMechE
Frazer–Nash Consultancy Limited, Dorking, UK

ABSTRACT

This paper describes two examples of multi-body dynamic problems recently assessed by Frazer-Nash Consultancy. Both examples have utilised GENDYN, a lumped mass solution code developed by Frazer-Nash Consultancy for the analysis of linear and non-linear dynamic problems. In each case, models of engineering systems are excited by a range of applied forces and displacements in order to determine their transient response.

GENDYN models are constructed by coupling masses and inertias together by resilient elements from a comprehensive library. These elements include both linear and non-linear characteristics e.g. gaps, friction, crushing, etc.

The two example applications are very different engineering problems in which dynamic simulation proved to be an extremely useful design and assessment tool. They are:

- The design of a rail buffer that produced acceptable jerk levels to passengers.
- Investigations into the roll stability of heavy goods vehicles.

1. RAIL BUFFER DESIGN

1.1 Problem definition

The objective of the first application of GENDYN was to design a railway buffer stop that met the following criteria.

- To stop trains of varying mass and velocity within a specified stopping distance.
- To limit the acceptable accelerations and jerks received by passengers to acceptable levels.

The basis of the design is shown in Figure 1. The figure shows the major elements of the system:

1. A multiple carriage train provides the energy to the system with fixed mass (m) and initial velocity (V).
2. It strikes a hydraulic buffer which is rigidly connected to:
3. A sliding buffer. As the buffer slides it collects:
4. A series of friction 'shoes' which are connected to, and slide along, the rails.

This system is further complicated by the following factors:

- The train can be up to eight carriages long with each carriage connected with hydraulic dampers / couplers.
- Both the inter carriage couplers and the hydraulic buffers (2, in the figure) exhibit highly non-linear behaviour.
- The friction force generated by the sliding 'shoes' is not constant and is inversely proportional to the distance travelled.
- The initial friction force of the 'shoes' is adjustable during commissioning.
- The performance of the shoes in service is extremely variable.

The bounding parameters of the problem can be solved with simple hand calculations using energy absorption equations. It is common sense which indicates that the maximum stopping distance will be based on the allowable acceleration (or deceleration). This means that the maximum stopping distance is found when the fastest train is fully loaded and the 'shoes' work at their minimum efficiency. The number and spacing of the shoes can then be calculated.

The other design parameter is when the lightest train hits the same buffer system with the shoes working at their maximum performance (highest friction). The same allowable values of acceleration and jerk have to be maintained.

Jerk is a term which is common in its use but little understood in its technical interpretation. Jerk is defined as the rate of change of acceleration. Whilst there is some published data linking acceleration to human tolerance, there is very little published data on jerk and its effects. However, it is known that unbraced passengers will fall and injure themselves at very low levels of acceleration, which on the face of it would be considered acceptable. The destabilising effect is due to the magnitude of the acceleration and the rate at which it varies (i.e. the jerk).

1.2 Model definition

The GENDYN model created is shown schematically in Figure 2.

The model in Figure 2 represents a five car train. As such there are five masses linked by four couplers between them (K1). The front car is linked to the mass of the buffer by another link (K2). The buffer mass is linked to the ground via a third link (K3).

The lumped masses, representing the cars and the buffer, are each assigned the appropriate mass for the assessment in hand (i.e. whether laden or unladen). The links, representing the carriage couplers (K1), are assigned properties representing their highly non-linear, hysteretic behaviour. This has been modelled in GENDYN using a user-defined spring link which uses a look-up table to determine the force associated with an element deflection. The hydraulic buffers, mounted on the front of the sled (K2), have been modelled as a user-defined spring, based on the manufacturers data. The effect of velocity has been modelled by use of a damping element.

The element representing the sliding shoes (K3) is again a user-defined force/distance element. The force displacement profile for this element is shown in Figure 3. The definition of this element was initially based on the manufacturers data (shown in dotted line).

After the initial results were studied, the profile was modified (solid line) so that the initial force is ramped up, via the careful positioning of the shoes, to reduce jerk. The profile reaching its initial maximum at point (A). The 'saw-tooth' profile is due to the sliding 'shoes'; the vertical 'jumps' in the profile represent the positions at which a pair of 'shoes' is engaged. After each 'jump' there is a gradual reduction in friction force as the 'shoes' wear. The force generated (A to B) is the force that would decelerate the minimum mass train at an allowable rate. It was calculated that the lightest train, at maximum impact velocity, would have stopped by point (B). At this point the force is further increased via the collection of more friction shoes. The positioning of these shoes was calculated so that only the heavier trains, travelling at faster speeds, would receive this force. Thus, all accelerations should remain within acceptable values.

This profile was considered to be the most efficient in terms of energy absorption, without exceeding nominal acceleration and jerk limits.

1.3 Results

The model was used initially to check the assumptions used in the scoping calculations. Once achieved, the model results were assessed in detail to determine displacement, velocity, acceleration and jerk profiles.

Initial results for a model of a seven car train are shown in Figures 4 and 5. Figure 4 shows the predicted acceleration traces for car 1 (car 1 being that which is impacting the buffer). It can be seen that the acceleration traces exhibit an oscillating acceleration about an average of approximately 2 m/s^2. The typical peak acceleration of this oscillation is approximately 5 m/s^2 and the frequency is 4 - 5 Hz.

If one assumes this oscillation is of sinusoidal form one can calculate the peak velocity of the vibration as approximately 0.08 m/s and the peak displacement as approximately 3 mm.

The jerk was calculated from the predicted acceleration and the jerk time histories for three of the cars in the consist are shown in Figure 5. It can be seen that the jerk predicted by the model is typically in excess of 200 m/s^3 .

The mean values of acceleration and jerk are unacceptably high. However, the peak values occur at a high frequency. As mentioned, the oscillation will only result in a small relative displacement of the floor (a few millimetres). Examination of ISO 2631:1985 (Reference 1) revealed that a typical man would reach a point of reduced work proficiency after an exposure of these levels of one hour. Thus, one can conclude that these oscillations would be potentially uncomfortable rather than a hazard to passenger safety.

Consequently, re-positioning of the 'shoes' (Figure 3 - solid line) was conducted to further reduce the mean acceleration and jerk to below 1.5 m/s^2 and 1.2 m/s^3 respectively, whilst maintaining or reducing the oscillatory element. This improvement could not have been achieved without the use of the dynamic model.

1.4 Conclusions

The following conclusions have been drawn from the modelling work:

- During a buffer impact event there are high frequency oscillations which occur along the length of the assembly.
- The oscillations may induce peak accelerations of up to 10 m/s^2 and jerks of over 300 m/s^3 at frequencies of 4 - 5 Hz. However, oscillations of this level should not compromise passenger safety.
- The main parameter, affecting passenger safety, is likely to be the average jerk and average deceleration of the assembly.
- The highest jerks occur in the lightest trains.

2. ROLL STABILITY OF ARTICULATED HEAVY GOODS VEHICLES

2.1 Introduction

This section outlines the application of GENDYN to assess the roll-over of heavy goods vehicles. The method developed has been used to investigate over thirty roll-over accidents and to influence the design of a number of HGVs. The proven consistency and accuracy of results has contributed to the resolution of several legal cases and has led to Frazer-Nash Consultancy, having a unique capability in the assessment of HGVs and their stability.

2.2 Problem definition

There are four primary factors affecting the stability of a vehicle whilst cornering:

- Vehicle centre of gravity height
- Vehicle track width
- Radius of turn
- Vehicle speed.

In addition to these there are the 'secondary' overturning effects of:

- The roll of the trailer
- The shifting of the cargo.

The roll is dependent on the suspension characteristics. When cornering, the centripetal force on the cargo causes the vehicle to roll about its 'roll centre' which, for practical purposes, is at the centre of the axle.

The shifting of the cargo centre of gravity outwards produces an additional overturning moment which contributes to a reduced roll-over speed. This is shown schematically in Figure 6. Similarly if a cargo slides outward on a flat bed, the centre of gravity is shifted with a consequential reduction in stability.

All of the above effects can be represented and tested with a tilt table and can be considered static or quasi-static. Consideration of the above factors may be adequate for the simple assessment of stability; for instance on a constant radius bend at a constant speed. However, in practice there are many other factors that will affect vehicle stability, such as:

- Sudden changes of direction
- Acceleration and braking
- Liquid sloshing (in the case of tankers)
- Fifth-wheel effects.

The fifth-wheel contributes two destabilising effects, the coupling at the fifth-wheel pivot is fairly free, allowing around 2 degrees of roll before stiffening up, with a maximum possible roll of about 3 degrees. Secondly, and more importantly is the reduction in stability due to high articulation angles. If the vehicle is turning sharply, stability reduces considerably as the trailer receives much reduced roll stability from the tractor.

The inclusion of these additional effects is complex and has been achieved using a GENDYN model, developed over several years.

2.3 Modelling strategy

A complex GENDYN model has been developed which is able to simulate all of the factors mentioned above. Any vehicle can then be realistically analysed under various conditions of turning, acceleration and deceleration, travelling over road irregularities, and with different kinds of suspension. A simplified diagram showing some of the features incorporated in a simulation is shown in Figure 7.

The model consists of 22 masses and inertias. The model incorporates all of the important motions of the vehicle and cargo including:

- Yaw of the steer axle
- Bounce and roll of the axle groups
- Bounce, sway, pitch, roll and yaw of the tractor
- Bounce, sway, pitch, roll and yaw of the trailer
- Sideways shifting of the cargo.

The various degrees of freedom are interconnected by 38 resilient elements that represent the tyres, suspension, fifth-wheel and liquid sloshing (in the case of tankers). Rigid cargo shifting is modelled using friction elements which allow the controlled sideways shift of the cargo mass when a stiction force is exceeded.

The tyres are modelled to incorporate the following physical characteristics:

- Non-linear vertical force / deflection characteristics that allows wheel lift when all vertical load is removed
- Cornering force that is the product of wheel load, tyre slip angle and tyre characteristics
- Self-aligning torque that is proportional to wheel load and tyre slip angle.

Suspensions are modelled using pairs of vertical linear spring and damper elements that are separated by a defined spring width. Resilient elements are also used to represent bump-stops and anti-roll bars.

The cargo is treated as a mass connected by a spring to the trailer bed. For liquid tanker models, the natural frequency of the 'sloshing' has been calculated to be approximately 0.5 Hz with about 15% of critical damping. This is for petroleum products in 'max' section tanks. The additional overturning moment that acts when the liquid centre of gravity moves away from the centre-line of the tank is also included in the model. It is taken as the product of the cargo weight, relative movement of the cargo and a correction factor that accounts for tank cross-section and ullage (in the case of tankers).

As the vehicle corners, the weight moves outboard of the vehicle centre-line. The secondary overturning moment which is caused by the body rolling about its roll-centre is modelled using destabilising cross-coupling elements which connect the body sway to the axle groups. The body overturning is the product of the vehicle weight and the difference between the body centre of gravity and roll centre height. The axle overturning is the product of the axle load and axle roll centre height.

Throughout the development of the simulation method, verification was achieved by comparing predicted results with data compiled from accidents and field trials. Initial results were very good with predicted roll-over speeds, consistently coinciding within 1 mph of tachograph readings.

2.4. Application of the method

By far the most common occurrence of roll-overs investigated by FNC concerns the negotiation of a roundabout. Figure 8 shows the roll angle and tyre forces from a typical simulation. It can be seen how the trailer has a near constant roll, due to the camber of the lead-in to the roundabout. At point 'A' the vehicle makes a left turn followed by a right turn at 'B'. The consequential increase in trailer roll from one side to the other can be seen. At point 'C' the trailer makes a further left turn to exit the roundabout. At this stage the vehicle is out of control and the combination of vehicle speed and the sequence of manoeuvres cause the vehicle to roll-over at 'D'. The only way of avoiding roll-over would be to straighten out between 'C' and 'D' in which case the vehicle would have run out of road.

The nearside wheel forces, for both the trailer and the tractor rear axle, are shown in Figure 9. As the trailer rolls between 'A' and 'C' there is a corresponding cycling of these forces. At 'C', there is a slight wheel hop before the force disappears (i.e. the wheel is off the ground). At this point the vehicle is 'balancing' on its offside wheels and roll-over results as the final left hand turn is negotiated.

The simulation method of assessment has also been used to investigate the effect of tyre deflation during a manoeuvre and to study the forces or 'feedback' experienced by the driver while conducting these manoeuvres.

2.5 Future developments

Having developed and refined the assessment method for HGVs, the model has the potential to be applied to other kinds of vehicles. For example, some pivot-steer loaders and heavy tractors have a tendency to tip over, particularly the high speed versions used by the military.

Like lorries, construction vehicles are generally large and heavy, and loss of directional control and roll-over accidents have occurred. Roll-over is also a major hazard in the agricultural sector and, eventhough farm tractors are generally not pivoted, turning on a steep incline has led to many roll-overs.

Frazer-Nash Consultancy is now applying similar analytical procedures to study both agricultural and military vehicles, including trailers. Safer designs and better operator understanding of the consequences of excessive speed should result.

3. REFERENCES

(1). BS 6841:1987 (ISO 2631-1) Guide to measurement and evaluation of human exposure to whole-body mechanical vibration and repeated shock.

© With Author 1998

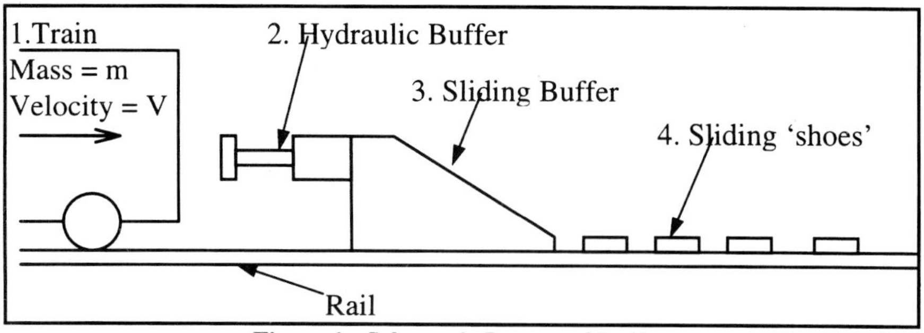

Figure 1 - Schematic Layout of Buffer

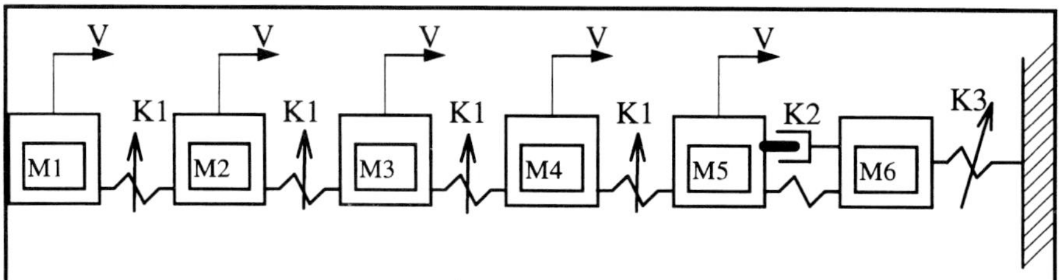

M1 to M5 are individual carriage masses.
M6 is the Buffer mass
V is the initial velocity at buffer imact.
K1 is the inter carriage coupler (force ∝ velocity, displacement).
K2 is the buffer hydraulic rams (force ∝ velocity, displacement).
K3 is the force distance profile of the 'shoes' (force ∝ displacement)

Figure 2 - GENDYN Model Representation

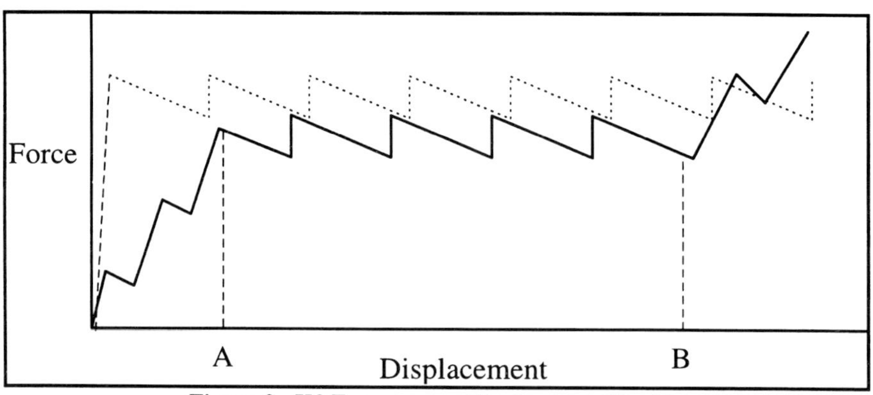

Figure 3 - K3 Force versus Displacement Profile

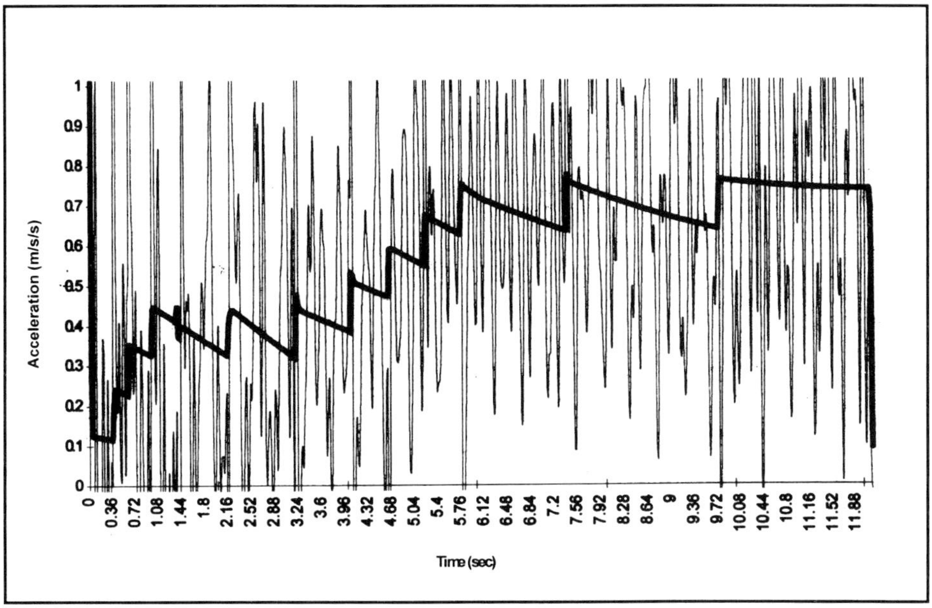

Figure 4 - Peak and **Average** Acceleration versus Time

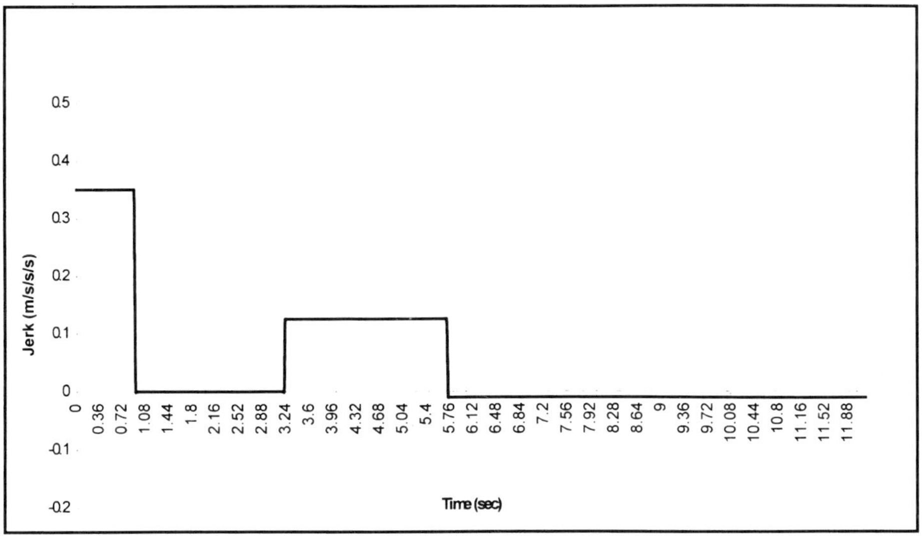

Figure 5 - Jerk versus Time

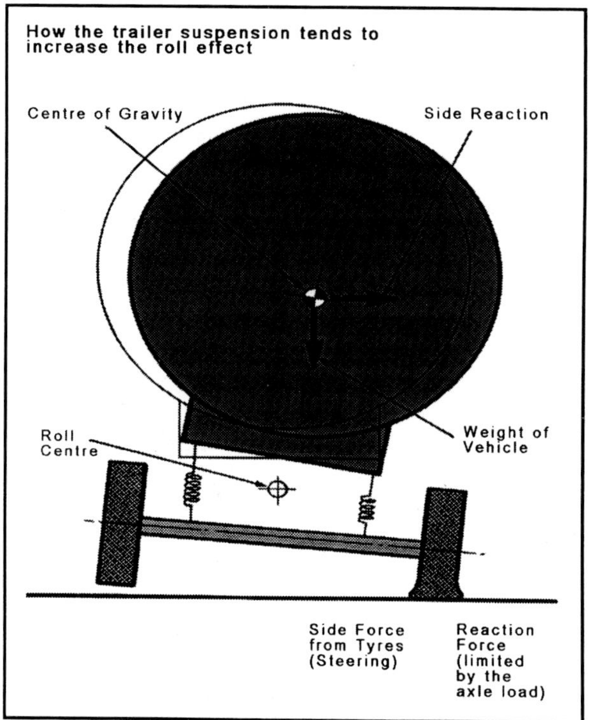

Figure 6 - Secondary Overturning Effects

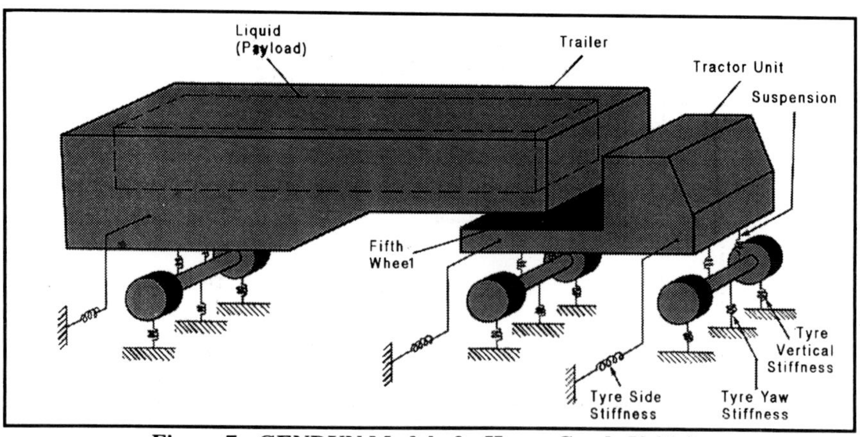

Figure 7 - GENDYN Model of a Heavy Goods Vehicle

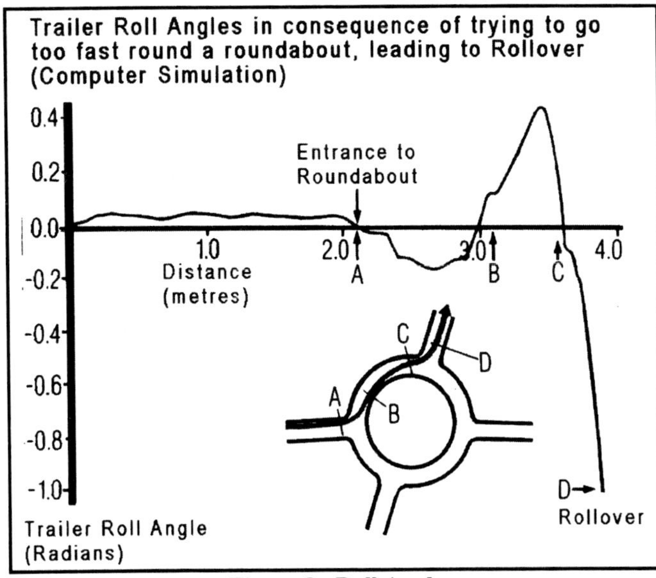

Figure 8 - Roll Angle

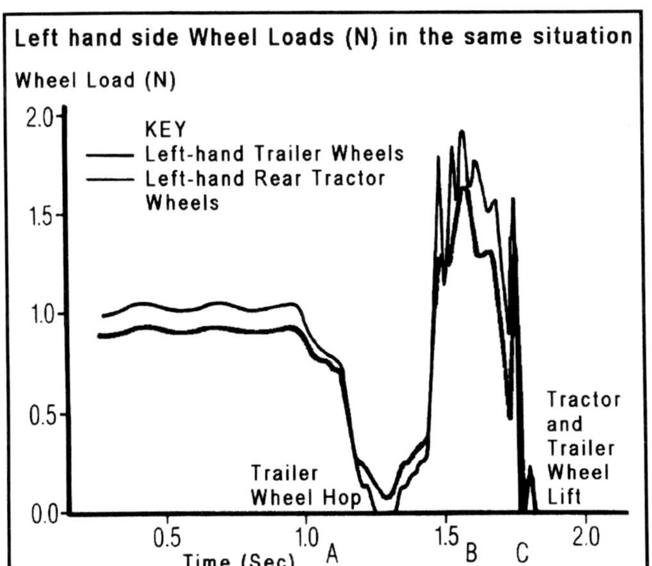

Figure 9 - Nearside Wheel Force

C553/001/98

An application of a powerful tool for analysis of mechanical systems containing rolling and spinning Hertzian point contacts

M TÖRMÄNEN and **L LINDKVIST**
Machine and Vehicle Design, Chalmers University of Technology, Göteborg, Sweden

ABSTRACT

This paper presents the use of a new method to analyse mechanical systems containing power-transmitting rolling and spinning Hertzian point contacts. For instance, mechanical traction-drive variators undergoing transient operating conditions. The technique used combines two efficient methods: Multi-body System (MBS) computer software that handles the dynamics of these system, and numerical procedures that treat the special traction motion relationships between the bodies in contact.

Several software packages for analysis of full dynamic systems are already available (e.g. ADAMS, DADS, MECHANICA). However, until now no efficient numerical procedures to solve the traction motion interactions have been available. Törmänen has recently developed new efficient numerical solutions that solve these relationships fast and with specified accuracy. By combining these solutions with an MBS-software, a powerful tool for analysis of mechanical traction drive variator is created.

The contact model used was based on Coulomb's dry friction and Hertzian pressure distribution. It can also be used to display the isothermal behaviour of contacts, where a solidified oil film is present, and it is general in the sense that it can handle slip perpendicular to the rolling direction.

To study the effects of transient changes of the speed ratio, this powerful tool is tested on the concept of a traction drive CVT from Torotrak Ltd. with a double-sided, full-toroidal variator. In this paper, it is demonstrated how the new tool could be used to predict theoretically the speed ratio of the mentioned CVT. In the future this will enable a theoretical refinement of control strategies for CVTs with mechanical traction drive variators.

Keywords: Rolling and Spinning Hertzian Point Contacts, Continuously Variable Transmission (CVT), Multi-Body Systems (MBS)

1 NOTATION

v_1, v_2, v_{rel} - translational velocities

$\omega_1, \omega_2, \omega_{rel}$ - rotational velocities

$e = [e_x, e_y, 0]$ - a vector from the theoretical centre of a contact ellipse to the speed pole

θ_x, θ_y - non-dimensional traction utilization factors

θ_φ - non-dimensional torque factor around the speed pole

F_x, F_y - tractive forces

F_{act} - actuator force

F - resulting tractive force

N - normal force at the point contact

M_e, M_z - tractive torque around speed pole and the theoretical centre, respectively

α - non-dimensional contact ovality

$\varepsilon_x, \varepsilon_y$ - non-dimensional speed pole location

a, b - length of the semi-axes of the contact ellipse

ξ, η - non-dimensional independent variables

ν - Poison's constant

E - Young's modulus

r, r_1, r_2 - principal radii of curvature

r', r'_1, r'_2 - principal radii of curvature

k - surface stiffness in the normal direction

δ - surface displacement in the normal direction

φ, φ' - orientation angles of the principal radii of curvature

$\xi_\vartheta, \eta_\vartheta, \psi_\vartheta$ - non-dimensional constants

$\Omega_{in}, \Omega_{out}, \Omega_{disc}$ - global velocities of the interacting rigid bodies

T_{in}, T_{out} - global torques

Δ - actuator displacement

γ - axial rotation angle for the disc actuator rod

β - disc caster angle

C553/001 © IMechE 1998

2 INTRODUCTION

Many mechanical systems contain power-transferring elliptical point contacts. This kind of contact is found between bodies that are pressed together and moved against one another. Therefore, the power is transferred from one body to the other by utilizing tractive forces. In this treatise, the relative motion that causes the tractive effort is considered as a combination of rotational and translational velocities, and the contact shape as a plane ellipse. This view agrees well with Hertzian contact theory. Therefore, these contacts are referred to here as *rolling and spinning Hertzian point contacts*.

One important category of applications, where this kind of contact is found, is mechanical traction drive variators. These could be used to design different concepts for Continuously Variable Transmissions (CVTs). The Torotrak vehicle transmission, see fig. 1, is one concept for a CVT with a mechanical variator. Even today, CVTs are used in some vehicle applications to achieve optimal engine operating conditions. Some of the advantages of using CVTs are increased driveability, reduced fuel consumption and better overall vehicle performance.

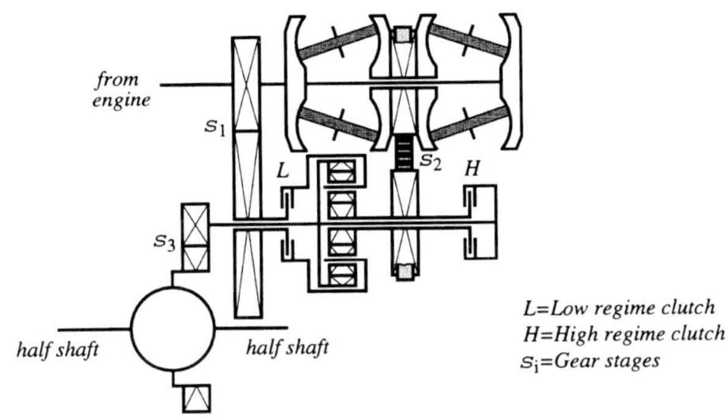

Figure 1. Torotrak front wheel drive CVT, c.f. (1).

The major requirement for all types of variators is power transfer while the speed ratio is being changed. For mechanical traction drive variators this is established by utilizing traction between pairs of rotating bodies. Thereby, the power is transferred through the point contacts by traction and the change of speed ratio is achieved by orientation of the bodies. Therefore, the surfaces of the mating bodies, the so called races, are usually rotational symmetrical. In applications, these races are also usually separated by a thin high-pressurized oil film with high traction coefficient to avoid wear and unnecessary heat-dissipation.

One way of describing the relationships between traction and motion is with a model introduced by Mägi (2). His approach is based on a concept of dry friction and is practical to use from the engineering point of view. It also covers those cases where a iso-thermal solidified lubricant is present as in mechanical variators. Until now, numerical contributions have only been efficient for some special cases of tractive motion where the operating conditions are quasi-stationary. Arbitrary operating conditions have been more time-dependent to evaluate and have, therefore, been out of reach for analysis of transient conditions (e.g. change of speed ratio). Recently, Törmänen has developed new efficient numerical solutions that make it possible to analyse such rolling and spinning contacts under arbitrary operating conditions. A more thorough description will be found in (3).

By combining an efficient contact subroutine with some package of computer software for

Multi-body System (MBS) dynamics, a powerful tool is created for analysis of complex mechanical systems containing rolling and spinning Hertzian point contacts. The MBS-software takes care of all system dynamics and the contact subroutine describes the force interactions in all power transferring point contacts. The subroutine must have high evaluation capacity, since the MBS-software calls it several times within each iteration for dynamic equilibrium. In such iterations the MBS-software provides the contact subroutine with the normal forces, velocities and geometrical properties of the races and the subroutine in turn provides the MBS-software with the necessary tractive forces. This is shown schematically in fig. 2.

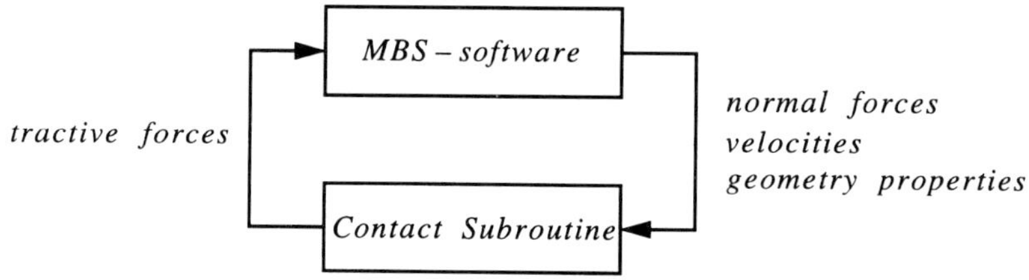

Figure 2. Communication between an MBS-software and the contact subroutine.

This kind of powerful tool enables fast product development processes, which is the fundamental requirement when minimizing costs and development time when optimizing complex products as the mentioned mechanical systems above.

3 CONTACT SUBROUTINE

3.1 Constitutive relationships

Fig. 3 exemplifies two bodies pressed together with the global velocities v_1, v_2, ω_1 and ω_2. The relative motion that causes the tractive effort between the two bodies is uniquely defined by one single point at the common tangential plane. This point is the instantaneous centre of the relative rotation and is here referred to as the speed pole. In the same manner the resulting traction can be defined with one single force F at a specific location (see fig. 3). This is the fundamental form of the pole concept that Mägi has presented in (2).

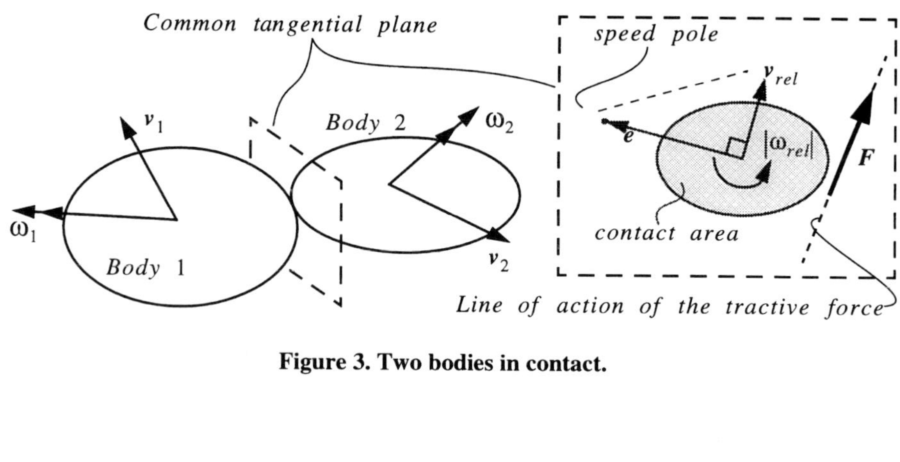

Figure 3. Two bodies in contact.

By applying vector algebra the location of the speed pole (e_x, e_y) can be established as:

$$e = [e_x, e_y, 0] = \frac{\omega_{rel} \times v_{rel}}{\omega_{rel} \cdot \omega_{rel}}, \qquad [1]$$

where the relative velocities are defined as

$$v_{rel} = v_1 - v_2 \text{ and } \omega_{rel} = \omega_1 - \omega_2. \qquad [2]$$

The constitutive relationships between traction and motion can then be described by the following three non-dimensional integrals, found in (2).

$$\theta_x = \frac{F_x}{\mu N} = \frac{3}{2\pi} \int\int_D \sqrt{1 - \xi^2 - \eta^2} \frac{\eta/\alpha - \varepsilon_y}{\sqrt{(\varepsilon_x - \alpha\xi)^2 + (\varepsilon_y - \eta/\alpha)^2}} d\xi d\eta, \qquad [3]$$

$$\theta_y = \frac{F_y}{\mu N} = \frac{3}{2\pi} \int\int_D \sqrt{1 - \xi^2 - \eta^2} \frac{\varepsilon_x - \alpha\xi}{\sqrt{(\varepsilon_x - \alpha\xi)^2 + (\varepsilon_y - \eta/\alpha)^2}} d\xi d\eta, \qquad [4]$$

$$\theta_\varphi = \frac{M_e}{\mu N \sqrt{ab}} = \frac{3}{2\pi} \int\int_D \sqrt{1 - \xi^2 - \eta^2} \sqrt{(\varepsilon_x - \alpha\xi)^2 + (\varepsilon_y - \eta/\alpha)^2} d\xi d\eta, \qquad [5]$$

$$D = \{(\xi, \eta): (\xi^2 + \eta^2 \le 1)\}. \qquad [6]$$

These integrals state the equilibrium at the speed pole and depend only on the non-dimensional speed pole location $(\varepsilon_x, \varepsilon_y)$ and the non-dimensional contact ovality (α), and are here referred to as the *Hertzian traction integrals*. Efficient solutions to these can be found in (3). The non-dimensional dependent and independent variables are defined in equation [7] below, where the variables a and b are the length of the semi-axes of the contact ellipse, c.f. fig. 4.

$$\varepsilon_x = \frac{e_x}{\sqrt{ab}}, \quad \varepsilon_y = \frac{e_y}{\sqrt{ab}}, \quad \alpha = \sqrt{\frac{a}{b}}, \quad \xi = \frac{x}{a}, \quad \eta = \frac{y}{b}. \qquad [7]$$

However, by simplicity it is better to define the tractive force interaction at the theoretical point of contact of the contact ellipse as it is shown in fig. 4 below. In this case the torque must be modified with the following equation:

$$M_z = M_e - F_y e_x + F_x e_y. \qquad [8]$$

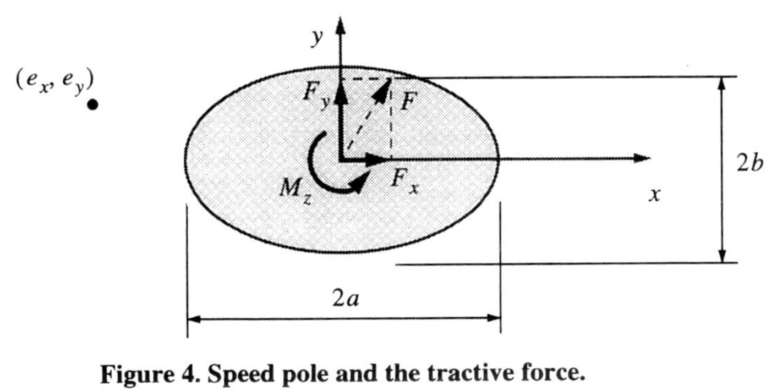

Figure 4. Speed pole and the tractive force.

3.2 Evaluation of the theoretical point of contact

Different methods can be used to find the theoretical point of contact between two bodies and is strongly dependent on the complexity of the race geometries. If the geometry of races are very simple (e.g. ball inside a sphere or ball on flat surface) the location of the point of contact can be found analytically. But, if the races are more complex (e.g. a torus inside a toroid), one way to evaluate the theoretical point of contact could be by using a subroutine that evaluates the shortest distance between the bodies and detects when this distance becomes zero.

3.3 Evaluation of contact parameters

The contact subroutine needs information about three contact parameters: the contact ovality α, the mean constant radius $\sqrt{ab}$; and the surface stiffness k in the direction of the surface normal. These parameters depend on the location of the theoretical centre and the orientation of the compressed bodies (i.e. the orientation of the four principal radii) and must, therefore, be evaluated every time the subroutine is called. According to Hertz, c.f. (4), the length of the semi-axes a and b can be evaluated from:

$$a = \xi_\vartheta \cdot \sqrt[3]{\frac{3(1-\upsilon^2)N}{E\left(\frac{1}{r}+\frac{1}{r'}\right)}}, \qquad b = \eta_\vartheta \cdot \sqrt[3]{\frac{3(1-\upsilon^2)N}{E\left(\frac{1}{r}+\frac{1}{r'}\right)}}, \tag{9}$$

and the surface stiffness coefficient k from:

$$N = k \cdot \delta^{3/2} = \left(\frac{E}{1.5(1-\upsilon^2)}\sqrt{\frac{2}{\left(\frac{1}{r}+\frac{1}{r'}\right)}}\left(\frac{\xi_\vartheta}{\psi_\vartheta}\right)^{3/2}\right)\delta^{3/2}. \tag{10}$$

The variables ξ_ϑ, η_ϑ, and ψ_ϑ can, thereafter, be constructed as spline functions from Table 1 by using the auxiliary angle defined in equation [11] as:

$$\vartheta = \arccos\left(\left(\frac{1}{r'}-\frac{1}{r}\right)\Big/\left(\frac{1}{r'}+\frac{1}{r}\right)\right). \tag{11}$$

Table 1. Non-dimensional constants.

ϑ	90°	80°	70°	60°	50°	40°	30°	20°	10°	0°
ξ_ϑ	1	1.128	1.284	1.486	1.754	2.136	2.731	3.778	6.612	∞
η_ϑ	1	0.893	0.802	0.717	0.641	0.567	0.493	0.408	0.319	0
ψ_ϑ	1	1.12	1.25	1.39	1.55	1.74	1.98	2.30	2.80	∞

A general way to calculate the radius r' and r is by using the following equations:

$$\frac{1}{r'}+\frac{1}{r} = \frac{1}{r'_1}+\frac{1}{r_1}+\frac{1}{r'_2}+\frac{1}{r_2}, \tag{12}$$

$$\frac{1}{r'}-\frac{1}{r} = \sqrt{\left(\frac{1}{r'_1}-\frac{1}{r_1}\right)^2+\left(\frac{1}{r'_2}-\frac{1}{r_2}\right)^2+2\left(\frac{1}{r'_1}-\frac{1}{r_1}\right)\left(\frac{1}{r'_2}-\frac{1}{r_2}\right)\cos(2\varphi)}, \tag{13}$$

C553/001 © IMechE 1998

$$\left(\frac{1}{r'}+\frac{1}{r}\right)\sin(2\varphi') = \left(\frac{1}{r'_1}-\frac{1}{r_1}\right)\sin(2\varphi), \qquad [14]$$

where r'_i and r_i are the smallest and the largest principal radii of the body i. The angle φ defines the relative orientation between the compressed bodies, and φ' defines the orientation of the contact ellipse relative to body 1.

3.4 Limitations

When using this subroutine the following assumptions must be fulfilled: the mating surfaces should be rigid in the direction of the surface tangent, but locally compressible in the direction of the surface normal according to the classical theory by Hertz; thc ideal Coulomb law of friction should be present over the entire contact area; and the contact area should be a plane ellipse. Typical cases where this subroutine cannot be used are: railway wheel contact, where the friction is not fully utilized in the whole contact; and bearings, where the geometrical properties are such that the contact area is not a planar ellipse. It should also be mentioned that the contact ovality should not be too small or too large to avoid line contacts.

4 MODEL DESCRIPTION

The main purpose of the studied MBS-model is to show the benefits of the proposed method to analyse rolling and spinning Herzian point contacts. Therefore, it is a quite simplified model of 1/6 of an ordinary Torotrak variator and consists only of one intermediate disc with actuator equipment, input shaft, and output shaft as it is shown in fig. 5. The actuator is modelled as a hydraulic cylinder and consists of a piston that can rotate and translate relative to a cylinder which in turn can rotate, in one direction, around a point connected to the housing as it is illustrated in fig. 6.

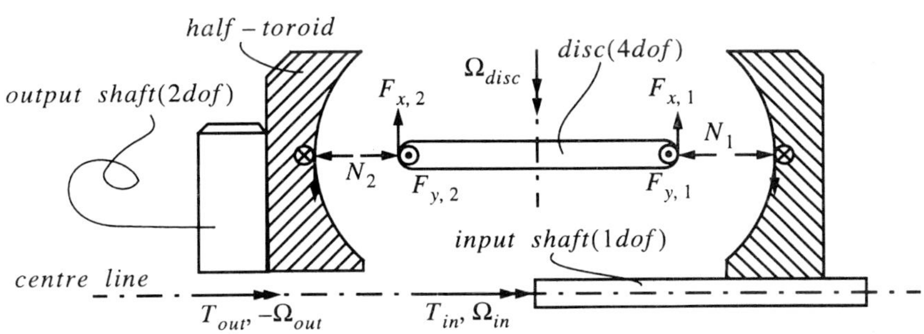

Figure 5. Graphical representation of the simplified Torotrak variator.

In this model the power-split function is not taken into account, but it could be simply built in with the ordinary MBS-elements for gears. The components that are used in the modelling are ordinary MBS-elements such as rigid bodies, joints, springs, dampers, forces, impact-modules etc. and a FORTRAN subroutine for the Herzian point contacts.

The input shaft has 1 dof (degree of freedom) which is the rotation Ω_{in}. The output shaft has 2 dof, the rotation Ω_{out} and translation along the output axis which makes it possible to apply a pre-load which gives the normal forces $N_{1,2}$. The disc has 1 dof, the rotation Ω_{disc} relative to the piston, but it has globally 4 dof. The piston has 2 dof relative to the cylinder, the rotation γ and translation Δ. Finally, the cylinder has 1 dof, rotation around the fixed point. Therefore, the simplified model has 7 dof. In the current simulations, however, the position of the piston relative to the cylinder, Δ, is restricted to follow a specified curve (shown in fig. 7) and consequently this model has only 6 dof. The angle β in fig. 6 defines the caster angle between the disc and the mating races.

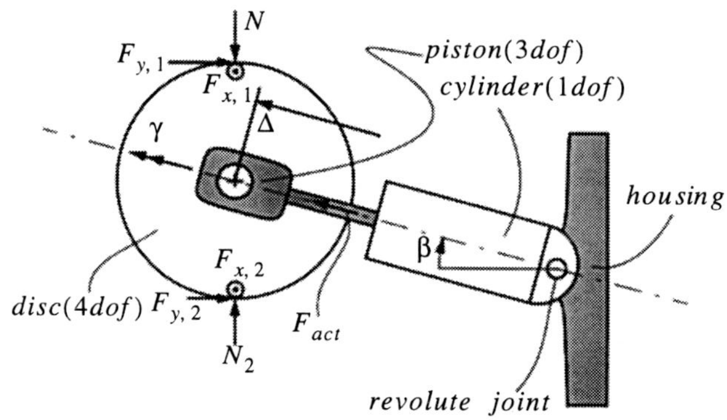

Figure 6. Disc actuator with the view from above.

The input shaft is driven by a constant torque and the reaction torque on the output shaft is generated by a viscous damper. The only losses that are considered are those that occur in the Hertzian point contacts. However, it is simple to add additional losses such as bearing losses, meshing losses, etc. by adding some MBS-elements in appropriate places.

The parameters of the model are chosen to be in the same order of magnitude as they would be in a real application (e.g. the inertia of the input shaft is about 1/6 of the inertia of an ordinary combustion engine).

5 SIMULATION RESULTS

The main objective of this study is the non-intuitive way to change the speed ratio. By moving the disc actuator forward or backward the disc will rotate around the actuator rod due to transversal tractive forces: $F_{x,1}$ and $F_{x,2}$. This in turn causes the change in the speed ratio.

The first plot, fig. 7, shows how the parameters: disc angle, speed ratio and actuator force vary with the transient piston motion. This plot also shows an important feature of this tool: the possibility to evaluate the time it takes for the system to become stationary after a transient change-over. The transient piston motion stops at point *a*, but first at point *d* the system becomes stationary. Point *b* and *c* indicate where the disc angle and the speed ratio become stationary.

The second plot, fig. 8, shows how much of the traction is used in different directions at each contact. The tractive forces are directly proportional to the utilization factors and the normal forces, c.f. equation [3] to [5]. In the same manner, the third plot (see fig. 9) shows how the different efficiencies vary within the same interval. It should be mentioned that the efficiencies are greater than one when the power-flow is reverse. This is due to retardation of the vehicle.

C553/001 © IMechE 1998

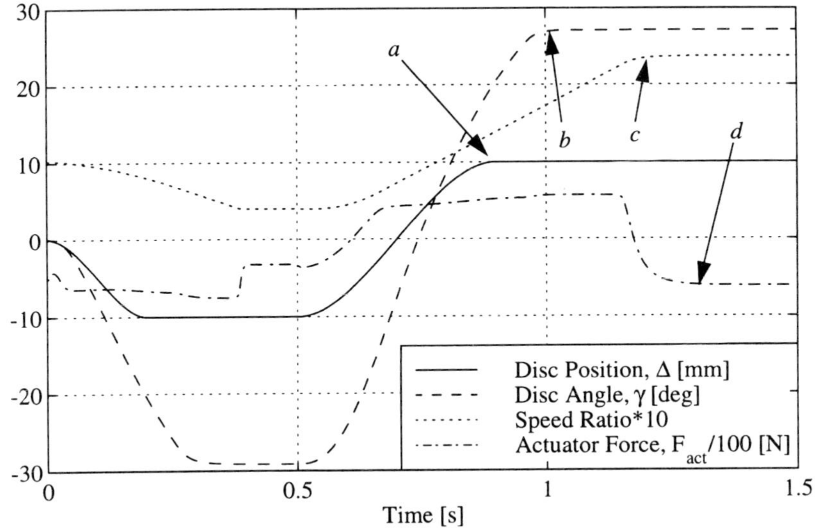

Figure 7. Disc position, disc angle, speed ratio and actuator force vs. time.

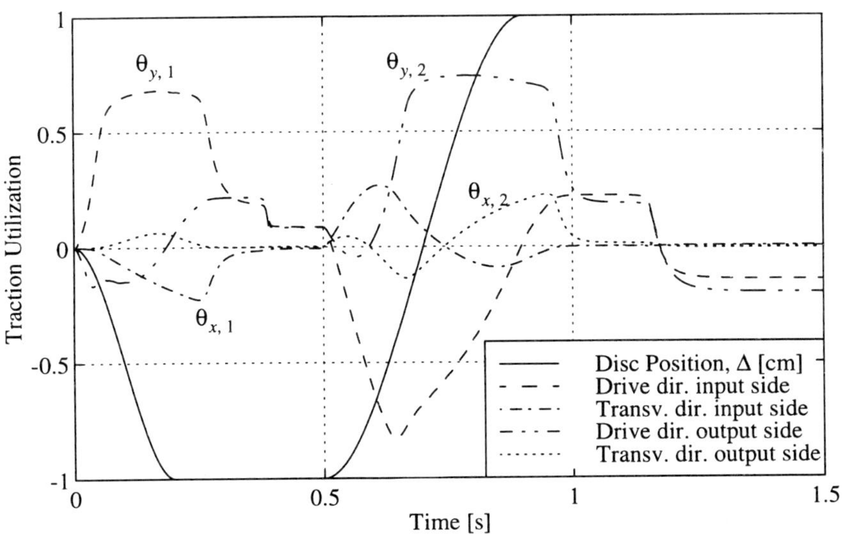

Figure 8. Traction utilization in all direction in all contacts and disc position vs. time.

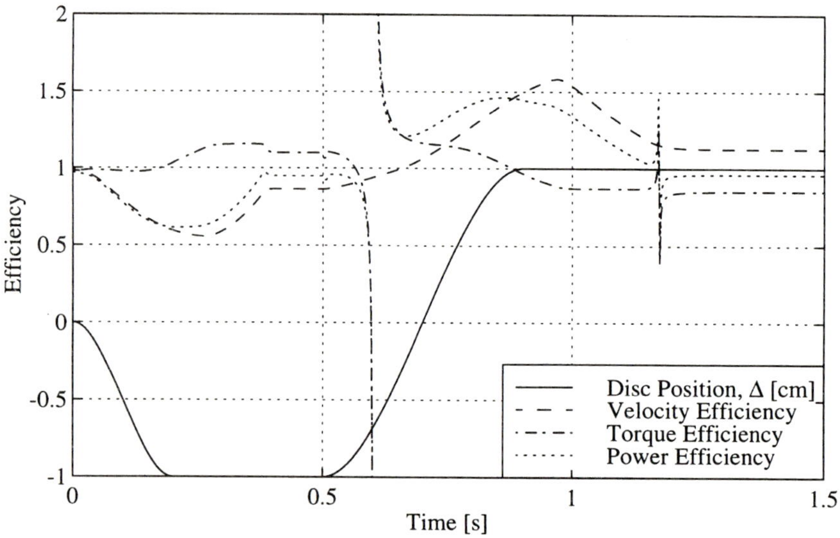

Figure 9. Torque efficiency, speed efficiency, overall efficiency and disc position vs. time.

6 DISCUSSION & CONCLUSION

The proposed model for simulation of toroidal variators can be used in various ways to improve the performance of this type of transmission concepts at transient conditions. In the early stages of the design process it can be used to compare different concepts. It can also be employed for parameter studies to tune the behaviour of a specific concept.

Another major field of application is the design of control strategies. Quite complicated control of the actuator is required to obtain the desired ratio, especially at transient driving conditions. Also the pre-load of the discs needs to be controlled in order to obtain the optimal total efficiency of the transmission without excessive wear. The control system could be modelled in some software for control design (e.g. ISI SystemBuild) and then this model could be connected to the MBS-model of the variator. These models can then run together to evaluate the control system. This would give a very powerful tool for the design of the control system.

7 REFERENCES

(1) Fellows, T. G., Greenwood, C. J., The Design and Development of an Experimental Traction Drive CVT for a 2.0 Litre FWD Passenger Car, SAE Paper 910408, February 1991.

(2) Mägi, M., On Friction at Elliptic Hertzian Contact Areas with Special Reference to Variable Drive Application, Lic. thesis, Göteborg, Sweden, 1964.

(3) Törmänen, M., Design of A Powerful Tool for Analysis of Mechanical Systems Containing Rolling and Spinning Hertzian Point Contacts, Lic. thesis, Göteborg, Sweden, 1998. (To be published)

(4) Dubbel, *Taschenbuch für den Maschinenbau*, Hrsg. von W. Beitz and K.-H. Küttner, page C36, 16th ed. Springer-Verlag New York Heidelberg Berlin, 1987.

C553/001 © IMechE 1998

Authors' Index

T

V

W

The Institution of Mechanical Engineers is a leading forum for the exchange of knowledge and expertise in the field of mechanical engineering.

A wide range of events is organized by the IMechE, to which all are welcome to attend. For further information about IMechE Conferences, Seminars, Workshops, and other events please contact us for further information.

Visit our website	www.imeche.org.uk
Telephone	+44 (0) 171 222 7899
Fax	+44 (0) 171 222 4557
Or write to	Conferences and Events Institution of Mechanical Engineers 1, Birdcage Walk London SW1H 9JJ

We look forward to seeing you at our future events.